PIMLICO

815

MAX PERUTZ AND
THE SECRET OF LIFE

Georgina Ferry is a former staff editor on *New Scientist*,
and contributor to Radio 4's *Science Now*. Her books
include the acclaimed biography *Dorothy Hodgkin:
A Life* (1998); *The Common Thread* (2002, with Sir
John Sulston) and *A Computer Called LEO* (2003).
She lives in Oxford.

Dorothy Hodgkin: A Life
The Common Thread: Science, Politics, Ethics and the Human Genome
(with John Sulston)
A Computer Called LEO: Lyons Teashops and the World's First
Office Computer

MAX PERUTZ AND
THE SECRET OF LIFE

GEORGINA FERRY

PIMLICO

Published by Pimlico 2008

4 6 8 10 9 7 5 3

Copyright © Georgina Ferry 2007

Georgina Ferry has asserted her right under the Copyright, Designs
and Patents Act 1988 to be identified as the author of this work

First published in Great Britain in 2007 by
Chatto & Windus

Pimlico
Random House, 20 Vauxhall Bridge Road,
London SW1V 2SA

www.randomhouse.co.uk

Addresses for companies within The Random House Group Limited can be found at:
www.randomhouse.co.uk/offices.htm

The Random House Group Limited Reg. No. 954009

A CIP catalogue record for this book
is available from the British Library

ISBN 9781845952198

The Random House Group Limited supports The Forest Stewardship
Council® (FSC®), the leading international forest-certification organisation.
Our books carrying the FSC label are printed on FSC®-certified paper.
FSC is the only forest-certification scheme supported by the leading
environmental organisations, including Greenpeace. Our
paper procurement policy can be found at
www.randomhouse.co.uk/environment

MIX
Paper | Supporting
responsible forestry
FSC® C018179

Printed and bound in Great Britain by Clays Ltd, St Ives plc

Contents

List of Illustrations

All supplied courtesy of the Perutz family, who own the copyright, unless otherwise stated

Plate section 1
Max's mother Dely Goldschmidt
Max's father Hugo Perutz
Max with Cilly Jetzfellner
Franz, Max and Lotte
Max aged fifteen
Max with Fritz Eirich and Pussy Gatzenburg
Students in organic chemistry, Vienna 1935 (© Dr Hans Friedmann, photo kindly supplied by Gerhard Pohl)
The Villa Perutz
Gerald Seligman, Philip Bowden and guide on the Jungfraujoch
Max and his mother
Max in 1941 (photo by Ramsey and Muspratt)
Gisela as a young woman
Max and Gisela's wedding in 1942
Gisela on holiday in Switzerland in 1946
Max with Vivien on the Jungfraujoch, 1948
Max with Robin in Austria, 1952
Camping holiday in Devon, mid-1950s
Max in ice cave on the Jungfraujoch, 1938
Max with inclinometer, Jungfraujoch, 1948
Return visit to the Jungfraujoch igloo, 1950 or 1951
Max during the Hirschegg protein workshop, 1968
Max's sister Lotte in Vermont
Family holiday at Feder, 1994
Felix Haurowitz
John Desmond Bernal
David Keilin
William Lawrence Bragg (© the Cavendish Laboratory, Cambridge)

Sketch of Max by Bragg (courtesy of the Laboratory of Molecular Biology)

Plate section 2

Max and his colleagues outside the Hut, late 1950s (courtesy of the Laboratory of Molecular Biology)

X-ray diffraction photograph of oxyhaemoglobin

'Visceral' model of myoglobin

Christmas card to Steffen and Primrose Peiser, 1959

Still from *Eye on Science*, 1960 (© BBC)

Nobel telegram, 1 November 1962

Max, Gisela and John Kendrew celebrate the announcement of the Nobel prize (© Keystone Press Agency)

Max with fellow Nobel prizewinners in Stockholm (© Svenskt Pressfoto/Photoshot)

Nobel medal

Max and Gisela at the Nobel ball (© Svenskt Pressfoto/Photoshot)

Max shows the Queen a model of DNA

Governing Board of the LMB (courtesy of the Laboratory of Molecular Biology)

Fred Sanger, Sydney Brenner and Max, 1980 (courtesy of the Laboratory of Molecular Biology)

The LMB building in 1962 (courtesy of the Laboratory of Molecular Biology)

Max with high-resolution model

Max in his office after his 'retirement'

Don Abraham and Max in the LMB canteen (courtesy of Don Abraham)

Kiyoshi Nagai with the haemoglobin model (courtesy of Kiyoshi Nagai)

Max on a beach in the US

Images from *The Souvenir Book of Crystal Designs*, produced for the Festival of Britain in 1951

Max greets Pope John Paul II, 1983 (© Fotografia Felici)

Max's favourite painting in the National Gallery, *The Agony in the Garden* (© National Gallery, London)

Max lecturing, with model of a protein chain

Plaque outside Max Perutz Lecture Theatre (courtesy of the Laboratory of Molecular Biology)

Preface and acknowledgements

'Many call me a famous scientist,' wrote Max Perutz wistfully to a friend, 'but few know what I am supposed to be famous for.' Such is the lot of scientists, even Nobel prizewinners, whose work lacks the element of public participation that brings automatic celebrity to artists, writers, politicians, actors and sportspeople. Posterity mattered to Max, and when he reached the end of his life without having had time to write his own biography, he called me to his bedside and asked if I would take the job on.

The scientific biographer faces a dilemma. Science is a collective enterprise and to focus on the lives of individual 'great men and women' can be misleading. Max Perutz's 70-year career in science began as others made the first attempts to understand life at the level of individual molecules, and ended as international databases were making the complete readout of the human genome available to all. His own twenty-two-year quest to reveal the structure of haemoglobin, the molecule that transports oxygen and makes blood red, was rewarded with a Nobel prize; yet he would be the first to admit that he could not have done it alone. His story is also the story of an international community of men and women dedicated to uncovering life's secrets.

At the same time the diverse personalities and experiences of individual scientists provide absorbing narratives of creative endeavour, and in that sense scientists are no different from artists, poets or politicians. Accordingly I do not think I need offer any apology for making Max the object of focus. His exile from his native Vienna and adoption of a new identity as an Englishman provide a powerful counterpoint to his efforts to bring a new branch of science into the mainstream. He saw science not as an abstract mystery for an exclusive club of initiates, but as integral to human civilisation and culture, and through his writing sought to share his own passion for enquiry with others. He was all the more engaging for not being a superman, but a human being with the usual quota of human frailties. Understanding Max Perutz may not help you understand protein molecules, but following the life of a scientist certainly tells you a great deal about the day-to-day practice of science.

When I went to see Max in the weeks before he died, to talk about the possibility of writing this book, I mentioned the enormous volume of personal letters and other unpublished papers that I had drawn on for my life of Dorothy Hodgkin some years previously. His face fell. I would not have the same luxury in his case, he said – he had kept hardly anything.

Max, however, reckoned without the conservationist tendencies of his family. In the months after his death, it transpired that his wife Gisela had kept every letter he had ever sent, secreted in unlikely places all around the house. His children and grandchildren likewise had preserved his often beautifully-written letters, full of his own doings but also his fatherly (and grandfatherly) encouragement or sympathy. At the same time, it turned out that Max himself had kept far more than he implied. In particular, he had recovered (after the recipients' deaths) the bulk of the letters he had written to the girlfriend of his teenage years in Vienna, Evelyn Baxter (later Machin), to his sister Lotte Perutz, and to his parents and parents-in-law.

I am indebted to Robin and Vivien Perutz for giving me unrestricted access to all the surviving private correspondence, and for allowing me to quote extensively from them. There are so many more choice passages I could have included, and I am delighted that Cold Spring Harbor Laboratory Press is to publish a selection of Max's letters edited by Vivien.

Max's correspondents included many distinguished figures whose papers have been formally archived, and I thank the following people and institutions for providing access to material and, where appropriate, allowing me to quote from it: the American Philosophical Society Library, Philadelphia (Valerie-Ann Lutz and Horace Freeland Judson himself) for the Judson archive; the Bodleian Library, Oxford (Colin Harris) for the papers of Dorothy Hodgkin, John Kendrew and David Phillips; the British Library Sound Archive; the Cambridge University Library (Adam Perkins) for the Bernal papers; Christie's, New York (Francis Wahlgren), where the Perutz papers now in the Venter Institute archive temporarily rested; the Cold Spring Harbor Laboratory Archives (Ludmila Pollock, Teresa Kruger and James D. Watson) for the James D. Watson Collection; the literary estate of Lord Dacre of Glanton (Blair Worden); the Lilly Library, University of Indiana, Bloomington, Indiana (Becky Cape, and Rachel Maranto for research assistance) for the Haurowitz papers; the Medical Research Council (Philip Toms); the MRC Laboratory of Molecular Biology (Annette Faux); National Archives, Kew, for the Habbakuk papers and MRC archives; Nuffield College, Oxford, for the Cherwell papers; Oregon State University, Special Collections, Ava Helen and Linus Pauling Papers; the Pontifical

Academy (Mgr Marcelo Sanchez-Sorondo); the Rockefeller Foundation Archives, Rockefeller Archive Center, Sleepy Hollow, New York (Charlotte Sturm and Ken Rose); the Royal Institution (Frank James) for the Bragg archives; the Royal Society; the University of Vienna Archives; the Wellcome Library.

Sadly I never had the opportunity to interview Max myself, but have benefited from interviews recorded by others, particularly those held by the British Library Sound Archive and the Imperial War Museum, and the transcripts of the interviews Horace Judson conducted with Max for his book *The Eighth Day of Creation*.

The recently-retired director of the LMB, Richard Henderson, played a crucial role in acting as a go-between when Max first thought of approaching me to take on this task. Henderson and his staff, especially Annette Faux and Michael Fuller, could not have been more accommodating during the book's long gestation. It was a delight to talk to or correspond with a large number of Max's former colleagues, friends and family: Don Abraham, Raymond Appleyard, Uli Arndt, Joyce Baldwin, Sir Hermann Bondi, Sydney Brenner, Andrew Brown, Maurizio Brunori, Christine Carpenter, Robin Carrell, Henry Chadwick, Jean-Pierre Changeux, Anne Corden, Carol Corillon, John Constant, Francis Crick, Tony Crowther, David Blow, Mark Bretscher, David Davies, Guy Dodson, Raymond Dwek, Manfred Eigen, Fritz Eirich, Sir Alan Fersht, John Finch, Michael Fuller, John Galloway, Marie-Alda Gilles-Gonzalez, Andrew Grace, Sir John Gurdon, Freddy Gutfreund, Roger Hanna, Peter Harper, Samir Hasnain, Richard Henderson, Chien Ho, Hugh Huxley, Vernon Ingram, Joy Fordham, Ken Holmes, Lindsay Johnson, Martin Karplus, Olga Kennard, Ann Kennedy, John Kilmartin, Sir Aaron Klug, Blaise Machin, Graeme Mitchison, Kiyoshi Nagai, Robert Olby, Steffen Peiser, Gerda Perutz, Gisela Perutz, Robin and Sue Perutz, Vivien Perutz, Gottfried Peloschek, Gretl Petziwal, Gerhard Pohl, Alex Rich, Matt Ridley, Daniela Rhodes, Michael Rossmann, Fred Sanger, David Sayre, Alan Schechter, Jon Sessler, Robert Shulman, Robert Silvers, Fritz Stern, Alice Frank Stock, Mac Stock, Sir John Meurig Thomas, Marion Turnovszky, James Watson, Meta Werner, Nancy Wexler, Bob Williams. Many others not formally interviewed told me stories about him. Sadly, a number of those in this list have not survived to see the book in print.

A brief effort on my part to learn German proved inadequate to the task of reading the young Max's closely handwritten letters. For assistance with translation I must thank Linde Davidson, Ronald Gray, Barbara Hott, Vivien Perutz, Daria von Esterházy, Alice Frank Stock and Mac Stock.

Marion Turnovszky, the only remaining close relative of Max's still to live in Vienna, gave me a wonderful insight into that city and arranged for me to meet three friends from Max's student days. Gottfried Peloschek, a classmate, took me to see Max's old school, the Theresianum. I am most grateful for their hospitality.

Max's acquaintance was international, and it was thanks to a generous grant from the Alfred P. Sloan Foundation that I was able to visit contacts in North America and continental Europe, and to meet most of the other expenses involved in researching this book. I have been greatly encouraged by the sensitive editing of Jenny Uglow at Chatto: her backing of this project from proposal to publication has been enormously important to me. Both she and my agent Felicity Bryan have been admirably patient.

Robin and Vivien Perutz have painstakingly read drafts at every stage: their input has been always supportive and never intrusive. I am grateful to many of my other interviewees for checking relevant passages, but must particularly mention Aaron Klug, Alice Frank Stock and Guy Dodson, who read large sections. Any faults that remain are entirely my own.

Georgina Ferry
February 2007

I

Scenes from a Vienna childhood

La science n'a pas de patrie, mais le savant doit en avoir une.
(*Science has no fatherland, but the scientist ought to have one.*)

Louis Pasteur

In 1936 Max Perutz arrived in Cambridge from Vienna as a hopeful 22-year-old graduate student, and he never left. He came full of eagerness to ask questions about the secrets of life, and stayed to found the research laboratory that would solve one of the biggest mysteries of all – how DNA, the physical basis of our genetic inheritance, encodes the instructions to make a living organism. His own research into the structure of proteins established the foundations of much of modern biology, and won him the ultimate honour of a Nobel prize. His creation, the Medical Research Council Laboratory of Molecular Biology, has garnered twelve Nobel prizes, a record unsurpassed by any other institution of a comparable size.

Yet, far from claiming to have brought glory to his adopted country and its ancient university, Max considered the gain to be all on his side. 'Had I stayed in my native Austria,' he wrote,

> even if there had been no Hitler, I could never have solved the problem of protein structure, or founded the Laboratory of Molecular Biology which became the envy of the scientific world. I would have lacked the means, I would not have found the outstanding teachers and colleagues, or learned scientific rigour; I would have lacked the stimulus, the role models, the tradition of attacking important problems, however difficult, that Cambridge provided. It was Cambridge that made me, and I am forever grateful.

It was perhaps not so surprising that Max, born in Vienna in 1914 of a Czech father and a Viennese mother, should have ended up British. The elegant apartment where he grew up shared a building with the British Embassy: No. 10 Richardgasse in the fashionable Third District. Opposite its nineteenth-century classical façade stood the Anglican Christ Church, a modest brick affair with a steeply pitched roof but no tower, the only place of worship in the Austro-Hungarian Empire permitted to conduct Church of England services. In addition to this accident of place, Max's father Hugo, the owner of a textile business, had served his apprenticeship in Manchester and was a confirmed Anglophile.

Of course it is romantic nonsense to suggest that these benign influences surrounding his birth and early childhood persuaded Max to renounce his Austrian homeland in favour of Britain. A passion for scientific enquiry first led him to make the move, before the twin evils of racial oppression and war made his choice irreversible. There can be no doubt, however, that when the moment came, Max was readier than many of his fellow refugees to begin a new life as a British subject.

A love of things English may have been in his blood, but science certainly was not. Max Ferdinand Perutz was born into a large family of successful industrialists. His Jewish ancestors came from a small town in Bohemia called Rakonitz (or Rakovnik) 60 km to the west of Prague. There his great-grandfather Josef Perutz had owned a draper's shop, and established a certain position for himself in the society of the town. Josef and his wife Charlotte had twelve children, many more than could be comfortably absorbed into the business. So, showing a streak of enterprise that was to resurface in the family again and again, the two eldest sons, Sigmund and Leopold, decided to head for Prague and establish themselves independently. While still in their twenties they founded Brüder Perutz in 1862 as a small shop in a quiet back street.

The Austro-Hungarian Empire, under its long-lived monarch Franz Josef, had been slow to catch up with the Industrial Revolution, but one of its few centres of industrial development was Bohemia, and its most rapidly developing industry was textiles. Mills producing wool, cotton and linen cloth proliferated in the industrial heartland of the Empire. The Perutz brothers worked hard and lived modestly, and, within a decade of their arrival in Prague, they made the jump from selling to manufacturing. They imported mechanised looms and spinning machines from England, set up factories in Bohemia, and sold their goods throughout Austro-Hungary. By the mid-1930s the company would own five factories, the heart of an international business. The modest shop was rapidly replaced by a succession of increasingly grand offices on the main boulevards of Prague, with subsidiary bureaus in Vienna and Budapest.

Sigmund and Leopold married two sisters, Mathilde and Clara Weiner, and had twelve children between them. Hugo Perutz, Max's father, was Leopold's third son and the tenth of the twelve cousins. There clearly wasn't room for all of them in the offices of Brüder Perutz, even after both Sigmund and Leopold had died in early middle age. While the two eldest sons from each family were groomed to take over the business, Hugo (having first been sent to England for training) was given a sum of money to set up on his own. At the dawn of the new century he bought a partnership in a cotton printing works in Böhmisch Leipa, a town in Sudetenland now known as Ceská Lípa in the Czech Republic. The firm had its head-quarters in Vienna, the imperial capital; as did many up-and-coming industrialists from Bohemia, Hugo took the opportunity to settle there.

In 1904 he met and married Adele Goldschmidt, known as Dely, herself the heiress to a considerable textile fortune. Dely's father Ferdinand Goldschmidt had been married twice; after his first wife ran off with a writer he married Thekla Ehrmann, a much younger woman from Frankfurt, and they had four daughters: Dely, Valerie (Valli), Alice and Anni. Ferdinand died young, leaving his weaving business to Egon, the only remaining child of his first marriage. On the outbreak of the First World War, Egon was called up as a reserve officer in the imperial army. While on his first mission to look for quarters in Hungary he was shot dead by mistake by a nervous border guard whose language he didn't understand.

Dely had adored her dashing half-brother, who was an accomplished rider, and in later years would describe him to her grandchildren as 'a paragon'. However, her sister Valli told a different story: on inheriting the business he had squandered its capital on extravagant purchases, including a stable of forty racehorses and a château for a discarded mistress. Egon's death hit Dely hard, though it was only the latest chapter in a history of tragedies involving her siblings. Another half-brother, Rudolf, fell ill and died in his teens, and a gifted but unhappy half-sister called Else committed suicide. Dely's younger sister Alice also died, in her early twenties, after throwing herself from a window, distraught over a thwarted love affair.

After Egon's death, Dely and her two remaining sisters Valli and Anni inherited equal shares in the factories. From the Perutz family's point of view, Dely's weaving concerns would complement Hugo's printing inter-ests most satisfactorily. For Hugo it was enough that she was beautiful and elegant, and he adored her. Dely wanted to live in the style of an aristo-crat, and Hugo, who was small, dark, quiet, hard-working and with perfect manners – though 'rather ugly' in Max's words – wanted nothing less for her. To run the household they employed five devoted domestic servants, and Dely furnished her home with style. When she came into her inher-itance in 1914 she commissioned a charming chalet-style country house – the Villa Perutz – in Reichenau, about 50 km south of Vienna and close to the mountains. Reichenau had been put on the map when the Emperor Franz-Josef built a hunting lodge there, and much of Viennese society decamped to breathe its mountain air in the summer months. Though she probably didn't realise it, the house was a shrewd investment: the rest of her cash became worthless in the inflation that followed the First World War.

Hugo and Dely's first child, Franz, was born in 1905, their daughter Lotte in 1909. Max, who arrived five years after Lotte in 1914, was very much an afterthought: family legend has it that the pregnancy was unwanted, and that Dely took various measures supposed to bring on a miscarriage. Max survived regardless, making his entry into a world that was about to vanish forever. In the final years of the reign of the Emperor Franz Josef, who had succeeded to the title in 1848, the Empire was a political mosaic of mutually antagonistic interests. There were divisions along national lines, particularly between German speakers who identified culturally with the newly-unified Germany, and the Slavs and Magyars to the east who wanted more autonomy. The social spectrum ran from peasant farmers, through small shopkeepers and artisans, to the increasingly numerous and powerful bourgeois industrialists and professionals, and those who adhered to the old imperial aristocracy. As the government gradually extended the franchise to more and more of these groups, around fifty political parties brokered constantly shifting coalitions. It was too unstable a system to survive the shock that was to come.

When Max was one month old, a Serbian nationalist assassinated Archduke Franz Ferdinand, the heir to the imperial throne, in Sarajevo, and Europe was plunged into four years of war. Austro-Hungary fought on the side of Germany against the British, French and Italians. While little fighting took place on Austrian soil, when defeat finally came in 1918 it left Austria at least as badly off as any of the other combatants. Franz Josef had died at the age of over ninety in 1916; his great-nephew Charles had succeeded him, but was forced to abandon the throne when Austria was declared a republic in 1918. At the Treaty of St Germain the victorious allies dismembered the former empire along more or less nationalist lines: Hungary and Czechoslovakia became independent, leaving Austria as a small country with a shattered economy and few natural resources. As Barbara Jelavich put it in her history of the country: 'The German-Austrian lands consisted of a capital city containing about a third of the population, an industry highly dependent on outside resources ... and a great deal of very beautiful scenery.'

Eighty per cent of the industrial equipment of the former empire was now in Czechoslovakia. The politics of the new republic remained volatile, with socialist, nationalist and national socialist factions struggling for supremacy. Perhaps nothing illustrates this crisis of identity better than the number of times the Perutz family had to change their address, while remaining in the same apartment. Originally Richardgasse after Richard Metternich, the son of the nineteenth-century statesman, their street became Jaurèsgasse in 1919 after Jean Jaurès, the French socialist murdered for opposing the First World War; then switched to Lustig-Preangasse from 1934 after a general in the nationalist militia, the *Heimwehr*; then Richtofengasse under the Nazi regime from 1938. (It reverted to Lustig-Preangasse in 1945, and finally changed back to Jaurèsgasse in 1947, but by that time the building was in ruins and the Perutzes long gone.)

The war took two of Max's uncles: soon after Egon's accidental shooting, Victor, a favourite younger brother of Hugo's, was mown down by Russian machine-gun fire with his cavalry regiment. However, Max's immediate family survived the war and its aftermath relatively unscathed. Hugo, though nearly forty, was called up in 1916 and posted to a mortar company on the Italian front. He returned uninjured and never spoke to Max about his experiences, although 'he once told my sister that he was pleased he never killed anybody'. True, the family must have had to endure the privations of rationing, and witnessed the country's decline from its former opulence to '*ein Trummerfeld*' – a (metaphorical) field of ruins. However, when food became scarce in Vienna – a contemporary of Max's remembers living almost exclusively on turnips from 1917 until 1920 – Dely took the family to live in Reichenau where she could barter clothes for food with the local farmers. Money, of course, was useless in those days of hyperinflation.

The country's economic difficulties hit the middle classes as hard as the workers. Many were left desperately trying to maintain appearances while unable to pay for even basic necessities, as memorably described by the historian Eric Hobsbawn in his autobiography. But while both Dely and Hugo lost all their savings as a result of inflation, the Perutz family businesses pursued a range of inventive strategies to remake their fortunes. Throughout the war, the Hungarian and Czech factories of Brüder Perutz supplied fabrics to the imperial army for uniforms, rucksacks and so on, and to hospitals treating the wounded for uniforms, stretchers and bandages. When supplies of cotton ran out, they made yarn from paper. When, at the war's end, they found their supplies of raw materials exhausted, they turned to importing ladies' fashions from France and Switzerland until they could get the factories going once more. In the face of great difficulties Brüder Perutz maintained an income; Hugo and Dely's firms were able to resume supplying goods to the devastated country, and, within a few years of the war's end, they were living as well as before.

Viennese society in the first decades of the twentieth century was full of contradictions for a family of Jewish origin. Most ambitious, middle-class Jews who moved to Vienna from the east left their religious practice behind in their grandparents' villages. Hugo and Dely were no exception – Max wrote that they 'never entered a synagogue'. Austrian law enshrined a degree of anti-Semitism, in that certain government posts were barred to Jews unless they converted to Catholicism. Yet among the educated middle class in cities such as Vienna and Prague, the distinction between Jew and Gentile meant little. Intermarriage was common, and expedient conversions frequent. When Hugo's cousin Franz died in 1918, for example, his eighteen-year-old son Felix converted to Catholicism at the same time as he took his father's place as a director of Brüder Perutz. He subsequently married Hilde Guszty, the Christian daughter of a former officer in the imperial army, and they brought up their children as Catholics. (Their daughter Marion had so little awareness of her Jewish heritage that, years later, she was baffled to be wished a happy new year in September by Jewish colleagues in New York.)

To remove any risk of discrimination (or so they thought), Hugo and Dely had Max baptised as a Catholic at the age of six, only waiting for the passing of the ageing grandmothers who still retained some attachment to the old religion and might have been upset. The older children, Franz and Lotte, were never baptised – Lotte gave her religion as 'Konfessionslos' (no religion) when she registered at Vienna University. Cilly Jetztfellner, the nanny Max's parents engaged to look after him, was a Bavarian Catholic, under whose gentle influence he 'became a very devout little boy' and began to trot off regularly to Mass and Confession.

Although Jews constituted no more than 12 per cent of the total population of central Vienna, they represented a far higher proportion of the urban middle class. People of Jewish descent made up at least a third and, in some instances, as many as three-quarters of those in liberal professions such as medicine, law and journalism. They were also well represented in commerce and industry. Their sons occupied around a third of the places at the prestigious academic high schools, the *Gymnasien*. They eagerly and easily adopted the cultural traditions of German humanism: from this population of assimilated Jews came many key contributors to literature, music, thought and the arts, including Arthur Schnitzler, Gustav Mahler, Gustav Klimt, Ludwig Wittgenstein and Sigmund Freud.

Hugo and Dely Perutz belonged to this moneyed, educated class, though as business people they tended to be more conservative in their politics than the liberal intelligentsia. They went to the theatre, concerts and the opera occasionally, owned a Steinway that was hardly ever played, and kept a library of books in several languages that they rarely had time to read. Others in the family were more creative: Dely's sister Valerie was a talented pianist, and the noted fantasy author Leo Perutz, a pioneer of what later became known as magic realism, was Hugo's first cousin. Max described his father as humane and cultured, fluent in English, French and Czech as well as German, but Hugo was mostly preoccupied with his work.

Dely, who also spoke good French and reasonable English, concerned herself with establishing a social position for herself and her family. She visited theatres and art galleries in the same spirit as she sought admission to the golf club and attended the Turkish baths: because it was what one did if one moved in the right circles. Once a week she would hold a 'jour' for her friends to come and sit in her elegant drawing room, drink coffee, eat cake and make conversation – though according to her cheeky youngest sister Anni, who infuriated Dely by writing down everything that was said at one of these occasions, the conversation was mostly about servants. She and her husband took a luxurious holiday once a year in St Moritz, while at home she honed her considerable skills as a golfer and tennis player. (One of the few possessions she took with her when she left Austria in 1938 was an elaborate ash tray she won in a golfing tournament.) 'My parents had wide interests which included almost everything except science,' concluded Max.

How, in this humanistic but relatively unintellectual environment, did Max conceive his life-long passion for the science of chemistry? His early education promised little. Before his first birthday he suffered from a severe bout of pneumonia, which recurred twice more before he was five. His earliest memory, he wrote, was 'the shock of being wrapped in cold, wet sheets to bring my fever down'. Although his birth had been unplanned he was never unloved, indeed his mother seems to have treated him with an indulgence denied to either of his older siblings. His childish coughs and colds inevitably resulted in his being kept at home with both mother and nanny in anxious attendance:

Lying in bed I looked forward to the visits of our doctor, the paediatrician Professor Knöpfelmacher, who tickled me with his beard when he listened to my chest and prescribed inhalations of a salty steam spray from a little boiler heated with an evil-smelling spirit flame.

When Max fell dangerously ill once more at the age of nine, Dely took Dr Knöpfelmacher's advice and sent him off with Cilly to the Semmering, a resort on a picturesque mountain pass about 90 km from Vienna. Probably they travelled on the scenic railway, completed in 1854, that leaves Vienna and winds up the mountains and over the pass to Mürzzuschlag, Max with his nose pressed against the window as he enjoyed the vertiginous views. They stayed for several weeks, benefiting from the clean mountain air at 1,000 metres. Max remembered tobogganing, gripped firmly in the arms of his nanny as they whizzed down the slopes. The treatment seemed to do the trick, as he returned in good health and did not suffer any later recurrence of the illness. When he went for a routine chest X-ray as an adult after the Second World War, a telltale shadow on his lungs suggested that he might have survived tuberculosis. He later realised that Knöpfelmacher may have privately made this diagnosis himself, but kept it from Dely to avoid causing her anxiety.

From the age of six, when his health allowed, Max attended a small private school, of which he claimed to have no recollection whatever. As a small and sickly child he found the rough and tumble of playground encounters intimidating, and he made few friends. At home he spent much of his time in the comforting company of his nanny, a circumstance to which Max later attributed his slow intellectual development: 'She wasn't very bright – she was a very good person, kind and [she] tried to keep me tidy, but she didn't stimulate me.'

Max's mother, meanwhile, was indulgent towards him yet remote, a gorgeous creature who would flit in and out of nursery life. A comment he made to an interviewer years later suggests that he yearned for more of her, and his father's, attention: 'If I had been more with my parents I might have been a bit brighter.'

Home life fell into two distinct phases. In Vienna Cilly shepherded the young Max through the city streets to and from school. Judging from his own hazy recollections, primary school was a threatening environment, where the possibility of a kicking in the playground was never far off. His mother made him take piano lessons though he had no interest in the instrument, and he gave up as soon as he 'felt independent enough' to resist her will. She also had him privately tutored in French and English, which he came to enjoy. His English teacher, Miss Rein, was a spinster from Hamburg who had worked as governess to the family of an English aristocrat married to an American heiress. Though she could never give Max a convincing English accent, she inspired him with a love of the language and its literature, and by his mid-teens he was an avid reader of English books.

At home in the spacious third-floor flat in the Jaurèsgasse, Max had the best books and toys that money could buy, but few friends to share them with. His parents were affectionate, but busy and distracted. Yet the strength of the bond between them gave him a sense of security. There were treats, such as visits to the theatre. Dely began his theatrical education when he was eight or nine years old by taking him to see plays by Johann Nestroy and Ferdinand Raimund, broad comedies from the nineteenth-century heyday of the Wiener Volkstheater. Max laughed so much he fell off his seat.

Vienna life, however comfortable and elegant, could never compete in Max's mind with his other life in Reichenau. Every summer Dely and the children travelled to the Villa Perutz, where Hugo would join them at weekends. During the years immediately after the war, when they lived there almost full-time, Max grew to love the house, the surrounding countryside, the greater freedom he was allowed and the opportunity to spend more time in his mother's company: 'When I woke up at six in the morning, during the week when my father was working in dusty, hot Vienna, I tiptoed into my mother's room and climbed into her bed to the comfort of her warmth and some nice stories.'

Villa Perutz had a large 'English garden' full of fruit and flowers. Dely retained a gardener, who lived in his own house in the grounds and provided fruit and vegetables for the kitchen. He had a huge dog called Lux, a cross between a St Bernard and a German Shepherd, that Max recalled with fondness: 'When I was out there, [the dog] became my permanent companion and close friend. I could sit for hours on the porch with my arms around him.' Boy and dog went on expeditions in search of bilberries or mushrooms in the woods; later Max would undertake much more ambitious forays into the surrounding mountains, but from an early age he was allowed plenty of freedom to range beyond the garden fence on his own.

In his written reminiscences, a picture emerges of Max as a solitary and self-contained child, but he was not wholly lacking in company. He was never close to his brother Franz, who left to study in Switzerland when Max was only nine, but his sister Lotte was another matter. Tall, blonde and athletic, she was as physically unlike her younger brother as it was possible to be. At the same time she was intelligent and independent, and perhaps provided the intellectual stimulation he missed from his mother. She was just near enough in age to be a companion to him, and they remained the closest of friends and confidants until her death in 2000.

One of very few first-hand accounts of his childhood by a third party greatly adds to the sketchy recollections left by Max himself, which focus almost exclusively on his blissful Reichenau summers and less than blissful schooldays. Evelyn Baxter was the niece of the Perutzes' near neighbours and close friends, Alfred and Mitzi Teller. Alfred was the architect who had designed the Villa Perutz; Mitzi and Charlotte, Evelyn's mother, came from another wealthy Jewish textile family. Possibly due to tensions in Charlotte's marriage to Evelyn's English father, Evelyn spent long periods staying with the Tellers, sometimes accompanied by her mother. The Tellers' own children, Willy and Lilli, were of an age with Franz and Lotte and spent a lot of time with them 'tearing up and down the street on bicycles, playing table-tennis or just hanging around'. At only ten years old Evelyn was not of much interest to her cousins – but she was thought to be 'a suitable friend' for Max, who was about the same age. She had first noticed the 'pale and delicate little boy' when they had briefly overlapped at the same elementary school.

> I went with him on sedate outings to the Wienerwald and to his flat for tea. Here several sets of double doors were opened for us to lay out rails, points, stations, turntables and signals for a great railway network. The rolling-stock was superb and I remember especially a Red Cross carriage with beds, patients, stretchers and orderlies. I played with Max in perfect accord.

Max doesn't mention Evelyn in his own accounts of his early childhood, although they met again in his teens and remained in touch ever after. The only friend of his own age to warrant a warm recollection was Franz, the son of the Reichenau cobbler, with whom he went bicycling or played croquet when it was fine, or played board games or table tennis when it was wet. 'He was always terribly nice to me,' Max recalled. When he tried to re-establish contact with Franz and his brother after the Second World War he discovered to his grief that both had been killed on the Russian front.

The arrival of the first autumn crocuses was a dreaded portent for Max – it meant that it was nearly time to go back to Vienna and start the school term. When he reached the age of ten, his parents enrolled him in the Theresianum, of the eleven *Gymnasien* in central Vienna the one with the smallest Jewish contingent and the strongest Catholic ethos. (The art historian Ernst Gombrich, five years older than Max, attended the same school – a coincidence they discovered in 1988 when they met at Buckingham Palace to receive the Order of Merit together.) According to Max, his parents chose the school because Knöpfelmacher advised them 'not to

coddle me but to ensure that I got plenty of fresh air and exercise'. The
Theresianum had been founded in 1746 by the Empress Maria Theresa, as
a military academy for the sons of the Austrian aristocracy. The school
occupied the buildings of a seventeenth-century royal palace, La Favorita,
whose interiors had been carefully preserved: the school library occupied
a former throne room, while every surface in the *Peregrinsaal* and the
Goldkabinett were lavishly decorated. It is still operating today, though it
takes four times as many students as in Max's day, and admits girls as well
as boys. After the First World War its survival was in doubt: it had lost
many of its Hungarian pupils, and two of its Moravian estates were now
part of Czechoslovakia. Between 1918 and the mid-1920s the school authorities and the government negotiated over its future, the school wanting to
retain some independence according to its founder's wishes, the Ministry
of Education wanting more control. By the time Max arrived it had become
part of the city's public education system, but it still charged fees and
retained a flavour of its aristocratic heritage.

The students were for the most part a wealthy and international crowd.
The class Max entered in 1924 included the sons of the Swiss and Bulgarian
ambassadors to Vienna, and the Austrian ambassador to Rome. 'There were
barons, counts and princes who cured me of snobbery for life,' wrote Max,
'because they were no brighter than the rest of us.' Half the students were,
like Max, boys from the surrounding districts who attended from 8 a.m.
until 1 p.m. each day, Monday to Saturday. The other half were boarders:
according to Max's classmate Gottfried Peloschek, who came from Lower
Austria, they tended to make friends among their number and did not mix
socially with the day boys.

Soon after Max started he failed a maths test. It turned out that he had
the wrong answers because he had copied the questions wrongly from the
blackboard: his eyesight was too poor to read from it properly. Astonishingly
no one had noticed before, although the evidence was all too obvious.
Max remembered as a small boy rushing joyfully up to a woman who was
walking in his street, thinking it was his mother, only to be overwhelmed
with embarrassment to discover it was a stranger. 'After that I would screw
up my eyes trying to recognise people, which made the other boys brick
and bully me for looking so stupid.'

He also recalled that he could never identify wild birds – they all looked
like sparrows to him. 'If my mother had been with me always she might
have spotted that earlier,' he told an interviewer. Once his teachers identified the problem, he acquired the further indignity (to a schoolboy) of
having to wear glasses.

Max was not stupid or unwilling to learn, but he wanted to do it on his own terms. He did not take to the school environment. The curriculum was similar to that of an English public school. He learned Latin and Greek during his first two years, but in 1927 the school's status changed from a *Gymnasium* to a *Realgymnasium*, and French replaced Greek. Literature, history, scripture (taught separately to Catholics, Protestants and Jews) and mathematics also featured prominently. For most of the eight years he spent there, Max regarded lessons as unremittingly boring, the examinations pointless, and his performance reflected his indifference: Peloschek would obligingly dig him in the ribs if he began to fall asleep in lessons. If he could get away with it, he read a book under the desk – often an English book – rather than listen to the teacher. The regular written tests in Latin or mathematics, to be completed against the clock, induced attacks of diarrhoea 'which cost me time and aroused the teacher's suspicion that I wanted to go out in order to cheat'. On the contrary, Max dreaded the stinking toilets even more than the expectation of his poor marks.

History lessons presented their own problems with such international students. After spending a lesson learning about Austrian victories over the Hungarians, Max heard from his Hungarian classmates that when they learned of the same battles in Hungary, the Hungarians were the victors. He quickly arrived at the conclusion that 'not everything I was taught was necessarily true', a valuable lesson for a future scientist. At the same time he felt the first stirrings of religious scepticism on realising that the Protestant and Catholic scripture classes were each taught that the other was destined for hell fire. Only one teacher won his respect from the start, the Latin master Dr Hrazky. He was also Max's form teacher throughout his school career, and he filled his lessons with excursions into philosophy and ancient history that managed to capture Max's wayward attention.

Small – he never grew taller than 5 feet 6 inches – bespectacled, and lacking either the physique or the inclination for ball games, Max initially made few friends. He became depressed to discover, as he thought, that he had 'no talent for anything, not for any of the academic subjects, not for drawing, music, acting, tennis, riding or football'. Only at the age of fourteen did he win a modicum of welcome attention, when the entire class signed a letter thanking him for gaining them several days' holiday by contracting scarlet fever. By his mid-teens, however, things had begun to improve. After-school and Sunday morning skating parties were a regular feature of the winter months, and Max skated well. His elder brother and sister were enthusiastic skiers, and his mother had introduced him to the sport when he was eleven. Considering his initially delicate constitution, he took to it with surprising ease. The need to keep up with his much older siblings soon turned him into a tough and competitive racer, and he also began to join them on their climbing expeditions in the summer. Physically he was transformed, as Evelyn Baxter immediately noticed when she paid another extended visit to the Tellers, five years after first meeting Max: 'It was difficult to see the delicate little boy in the now sturdy fifteen-year-old, already a keen mountaineer and skier.'

He was also old enough to experience some of the social life of young Vienna. Evelyn recorded an occasion that shows how little the economic turmoil of the time – the Great Crash had happened a year before, and the ensuing crisis lasted until 1937 – affected the lifestyle in the Jaurèsgasse:

> During this stay his parents gave a ball for their two older children and I was invited, to keep Max company. Everything was lavish and splendid: the food rich and imaginatively Viennese and there were wonderful concoctions with birds and flowers, made of ice-cream. Max and I enjoyed ourselves in a half grown-up, half childish way. We danced, ran around greeting the dancers and joking with them, or sat in a corner nibbling and making remarks. The highlight was a cotillion. The young men were called into another room and given posies of flowers and the girls pink muslin bags with drawstrings and our names embroidered on them. In them were ribbon favours to pin on the young men. The band played and the men paraded in and gave their posies to their chosen girls. Of course I got Max's and pinned my favour on his lapel, but I was also given others, and soon danced off with my bag full of flowers. There was a happy family feeling at this ball as all the guests knew each other at school, through their families, or from skiing or mountaineering together. Everyone was lively but not rowdy notwithstanding the good Austrian wine.

It was not just at home that Max was beginning to make an impression. His schoolmates were struck by his skill on the ten-day skiing camps organised by the Socialist administration of the City of Vienna each February. His status soared when at the age of sixteen he was a member of the three-man team that brought home the high school trophy. He modestly attributed their success to the third competitor Walter Innerebner, a 'fearless Tyrolean' who had run the entire course without turns and got down in record time. Nevertheless 'for the first time in my life I was treated with a certain degree of respect'.

Academically things were looking up as well. When Max reached the age of sixteen, the school leavened the diet of classics, scripture and German literature with the first chemistry classes:

> The chemistry teacher also taught at the technical university in Vienna; he was of a calibre quite above the others, and he really interested me in chemistry . . . He also organised practical classes once a week in the afternoon, mixing chemicals, all those lovely colours really fascinated me.

Max was impressed that this teacher's experiments always worked, while the physics teacher's – dreary demonstrations with pendulums – often did not. The chemistry teacher was Dr Arthur Praetorius, who lived long enough to see the sleepy schoolboy he had inspired win the Nobel prize. With the stimulus of this new interest, Max pulled himself together and achieved a reasonably creditable result in his school leaving exam, the *Matura*.

At about the same time, a couple of boys arrived in his class who seemed to have a sympathetic outlook, and they became his lifelong friends. One was René Jaeger, the son of a Swiss diplomat, who shared Max's love of the mountains. The other was Werner Weissel, who joined him in debates on literature and ideas and sat next to him throughout his last two years at school. While Max did not climb regularly with Jaeger until much later in his life, Weissel and his brother, together with another classmate Titi Hoefft, were his constant companions on the skiing and mountaineering trips he took in every holiday from his mid-teens until he left Austria in 1936. Years later Max's daughter Vivien remembered that: '[H]e was more animated with Werner than with anyone else and the conversation would fly from literature, to history, politics, science and medicine, always interspersed with jokes which with my bad German I could never understand.'

Weissel was a key member of a group of boys and girls that Max mixed with, sometimes shyly and a little on the edge of things, as he grew to manhood. Among the girls were the Steyrers, Elisabeth, Gerda and their younger twin sisters, who also had a house in Reichenau. Franz, now returned from Switzerland and working in the family business, seems to have been the main attraction in the Perutz household (years later he made his second marriage to Gerda), but Max enjoyed skating, card games and trips to the cinema with the Steyrer sisters. All his life he had been good at occupying himself on his own, with books, walks, and his new passion, photography: now he was learning how to mix. However his mother's efforts to get him to mingle with the 'right people' were largely unsuccessful:

> In old Austria advancement used to depend upon 'Protektion', by which was meant something similar to the old boy system in England: moving

in the right circles, being known and having the good will of people in high places. To have 'Protektion' as a young man you had to dance well and be good at tennis and bridge. I disappointed my mother by being hopeless at all three of these arts. Neither skiing nor rock climbing got you into high society.

In the spring of 1932, Evelyn Baxter made her final visit to Vienna. She had grown into an intense young woman, destined for Oxford, who spent her private moments writing poetry. Finding herself spurned by a former schoolfellow who had turned into a society beauty, she remembered her childhood playmate:

Very happily for me there was someone whose friendship had stronger roots: Max Perutz was a keen photographer and was in the habit of coming most days after school to use the darkroom which had been arranged in the upper floor of our [i.e. the Tellers'] flat. He always looked for me and soon enlisted me as his assistant and taught me how to develop prints and make enlargements . . . we soon found many interests to share. His true Austrian sense of humour accorded well with my English one and he teased me gently when I was too earnest and took my vocation as a writer too seriously. Our friendship grew naturally and affectionately.

Max invited Evelyn to spend the last weeks of her visit in Reichenau, where they passed idyllic days swimming and walking. He even took her, a complete novice, on an adventurous day-long mountain climbing expedition during which they had to crouch together in a bivouac tent while a sudden summer storm passed overhead. When he came to see her off at the station, he was so reluctant to make his farewells that he failed to get off the train in time and travelled with her to the next small station: 'We laughed as we parted and he ran along the line waving until he was out of sight. I felt sadness at parting with him.'

Max wrote to Evelyn almost as soon as she had left, and they continued to correspond frequently during the following four years. Almost all of Max's letters have survived, though sadly not Evelyn's. On his side, a childhood friendship had grown into first love. The letters are not typical love letters, however: they are mostly long and chatty, sometimes boastful, sometimes self-deprecating, sometimes serious and sometimes boyishly humorous. When she is dilatory about replying, they turn petulant and fretful: Max seems to have wanted desperately to impress Evelyn and keep her attention. In her turn she saw him as a kindred spirit, but not a prospect for romance. Max might have been mortified to know that in a letter to her mother she described him as a 'killing little man' and because of his willingness to develop photographs, a 'useful person to know'. Happily for his biographer, the correspondence did not end when, a little over a year after they parted, he heard that she had become engaged to Max Machin.

> Half an hour ago I was told that you have got engaged; I can't say I'm delighted about it – apparently it is the custom to write to say so in letters of congratulation – but I hope you will believe me that I wish you all the happiness imaginable . . . Of course I won't go in for tragic gestures as I'm not the sort to; I can't write to you that I can't live without you, because I've seen I can. I've realised that it's better to be unhappy in love than not to love at all . . .

While his love life had got off to an unfulfilled start, Max's academic career was showing immediate promise. Having enjoyed Dr Praetorius's lessons so much, he decided to choose chemistry for his degree, taking the advice of one of his father's friends, a 'very clever man' called Dr Strauss:

> He said that there's no point today in doing a programme for 4 or 5 years and that I should study whatever I find most interesting; I think he is right and I will therefore study chemistry because it's really pointless in today's world to rack my brain over what I'm going to do with it when I finish. Mind you I only decided while writing this letter and must still talk it over with my parents.

Chemistry was not part of the parental plan, however. Max's brother Franz had been sent to Zurich to study engineering, to prepare him for a technical role in the family firm; Max was destined to study law, to take over the business side of things. But Fritz Eirich, an older friend from his skiing circles who taught chemistry at the University of Vienna, successfully talked Hugo round. Max entered the university in October 1932, embarking on a seven-year programme in chemistry that should have concluded with a doctorate in 1939.

The *Nationale*, or registration form, that he filled in at the beginning of the first semester shows that he signed up for classes in chemistry, experimental physics, psychology and chemistry practicals. For seven of the eight semesters of his four-year taught course, he took practical classes weekly with Professor Ernst Spaeth, head of the second of the university's two chemistry institutes, which specialised in organic chemistry. Spaeth's research on the synthesis of compounds derived from plants had won him an international reputation, attracting research students from far afield:

from 1929–1931 he had supervised the doctoral work of the black American chemist Percy Julian, who arrived with impressive quantities of glassware and went on to make a number of discoveries in the field after his return to the US. Spaeth was a hard taskmaster but taught with flair, and Max considered himself fortunate in the quality of instruction he received.

The stars of science, in the first three decades of the twentieth century, were almost all in Germany. Since the reunification of the country in the nineteenth century, science had been at the forefront of a confident social and economic development. The new Kaiser Wilhelm Institutes of chemistry, physics and medicine gave the stamp of royal prestige to scientific research. Most of these institutes were in Berlin, which attracted a galaxy of leading physicists including Max Planck, Albert Einstein and Erwin Schroedinger. Vienna had also had its share of luminaries, including the physicist Ludwig Boltzmann and the physiologist Karl Landsteiner, who discovered blood groups. The Austrian universities shared with their German counterparts a tradition of rigorous research that was more often linked to industrial development than in British universities. While the most able Austrian students, such as Schroedinger, tended to gravitate to Germany, the science departments of the University of Vienna undoubtedly offered teaching and research of an international standard.

Max's whole attitude to academic work was transformed by his introduction to university life, and by being able to pursue a subject he found absorbing.

> I who had never worked in my life began to work like a madman. Already on day two I had a 14-hour day. Now I go to the lab at 8, work without a lunchbreak, come home at 6 and work again until 10.

He settled down to his first four years (which had to be passed before students could proceed to doctoral work), taught absolutely by the book – or rather two books. The 759 pages of Karl Hoffman's *Inorganic Chemistry* and the 866 pages of Paul Karrer's *Organic Chemistry* each had to be more or less memorised. He regarded this as a feat of endurance 'which gave me a certain sporting satisfaction, like walking from Land's End to John O'Groat's'. What the course did not provide was any explanation of why matter had the properties that it did. Why was diamond hard and graphite soft, when both are made of carbon? The basic courses in organic and inorganic chemistry did not begin to answer these questions: they were taught as empirical enquiry with very little theoretical background. Students worked through a series of practical experiments, first qualitative and then quantitative, testing samples day after day in the laboratory to isolate their components. Once, when carrying out an analysis of coal for a friend during the vacation (he was promised twenty schillings to do it), Max managed to blow up the technician's laboratory and narrowly escaped blinding when he was rushed to the hospital casualty department to have glass splinters removed from his eyes. It seems to have been an exceptional incident, however: in general he proved competent at lab work and always enjoyed doing it.

From a social point of view he confessed himself to be 'agreeably disappointed' by the chemistry department.

I imagined that the atmosphere would be terribly aggressive and tense. The atmosphere is smelly, but otherwise very friendly. I have never seen anybody fight; Jews, swastikas [Nazis], Christian Socialists and Socialists cooperate as good friends (so far); they wear the same uniform, which is a white lab coat, under which political identity and convictions seem hidden. If you ever hear a mad rumpus it is at most a waterfight in the lab which happens when the technician is absent. The whole environment is most agreeable and informal [gemütlich] . . . I prefer it to school, particularly because you can come and go as you wish from eight in the morning to six in the evening . . . There are all kinds of girls there from small and dark and fat to tall and blonde and slim and pretty.

After his all-boys school, Max greatly enjoyed having female classmates. A photograph of the organic chemistry lab from 1935 shows fourteen male students and five women, not a bad ratio for the times. He shared his practical bench throughout the four years with Gretl Schloegl, who was blonde and very attractive. She remembers that they were good friends: she would sometimes come out to Reichenau at weekends, where he compelled her to climb the nearest mountain with him as he had with Evelyn. 'It was horrible, but I did it,' she says. Clearly Max had yet to learn that forcing his girlfriends to acquire skills he thought essential was not the way to win their hearts. Nevertheless, his friendship with Gretl survived his realisation that it would never be anything more:

Originally I wasn't in love with her but when my friends Wildstrehel and Weissel got passionate about her, jealousy caught me and I fell in love with her. Now I stand here poor idiot and am a function of her moods. And depending on whether I am seeing her today, whether she's been with me or with my friends, I'm either very happy or sad to death. Anyway it crystallised itself and she fell in love with Weissel though she likes me very much but what can I do with that – my usual fate.

Dancing, parties, theatre and cinema certainly featured in Max's off-duty hours, though one friend, Meta Steinschneider, remembered that he was 'too bashful to dance'. (As she was almost six feet tall and very outgoing, it was perhaps not surprising.) Outdoor pursuits, and talking to very close friends, seemed to be his favourite activities. In the summer of 1933, aged just nineteen, he went on a month-long expedition to the Arctic with Werner Weissel, who was training to be a doctor, and a fellow chemistry student, Heini Granichstaedten. By train, bus, hitchhiking and on foot they toured northern Finland and Norway with their tent, knocking at the doors of incurious Lapps to ask for food and water. Finally they took the annual supply ship to the meteorological station on Jan Mayen, a small island 600 km north of Iceland.

Arriving by rocket on the moon could hardly be more sensational than this volcanic island. The earth is coal-black sand out of which small black or red pointed stone slopes rise; the wind draws wild shapes in the sand. Further off there are small mountains which in part have bright red rock walls, in part are overgrown with bright green moss ... Above it all, an extinct volcano of 2300 m rises that is completely covered with glaciers; the glaciers reach right down to the sea. You can't compare the landscape with anything, as it's quite unique in the world.

During this trip Max fell out with Weissel, who he said 'got on his nerves' and was a 'terrible swine' – eventually Weissel went home ahead of the others. The rift was only temporary, however, and their friendship survived. One of the bonds between them was their hatred of the Nazis. In April 1933 Max had written to Evelyn:

You can't imagine how the Hitler psychosis has also hit all young people here; there are only a few exceptions who can still see clearly and can see how insane and absurd this movement is. Wiltschko, my old friend, has regrettably become a member of the SS, only Weissel and Hoefft are still with me.

This is only one of many letters that reveal Max's acute awareness of the social, political and religious divisions in Viennese society. In later life he told interviewers, and even gave his family to understand, that he was 'not terribly interested in politics' as a young man, occupied as he was with skiing or mountain-climbing. On the contrary, he followed what was going on closely, though he gave his own political allegiance neither to right nor to left and regarded both with equal suspicion. After Hitler came to power in Germany on 30 January 1933, Max told Evelyn that he was reading three or four newspapers every day in order to keep up with events, pumped her for news of opinion in England, and kept up a lively stream of comment on the times he and his family were living through:

> My parents are terribly upset about the events in Germany, especially my father. He grew up in a world of material values, while today finan-cial agreements, laws, property and so-called secure positions and posses-sions have lost any kind of permanence; he's disturbed that law and rules no longer exist . . . I never believed in it, I'm not surprised.

Dely's sister Valli was married to a distinguished Bavarian Jewish judge in Munich, August Frank. On Hitler's accession he immediately lost his job, though because of his long service and high rank he was given a pension. Prudently he sent Max's cousins Richard and Alice to schools in England and Switzerland. (He and his wife somehow survived in Munich until three days before war broke out in 1939, when they escaped to England.) The Perutz family knew that many other German Jews were not so lucky. Yet they clung to the hope that the Austrian Chancellor, Engelbert Dollfuss, would prevent the Nazis coming to power in Austria. Dollfuss did indeed try to stifle both Socialists and Nazis by suspending parliament in March 1933 and ruling by decree, while negotiating secretly with Mussolini for help in staying independent of Germany. The situa-tion remained highly volatile: the university had to be closed for a fort-night at the end of May 1933 after Dollfuss ordered the police to break up a brawl between Nazi and Catholic students that erupted at a veterans' commemoration.

In February 1934 a small war broke out between the Social Democratic Workers' Party, who had fortified and armed the huge workers' apartment blocks built at strategic points around the city after the First World War, and government forces including the Nationalist Home Guard (*Heimwehr*). Shooting went on for three days: Max accurately gave the real death toll as 200–300, not the 2,000–3,000 that was widely reported, and expressed sympathy for Dollfuss's position.

> Actually the government here finds itself in a predicament; it has been forced to resort to these measures by the extreme methods of Nazi prop-aganda (bombs, assassinations etc.) and is now challenged by an intol-erable situation (an opposition party arms itself from taxpayers' money against the government). In our country it seems a democracy is impos-sible. That Dollfuss has forfeited English and French sympathy is very distressing of course.

Dollfuss was assassinated by Nazis in July the same year: for once Max heard the news late, as he was away on a climbing expedition with Lotte.

Max's lack of political affiliation did not mean he was without convictions. He was passionately opposed to the advance of fascism, and to the militarism that went with it. His Catholic faith finally evaporated when, despite a night of fervent prayers that they should not do so, the Italians attacked Abyssinia in October 1935 and committed terrible atrocities. Newspaper reports of the Pope blessing the troops as they left put paid to any notion he might have had of papal infallibility.

To what extent did he feel personally threatened? Although the atmosphere in the chemistry lab was relatively apolitical, 'not because of a lack of Nazis, but because of a lack of Jews', the same was not true elsewhere in the University of Vienna. Up to a third of the students were Jewish, and as many as 40 per cent of the faculty by descent (to reach the most senior posts, many academics had to convert to Catholicism). Yet the prevailing ethos among the student body was not that of the liberal bourgeoisie among whom Max had grown up. Anti-Semitism was fashionable, and ugly. Max felt safe among his fellow chemistry students, but not elsewhere: 'Nazi thugs permanently patrolled the main University building; I knew if I were to venture there I would be taking my life in my hands, as they might recognise me by my Jewish looks.'

His fears were not unfounded. In his first term a gang of fifteen youths burst into Lotte's English class at the university and viciously beat up a young Jewish man who had an artificial hand, so that Lotte's papers, three benches away, were splashed with blood. 'At the University there are similar incidents and punch-ups every day,' he told Evelyn. There are ample hints throughout the correspondence that whether for political or economic reasons, he saw his future in Austria as extremely uncertain. As early as 1933 his family had made plans to flee to Prague if Austria fell to the Nazis.

Yet in later life Max always insisted that it was neither the precarious state of his father's business nor the anti-Semitism that drove him to leave Vienna, it was his scientific ambition: he had decided to make chemistry his career. His dutiful efforts to learn Czech and find out more about his father's company during his undergraduate years had only confirmed his resolve to have nothing to do with the family business. He expressed his anxiety about his future with a characteristic combination of arrogance, prescience and self-deprecating humour in a letter he wrote late one night in the autumn of 1933:

My father's business is going very badly . . . I turn over and over in my mind feeble plans to avoid becoming a textile industrialist in a Czech village, having to forget chemistry and then when the factory fails being left with nothing. On the other hand as a chemist I can hardly earn enough to live on before I am 30 years old. Fortunately there is time – one has to consider what will be lost to researching humanity if I do not concentrate on science and instead devote myself to a business career. It would be very sad if the Nobel prize had to be awarded to somebody not worthy of it.

It was a joke, of course, but the old adage about true words spoken in jest holds here: if Max was to be a scientist, then he wanted to be one of the best, and never doubted that he could be. His performance as an undergraduate had attracted attention. Advancement in the University of Vienna was governed by a rigid system of patronage. Ernst Spaeth, head of the organic chemistry institute, whose practical classes Max attended for three hours a day throughout almost all of his undergraduate career, had invited him to be his PhD student, the first step in the long climb up the Germanic academic ladder. However, Max had seen how Spaeth's graduate students worked. They were given instructions each morning, and the great man made his rounds each evening to check that his orders had been carried out. It was all very hierarchical, and not at all what Max wanted.

In the summer of 1935, towards the end of his third year, Max took an introductory course with the protein chemist Friedrich von Wessely, and for the first time came across the idea that the tools of the chemist could be applied to problems in biology. Many of Wessely's examples came from the work of Sir Frederick Gowland Hopkins at Cambridge. Hopkins was the first to hold the chair of biochemistry there, and became the founding director of the Sir William Dunn Institute of Biochemistry, established in 1925. He had trained as a chemist, and came to Cambridge as a young lecturer in 1898 to develop chemical approaches to the study of physiology. The greatest discovery he made, for which he won the Nobel prize for physiology in 1929, was of the essential role of vitamins in the diet. He also made a number of other studies of the chemistry of proteins and specifically of enzymes, the specialist proteins that drive all the chemical reactions in a living organism. Nothing like that was going on in Vienna, and Max set his heart on working with Hopkins – but how could he, a mere undergraduate, get to Cambridge? That summer he poured out his feelings to his climbing friend Fritz Eirich, who had helped to win his father round to the idea of Max studying chemistry.

Eirich was working as a research assistant to Hermann Mark, the distinguished physical chemist and head of the inorganic and physical chemistry institute. Mark was a heroic figure to the students: not only had he played soccer for Austria and been heavily decorated for his valiant action on several fronts during the First World War, but, unlike Spaeth, he was cheerful, informal and even joined the students in skiing parties. He was also a formidable and innovative scientist, specialising in the chemistry of polymers: he had spent his early career in Berlin, the Mecca of physical science at the time, and had also worked in industry, as head of a research lab on synthetic fibres, for IG Farbenindustrie. He returned to the University of Vienna as professor in October 1932, just as Max enrolled as a student: Max attended his lectures from his second year onwards, and found the work a revelation:

> In chemistry you simply get to know what happens, learn the methods; physical chemistry teaches understanding of what happens. It's a wonderful feeling to penetrate ever more deeply the reason for and mechanism of what happens, to see how a theory by a genius allows you to grasp a dozen unexplained events mathematically in one go. In short the six weeks have profoundly satisfied my boundless curiosity about the why and wherefore of things.

As it happened, both Mark and Eirich were planning a trip to England that September, and would be visiting Cambridge together for a conference organised by the Faraday Society. Eirich arranged for Max to meet Mark, so that Max could ask Mark for a recommendation to Hopkins. Mark and Eirich duly set off for England but, so the story goes, Mark had an uncharacteristic lapse of memory and forgot to speak to Hopkins. However, during the same trip they visited the Cavendish Laboratory, hoping to speak to the physicist and crystallographer J.D. Bernal. He was away, but they had a meeting with a lecturer in his department, Peter Wooster, as Eirich remembered: 'Peter Wooster complained how hard it was to get graduate students who could pay for themselves. Putting two and two together, Mark said, "I have a graduate student for you," and that was Max.'

John Desmond Bernal headed the crystallographic lab of the Cavendish Laboratory, the university department of physics then presided over by the atomic physicist Ernest Rutherford. A physicist himself, he had assistants trained in both physics and chemistry, and was using the technique of X-ray crystallography to investigate the three-dimensional structure of biological molecules such as cholesterol, viruses and proteins. It sounded a promising opening, but when Max heard what Mark proposed he was very doubtful. He knew nothing about X-ray crystallography, and precious little about biology. 'Never mind, my boy,' replied Mark cheerfully, 'you'll soon learn', and wrote to Bernal on Max's behalf.

> I would like to ask you to-day, whether you could take in your laboratories a young Viennese student, who wants to make his doctor-dissertation in England. With this work he would like to make [i.e., complete] his degree in Vienna and I would be quite in agreement with his plan ... His name is Max Perutz and he would like to work especially in the field of X-ray-analysis and cristal-structure [sic] ...

Bernal appears to have asked for no more in the way of references, and accepted Max from the following October. The idea of going to Cambridge had so taken root in Max's mind that, with little more by way of introduction on either side, he agreed to go and work with Bernal, even though his desire to work on X-ray analysis was perhaps not as fully developed as Mark had suggested.

For his parents, the decision was a further disappointment. They still saw Max's future in the family business, and were very reluctant to let him leave Vienna for two years to study a subject with no apparent connection to the textile trade. Once more Max brought Fritz Eirich in as mediator. Eirich took the line that an English PhD would be a valuable asset to the business, but Hugo was not easily won over. After a couple of hours of intense discussion, however, he conceded – 'he was a reasonable man', says Eirich – and provided Max with £500, to last for two years, which he sent to his agent in London. He also saw to it that Max was kitted out with the accoutrements of an English gentleman: a bowler hat, white tie and tails for formal dinners, smart grey suit for work and tweed plus-fours for recreation.

Max learned that to be accepted as a graduate student at Cambridge he would have to be admitted to one of its colleges, whose historic and beautiful buildings line the banks of the River Cam and whose towers and courts give the city its distinctive character. So he wrote off to all the best known, including King's, Trinity and St John's. One by one, they all turned him down: Bernal's acceptance proved not to be a strong enough recommendation without further evidence of his ability. But he had made up his mind and was not to be deterred. He decided to go to Cambridge anyway, and sort out the college problem when he arrived. He had enough to think about with his final examination looming: it included a viva voce before his professors, during which he could be asked a question on anything in the syllabus, which essentially included the whole of chemistry, organic and inorganic. One of his classmates emerged from this ordeal distraught – she had failed for not knowing the different forms of silica. Max had time to mug up the answer before going in to his examination, and sure enough he was asked the same question. '[The professor] beamed when I could recite it all', he later recalled. He was duly awarded his *Absolutorum*, which qualified him to stay on another three years and take a PhD. It was, however, a purely internal exam: it did not give him a degree or diploma, the currency normally needed to move from place to place in an academic career. Vienna did not confer bachelor's degrees: as far as the university was concerned, he was simply taking the slightly unorthodox step of going to Cambridge to complete his seven years' work for the *Doctorandum*. He was going, therefore, purely on the strength of Mark's contact with Bernal – a classic example of the Viennese *Protektion* he had affected to despise.

'It was Cambridge that made me'

L'art du chercheur, c'est d'abord de se trouver un bon patron.
(*The art of the researcher is first of all to find himself a good boss.*)

André Lwoff

For Max to leave Vienna when he did could be seen as a wise, even prescient move. In 1936 the *Anschluss*, Hitler's annexation of Austria, was only two years away, yet the complacent Viennese remained reluctant to accept that they too might experience the ruthless expulsion of Jews from cultural life that was already taking place in Germany.

In June 1936, the central figure in the Vienna Circle of philosophers, Moritz Schlick (whose lectures Max had attended), was murdered on the main campus by a student he had failed. Schlick was not Jewish, but he had criticised Nazism in his lectures, and within weeks of his death Nazi sympathisers published articles excusing the murder on the grounds that his philosophy had inflamed the patriotic student. Soon afterwards most of the other members of the Circle left for the US. As early as 1933, Max had feared that the Nazis would come to power in Austria. He always insisted, however, that it was science alone that propelled him to Cambridge. He was not so much fleeing Austria as being drawn, as though by a magnet, to a place where (so he believed) real science was possible.

In a talk he gave in April 1993, Max retold the story of his arrival as a fairy tale:

> In 1936 I left my hometown of Vienna, Austria, for Cambridge, England, to seek the Great Sage. I asked him, 'How can I solve the secret of life?' He replied, 'The secret of life lies in the structure of proteins, and X-ray crystallography is the only way to solve it.'

'Sage', John Desmond Bernal, fictionalised by the author C.P. Snow as the character Constantine in his novel *The Search*, was one of the most charismatic and controversial figures of twentieth-century Britain. He had a life-long influence on Max, as he did on almost everyone he met:

He was the most incredibly magnetic and interesting and brilliant character I'd ever met. In my native Austria I never knew that anybody like that existed. His conversation was tremendous: you could listen to him for hours, he would pour out this continuous stream of original ideas and interesting knowledge such as I'd never seen anybody produce . . . So I counted myself very lucky that I had got that fantastic man as my supervisor.

Born in Ireland in 1901, Bernal discovered Marxism and rejected Catholicism while an undergraduate in physics at Cambridge from 1919 to 1923. He went on to combine original and multidisciplinary scientific research with the development of a radical philosophy of the place of science in society, and of the structure of society more generally. His knowledge of literature, the arts, history and politics was encyclopaedic: his nickname, 'Sage', used by almost all his closest associates, was only minimally ironic. His shockingly unconventional private life, involving serial and simultaneous liaisons with numerous women, was possibly less of an obstacle to his acceptance in Cambridge than his ultra-left-wing politics, but neither handicapped his scientific career to a significant extent. He very rarely completed a piece of scientific research himself, but provided an endless supply of ideas to colleagues who followed them up while their supervisor headed off to sit on yet another international committee of scientists and socialists.

Bernal opened a fresh chapter in biology by using a new tool – X-ray crystallography, which had been developed by physicists and adopted by chemists – to investigate the fundamental components of life. Today it is a commonplace that all biology is really chemistry when you get down to how molecules interact, and indeed that all chemistry is really physics when you get down to the subatomic level. In the 1930s it was a much more radical idea, and when Bernal attempted to found a large, interdisciplinary institute that would apply physics and chemistry to biological problems, he failed to attract financial support. The principle took root, however, notably with Warren Weaver who headed the natural sciences division of the extremely wealthy Rockefeller Foundation in the United States from 1932 to 1955. While he would not commit himself to funding an institute, Weaver, who is often credited with the first use of the term 'molecular biology', proved highly receptive to applications for funds from British scientists for small grants to pursue research along these lines. Almost all of those funded in this way were former students or colleagues of Bernal's.

Since the mid-nineteenth century scientists had used a variety of physical techniques to analyse biological material: spectroscopy, for example, was a method of analysing the chemical content of a substance by looking at the pattern of light and dark bands obtained as the different components of the substance absorbed light at different wavelengths. X-ray crystallography was a more recent technique that promised to reveal not the ingredients of a chemical substance, but the way its component atoms were arranged into a three-dimensional structure. Crudely, it was the difference between knowing on the one hand that a key is made of steel, and on the other that it has a smooth shaft with a ring for grasping at one end, and an intricately-cut plate at the other that matches a particular lock.

Everything, whether solid, liquid or gas, is made of atoms, the smallest possible units of elements such as carbon, oxygen or iron. Most everyday materials are compounds of two or more elements. Compounds are made up of molecules, in which two or more atoms are held together by chemical bonds. The properties of materials depend not just on the elements that compose them, but on the way the atoms of those elements bond together to form three-dimensional structures. As Max would come to learn, structure held the key to many of the physical world's most intriguing secrets:

> Why water boils at 100° [C] and methane at −161°, why blood is red and grass is green, why diamond is hard and wax is soft, why graphite writes on paper and silk is strong, why glaciers flow and iron gets hard when you hammer it, how muscles contract, how sunlight makes plants grow and how living organisms have been able to evolve into ever more complex forms . . . The answers to all these problems have come from structural analysis.

Most of the molecules that are specific to living things contain dozens to thousands of atoms. Each molecule has a particular job to do, from hormones that carry messages in the blood to enzymes that digest food and drive the body's biochemistry. Bernal believed that just as the structure of an eye or a limb was intimately related to the way it worked, the exquisitely-balanced mechanism of the living body would turn out to depend on the three-dimensional shapes of its molecules.

The technique of X-ray crystallography had been invented little more than twenty years previously. The distinguished German professor Max von Laue, then at the University of Munich, was engaged in a debate with other German physicists about the nature of X-rays, discovered by Wilhelm Roentgen in 1895. Were they highly energetic particles, or were they waves of electromagnetic radiation, like light? If they behaved as waves, then like light they should form a diffraction pattern of light and dark bars or spots when passed through suitably narrow slits or holes. The snag was that the wavelength of X-rays was so short that no slits could possibly be fine enough to test the theory. After discussing a problem in the PhD thesis of a student, Paul Ewald, it occurred to von Laue that the spaces between the atoms in a crystal might be just the right size to diffract X-rays.

Crystals grow when atoms or molecules accumulate in a perfectly orderly fashion, repeating the same fundamental element or 'unit cell' in three dimensions just as the flowers on a wallpaper pattern repeat in two dimensions. Von Laue believed that the lattice of atoms in crystals of simple compounds might diffract X-ray beams in the same way that a pattern of fine slits diffracts light. He suggested to his colleagues Walter Friedrich and Paul Knipping that they try the experiment with a crystal of copper sulphate. In 1912 they triumphantly published their result, showing that the beam of X-rays formed a beautiful interference pattern of bright spots on a photographic plate after it passed through the crystal.

A British scientist, William Lawrence Bragg, saw that this result could tell you something about the way that the atoms were arranged in the crystal. Bragg was born in Adelaide in 1890, where his father William Henry Bragg was professor of mathematics and physics. The family returned to the UK in 1909 when W.H. Bragg was appointed to the physics chair at Leeds, and his son went up to Trinity College Cambridge to study physics at the Cavendish Laboratory. When he read about the Munich experiment in 1912, he immediately worked out a simple mathematical rule, now known as Bragg's Law, that opened up the possibility of using crystallography to find the structure of molecules.

Bragg envisaged the atoms arranged in layers in the crystal, like apples in a greengrocer's box. The atoms in each layer would reflect incoming X-rays like a mirror. Some rays would pass through the first few layers and be reflected from deeper ones. If the reflections from different layers were in phase, with the peaks and troughs of their waves aligned, they would create stronger spots on the photographic plate. Whether or not they were in phase would depend on the distance between the layers, the wavelength of the X-rays and their angle of attack. Bragg realised that if you knew the intensity of the spots, the wavelength of the X-rays and the angle of their approach, you could – in principle – work out the positions of the layers of atoms. With his father he went on to demonstrate this relationship in the case of sodium chloride, showing that the cubic crystals were made up of a regular lattice of alternating sodium and chlorine atoms. Even more spectacularly, they showed how the carbon atoms in diamond were arranged, resolving a dispute that went back to the previous century. Three years later, while on active service in France and aged only twenty-five, the younger Bragg heard that he and his father had jointly been awarded the 1915 Nobel prize for physics (von Laue had won the same prize the previous year). He remains the youngest person ever to have received the award.

In 1923 Bragg *père* – now Sir William – became Director of the Royal Institution in London, where he continued to apply the now established technique to a range of simple, inorganic compounds, with just a few atoms in each molecule arranged in a regular fashion. At the same time, Desmond Bernal, who had impressed some (though not all) of his tutors with an extraordinary undergraduate thesis on the theoretical background to crystallography, came to work as Bragg's research student.

Bernal was working on the structure of molecules in relatively simple substances including graphite and metals, but in the same lab William Astbury had begun to turn his attention to the building blocks of living things. In the late 1920s the idea was beginning to circulate that the molecules in living things might have a definite structure, just as salt crystals do. Some biological materials have a structure that is almost as orderly as a crystal. The best examples are natural fibres such as wool and hair. Astbury had managed to get an X-ray pattern from a fibre of alpha-keratin, the main ingredient in hair, which, although blurred, was the first in a biological material to reveal the regular, repeating pattern that betrayed an orderly molecular structure.

In 1927 Bernal moved back from the Royal Institution to the physics department at Cambridge as the first lecturer in structural crystallography. His lab was based in outbuildings attached to the Cavendish Laboratory, renowned all over the world for the work of physicists J.J. Thomson and Ernest Rutherford on the atomic nucleus. The following summer, Bernal took a European tour that included Hermann Mark's lab at IG Farben in Ludwigshafen. He found that Mark had taken successful X-ray photographs of fibres of another natural product, cellulose, the starchy material that makes up plant cell walls. He became more and more taken with the idea that the secrets of life might yield to the penetration of X-rays. On his return, he discovered that Astbury was moving from the Royal Institution to the Textile Physics department at Leeds University. They agreed between them that Astbury would pursue studies of natural fibres such as hair and wool, while Bernal would attempt to study single crystals of compounds that occurred naturally in living things.

Each of the biological molecules Bernal wanted to study was highly complex, with up to several thousand atoms and no clue as to how these were arranged. Since the nineteenth century biochemists had known that if you made a solution of a pure biological compound – urea, for example, which helps to rid the body of nitrogen – it would form crystals just as a salt solution forms crystals as the water evaporates. The characteristic shapes of each compound's crystal were indirect evidence that the molecules that composed it also had a characteristic shape. Any doubters were silenced in 1932 when Bernal ended an international controversy about a family of steroid-based compounds called sterols (including vitamin D and cholesterol) by publishing evidence for their structure based on X-ray diffraction data from single crystals.

In 1934 he took the technique into a different league by obtaining the first X-ray photograph of a single crystal of a protein, the digestive enzyme pepsin. Proteins are the skilled workers of the living organism: there are hundreds of thousands of different ones, each with its own specialised role. They include structural components of the body such as collagen in skin and bone and keratin in hair; hormones such as insulin and growth hormone that carry messages in the blood; and, most intriguing of all, the huge family of enzymes such as lactase or pepsin that act as catalysts to drive every aspect of the body's biochemistry. They are large molecules, with between a thousand and tens of thousands of atoms. (To be clear about the scales involved, however, remember that on average *each cell* in our bodies, itself too small to see, contains around 100 *million* such molecules.)

A protein molecule consists of one or more chains of 150–1,000 small units called amino acids, each made up of atoms of carbon, nitrogen, oxygen and hydrogen (two of them also have an atom of sulphur). There are twenty or so different amino acids: just as a sequence chosen from the twenty-six letters of the alphabet spells a meaningful word or phrase, the order of the amino acids specifies a particular protein. By the 1930s this much was known about the composition of proteins, but important questions remained unanswered. There was no method of reading the *sequence* of amino acids in any individual protein chain, for example, so it was unknown whether this sequence was fixed or random. Furthermore, there was the question of how the chains arranged themselves. They clearly folded themselves into globular shapes, but did each have a unique, three-dimensional structure, or did they just bunch up randomly like a skein of wool? Bernal had a hunch that there was a unique structure, and that this would hold the key to how each protein worked. If he was right, then revealing this structure would be of crucial importance not only for its scientific interest but for applications in medicine and industry. Moreover at the time, many thought that proteins held the key to heredity – the secret of life itself.

Nucleic acids such as DNA (deoxyribonucleic acid), although they had been identified in the nineteenth century, were widely dismissed as too simple to encode the rich variety of living things.

When he began to work with the pepsin crystals, Bernal made a crucial discovery. In order to take a successful X-ray photograph of the crystals, he had to keep them wet: if he allowed them to dry out they lost their crystalline structure and gave no sharp spots. He inserted a tiny crystal into a fine glass tube, along with some of the solution from which it had crystallised, placed the tube in front of the X-ray beam and arranged his photographic film to catch the diffracted X-rays. The results were a crystallographer's dream: when he developed the film, he found a dense pattern of regular spots that revealed an orderly crystal structure. He confirmed the finding with his PhD student Dorothy Crowfoot (later Dorothy Hodgkin), and their 1934 paper on pepsin in the journal *Nature* marks the beginning of protein crystallography. The following year, having returned to Oxford and started her own lab, Crowfoot published X-ray photographs of the protein hormone insulin.

Proteins were becoming one of the hottest topics in science. The pepsin and insulin results were far from solutions of the structures: although no one knew it at the time (perhaps fortunately), protein structures were too complex to solve with the crude X-ray apparatus then available, and without electronic computers to analyse the data. But if one were optimistic enough, solving the structure of proteins could seem a realistic goal, given time and technological advance.

Max arrived in Cambridge in early October 1936 knowing nothing about crystallography and very little about proteins. Nor was Bernal there to greet him when he first turned up at the lab in his new grey suit. Instead he was confronted by three suspicious young men in shabby tweed jackets and creased flannel trousers. One of them, a dark, stocky, pugnacious American, looked him up and down and demanded to know his religion. Max was taken aback: his father had impressed upon him that the British were very reserved and that he must not ask personal questions. He replied cautiously that he was a Catholic. 'Don't you know the Pope's a bloody murderer?' came the response.

> It took me some time to realise that my questioners were all fanatical communists and on the side of the Republicans in the Spanish Civil War, and because of the Pope's support for Franco [he] was equally guilty in their eyes.

The American was Isidore Fankuchen, the atheist son of a Jewish Rabbi in Brooklyn, a physicist who worked as research assistant to Bernal. Left-wing politics and science were deeply intertwined in Bernal's circle: almost the only other scientific group in Cambridge where similar political beliefs predominated was the Biochemistry Laboratory headed by Gowland Hopkins, so discussions about the politics of the left segued naturally into discussions about the molecular basis of life.

No one seemed to be expecting Max, but eventually Peter Wooster appeared and showed him round. Wooster was at that time chairman of the Cambridge Scientists Anti-War Group (CSAWG) and General Secretary of the Association of Scientific Workers. CSAWG had been formed in 1932, but took a much higher profile after the outbreak of the Spanish Civil War in 1936, campaigning for the anti-fascist cause and carrying out experiments on such civil defence questions as how to make rooms gas-proof. The Association of Scientific Workers was a trade union for scientists and technicians. Bernal was a leading light in both organisations. Max quickly accepted the fact that he had landed in a hotbed of socialism, but did not personally become involved. His experience of Austrian politics had left him distrustful of both right and left, and anyway he had come to Cambridge for science, not politics. He was not disappointed. Fankuchen proved that his bark was worse than his bite, kindly teaching him the basics of practical crystallography. Wooster helped to orient him in the wider university, giving him advice about the lectures and courses he should attend (nothing was compulsory for a Cambridge PhD).

Wooster also solved Max's problem of acquiring a college affiliation. When he heard that Max had been turned down by some of them, he said, 'Why not go to my college, Peterhouse? It has the best food.' Founded in 1284, Peterhouse is the oldest college in Cambridge, and one of the smallest: some of its medieval buildings still survive, with elegant additions in a range of architectural styles and periods. In 1936 it had only eleven fellows and just over 100 students. Wooster provided Max with an introduction to the tutor responsible for graduate students, the taciturn mathematician Charles Burkill. Max went to see him, clutching the certificate the University of Vienna had given him to prove he had passed his *Absolutorum*. Possibly unbeknown to Max, Bernal had also written in support:

I realise that this application comes at a very late date but when Mr Perutz first arranged to work in this laboratory it was not then understood that he would have to become a member of the University ...

[Some economy with the truth is evident here.] He has come to me with recommendations from Professor Mark and I had already arranged for him to work in the laboratory . . . I therefore hope that you will be able to take him in the College from the beginning of term.

With no more than this letter and a brief interview, Burkill accepted Max and thereby gave him the formal status he needed as a member of Cambridge University. The College required him to take his meals in the college hall four or five times each week; a college room might have been available but Max regarded it as too expensive and took a modest rented room in a private house, 35 Owlstone Road.

By the end of the first week Max decided he liked Cambridge. Like every visitor, he was bowled over by the beauty of the buildings, and impressed that they were still used for the purpose for which they were built. He was also struck that even though he was a lowly student, everyone treated him with kindness and courtesy, from his academic superiors such as Wooster and Burkill, to the laboratory technicians and college servants. Students in Vienna had been accorded much less respect. The science, although he had hardly begun, promised to be exciting. He told himself that this was the place he wanted to stay.

Cambridge in the 1930s offered a curious study in contrasts. On the one hand, the ancient colleges perpetuated a system of patronage in which acceptance depended critically on one's being 'of the right sort' – having been a Cambridge undergraduate counted for a great deal in this respect. The head of the Cavendish Laboratory Ernest Rutherford, for example, an ebullient physicist from New Zealand, had been unable to get a Cambridge fellowship on completing his PhD despite his obvious ability, but easily obtained chairs at McGill University in Canada and subsequently at Manchester University. Only in 1920 did he return to Cambridge to succeed J.J. Thomson as Cavendish Professor of Physics, by which time he had won a Nobel prize for his work on radioactivity and been knighted.

On the other hand, the research environment fostered by men such as Hopkins and Rutherford, both of whom had risen purely on ability and without any family or old school connections, was highly egalitarian and richly productive. Both offered space to German Jews in the early 1930s, including the future Nobel laureates Hans Krebs and Ernst Chain, and the physicist Max Born. The critical difference between Cambridge and the European institutions the refugee scientists left behind was in the freedom given to young scientists to pursue their own ideas.

This was precisely the reason that Max had made the journey from

Vienna. But what was he to pursue? Rather to his disgust, the first job Bernal gave him to do involved not biological molecules but some 'horrible minerals' – including a metallic compound called iron rhodonite collected from slag heaps. Its crystals were not particularly suitable for a beginner, lacking the symmetry that makes the task of crystal analysis easier, but protein crystals were often difficult to grow, and Bernal had none on hand at that time.

Max had enough to do getting the hang of the equipment, which was extremely primitive. Bernal made him work with an X-ray generator of the type known as a gas tube. Essentially this was a glass cathode ray tube that emitted X-rays at the right wavelength if you connected it to a high tension electricity supply and created a vacuum in the tube. A vacuum pump evacuated the air, and there was a valve you could adjust to let air back in. The trick was to get the balance of the two just right so that the X-rays were neither too 'hard' nor too 'soft', in the parlance of the lab. Max spent his time 'sitting at this beastly tube adjusting the vacuum and the leak to make it run'. There were more modern, sealed Phillips tubes in the lab, but they were allocated to senior workers and Max had to make do. No one mentioned safety: there was nothing to shield users from the X-rays emitted by the tube. 'I must have received a large dose of radiation,' he later remarked casually, 'but it didn't do me any permanent harm.' Much more dangerous, in fact, was the high tension supply, which ran on fine wires across the ceiling of the dilapidated hut where Bernal's group worked. Bernal himself had received a shock that threw him across the room not long before, an experience Max was to survive three years later.

It didn't take long to finish the work on the iron rhodonite crystals. Max wrote up his preliminary findings and published them in the *Mineralogical Magazine*, the journal of the Mineralogical Society. At this stage of his career his enthusiasm rather outweighed his attention to detail:

> I wanted to work like Bernal: I was very slapdash and never checked anything. So I wrote up the work on iron rhodonite and never checked it, and Bernal sent it to the *Mineralogical Magazine* and I got it back from the referees with a whole lot of corrections of elementary arithmetical errors.

Still there was no prospect of an important biological molecule to study, but soon a new project was occupying all Max's time, one that had very little to do with X-ray crystallography. It came about because the table set aside for graduate students in the dining hall at Peterhouse was already full up when Max first went in to lunch. At the next table sat a group of young army officers, Royal Engineers who were sent to Cambridge for two years after their initial military training to broaden their outlook. They had a spare seat, and invited Max to join them, perhaps out of curiosity as much as friendliness. One of the group, John Constant, remembers that Max immediately stood out as 'foreign' in his looks and dress, and foreign students were a comparative rarity in those days. As they got to know him better, they discovered many other ways in which Max differed from the typical Cambridge science student, chiefly for his wide knowledge of literature and the arts, his appreciation of classical music, and his readiness to engage in philosophical discussion. He appeared to have no interest in sport (other than skiing), or in drinking, which were the main preoccupations of the engineers' circle.

Constant and Max found unexpected common ground, because Constant had spent several months in Austria in 1934, between school and military academy, and had witnessed the suppression of both Socialist and Nazi uprisings. They walked and talked in the Botanic Garden, Max leaving his friend in no doubt of his hatred of Hitler and despair at the inevitability of Austria's fall to Nazism. However, it was another of the group, John Carter, who gave Max his next scientific opportunity. Carter told Max of a discovery his father had made in the red cliffs near Budleigh Salterton in South Devon where they lived. Embedded in the Permian clay were roughly spherical nodules a few inches across, surrounded by 'haloes' where the red colour of the clay had been bleached away. Suspecting that the bleaching might be caused by natural radioactivity, G.E.L. Carter cut one of these nodules in half and exposed it to photographic paper for two weeks. To his great excitement, the imprint left no doubt that the nodules were radioactive, the radioactivity being strongest along the boundary between the dark, irregularly-shaped central core and surrounding clay. He also commissioned a chemical study from 'Miss Bennet at the Imperial Institute, London', who found levels of the metal vanadium that were more than a hundred times higher than in the surrounding sedimentary rocks. John Carter thought Max might be able to find the answers to more questions than his father, an amateur scientist retired from the Indian Civil Service, had the resources to pursue. How did the nodules form? What was the source of the radioactivity? So, when the engineers went down to Devon during the vacation for military exercises, he invited Max to visit and collect some of the nodules.

Max began by repeating Carter senior's observation that the cut surface of one of the nodules blackened photographic film. He decided to measure the strength of the radiation and approached the Cavendish's 'formidable' lab steward Mr Lincoln to borrow a Geiger counter. Lincoln, however, refused to give such a precious instrument to a first-year graduate student. Instead, he gave him an instrument called a gold leaf electroscope, which had been used a decade earlier to measure the relative strengths of radio-activity in samples of radium. When Max tried this, he became wildly excited: the rate of decay he recorded did not correspond to any known radioactive element! Could he have discovered a new chemical element in his first year? He rushed to tell Bernal, and Bernal told Rutherford. At last given a Geiger counter to check his result, however, Max was left 'sadly deflated': the rate of decay matched that of radium, which he already knew was there.

Max did not give up: he still wanted to know how the radium and the vanadium had got there. So he set about a series of chemical analyses, but learned nothing new. Carter had suggested that the nodules might contain fossils, as some living organisms included traces of vanadium, but Max found that the 'complex external structure [of the core]with fins and grooves, channels and pustules does not show any resemblance to the structure of a fossil'. At the end of his exhaustive series of studies, he had merely made a more thorough description of the nodules than Carter had managed, rather than developing a new understanding of this curious phenomenon.

Bernal was never less than encouraging, and perhaps led Max to believe his work was more important than it really was. He invited him to prepare an exhibit on the nodules for the forthcoming Royal Society Conversazione, to be held at the beginning of May 1937. The Royal Society in London is Britain's scientific academy, founded on the restoration of the monarchy in 1660. The Conversazione is a formal evening party, open to members and their guests, with an exhibition of the latest scientific discoveries, curiosities or expeditions. Max was thrilled. He had been in Britain less than a year, and already he was exhibiting his work at Burlington House to the most distinguished scientists in the country! He put on his white tie and tails, and invited a 'pretty Austrian girl' (probably Meta Steinschneider) to accompany him. Bernal had warned him that many of the more elderly Fellows might know nothing about radioactivity, and that he should keep his explanations simple. So when a 'dignified elderly gentleman' hove into view, Max embarked on his carefully prepared speech, explaining everything in the most elementary terms possible. His patient listener turned out to be Friedrich Paneth, an Austrian-born chemist (aged just fifty!) who had moved to England from Germany in 1933 and was one of the pioneers of research into radioactivity. Max's euphoria instantly evaporated.

Bernal also encouraged Max to write the work up for publication and undertook to submit it to the *Proceedings of the Royal Society*, a first-rank journal. Clearly Max had still to learn the lesson of the response to his paper on iron rhodonite. Sir Frank Smith, the Secretary of the Royal Society, left Bernal in no doubt that he thought he was wasting the referees' time with Max's paper, and suggested he read it 'very carefully' himself. The anonymous referee had said it was a 'confused account' adding very little to Carter's earlier paper, that it was full of minor errors, and that the author ('an Austrian') needed help with his English. By this time it was October 1937 and Max was busy with other projects, but he corrected the paper and Bernal resubmitted it. The Society responded in April 1938 that the referees were 'not enthusiastic' and that it still contained errors, but they were prepared to accept it if Bernal personally were to revise it again, the author himself appearing to be 'distinctly careless in detail'. Bernal advised Max to withdraw the paper 'now that you have established the fact that it was worth taking', and submit it elsewhere. Meanwhile, he had sent it to Lawrence Bragg: Bragg's response was dismissive, and decisive:

> I feel sure that the paper is not one for the Royal. I was rather disappointed in it; it seemed to me that it was too discursive and superficial. The whole thing has a sort of 'Gilbert White Natural History of Selbourne' flavour. I think the right place for the paper is the *Mineralogical Magazine*, but I expect they would like to see it shortened considerably before they take it.

Bernal did not tell Max of Bragg's opinion until two months later, and then only that he thought it unsuitable for the *Proceedings*. It seems unlikely that Max ever knew the full extent of Bragg's criticisms. At his tutor Charles Burkill's suggestion, he submitted the paper for his college student essay prize in physics, and was jointly declared the winner, pocketing the grand sum of £9. Buoyed by this success, he eventually succeeded in getting the paper published in 1939, in the respectable Leipzig-based *Mineralogische und Petrographische Mitteilungen*. (It still contains some typographical errors.) Years later he bemoaned the fact that the outbreak of war meant that his paper was ignored. Publishing in a German journal at such a juncture (although in English) probably did not help. He retained fond memories of this early piece of research, making it the subject of his speech at an eightieth-birthday dinner in his honour at Peterhouse, with the ironic title 'My First Great Discovery'.

Was the whole nodules study a waste of time? As a contribution to the great body of scientific knowledge, perhaps: but as an exercise in learning how to undertake and present a complex, multidisciplinary piece of research, it had some value. It forced Max to make contact with senior scientists outside Bernal's group in order to obtain equipment and appropriate laboratory space; it got him through the doors of the Royal Society, which must (if such a thing were possible) have increased his ambition to succeed; and it taught him a hard lesson in the combination of original thinking and solid evidence you need to satisfy your scientific peers. Most of all, he found the work enjoyable and absorbing, unlike his efforts with the gas tube and the 'horrible minerals'. He was more convinced than ever that if only he could find the right problem to work on, science was the career he wanted.

Another unexpected opportunity presented itself in the spring of 1937. An independent glaciologist called Gerald Seligman arrived in the lab, looking for someone who not only knew about crystals, but who could also ski. He was planning an expedition to the Jungfraujoch Research Station in Switzerland in the summer of 1938 to study the formation of glaciers. He was offering to cover all costs and include a generous allowance for expenses. Max, with Bernal's blessing, immediately volunteered (he was probably the only person in Cambridge with both qualifications). The damp, lowland atmosphere of Cambridge occasionally got him down, and the chance to spend the following summer up a mountain at no cost to himself was too good to miss. Scarcely able to believe his good fortune in finding someone so well qualified, Seligman hired Max at once. He also hired a physical chemistry student called Tom Hughes, who was to help Max with the glaciology work and also to study the physics of skiing. Now, in addition to his lab work, Max had to attend planning meetings of the Glaciological Society and write a research plan, which again kept him busy and extended his circle of scientific acquaintance.

In July 1937 he went home to Austria for a climbing holiday with Lotte and the old crowd. They went up to the Wilder Kaiser mountains near Kitzbühel, but to Max's intense disappointment heavy rain set in, day after day, and they had to give up and go home. On the train back to Vienna he brooded on his future. He had been in Cambridge a whole year and had learned something about crystallography, but nothing had come up that held any promise as a PhD thesis topic, still less an important piece of work that would launch him on a research career. He was determined to tackle a biological problem, as firmly convinced as ever that chemistry was poised to answer some of the really important questions. But where to begin?

As his parents updated him on the family news, he remembered that his cousin Gina was married to a professor of biochemistry, Felix Haurowitz, at the German University in Prague. He decided to visit him and ask his advice. In accordance with the formalities of Viennese society, he got his father to have a word with Felix while on one of his business trips to the Czech city, before writing himself on 1 August:

> It has emerged during my stay in Cambridge that there is hardly anyone who knows both crystallography and organic chemistry, and that a combination of the two would have good prospects for success. Second, no one has done a crystallographic analysis of a compound from the haemin area, and that such an analysis looks rather promising, if it was possible to find two isomorphous, well crystallised porphyrins. As I don't know anyone in England who even touched those compounds, I would value a meeting with you very much.

A few days later, he boarded the train for Prague.

Felix Haurowitz was Max's senior by eighteen years. A native of Prague, he had trained as a doctor after seeing active service in the Austro-Hungarian army during the First World War, but soon devoted himself to research in physiological chemistry. He corresponded regularly with an international circle of distinguished biochemists and was both widely respected and well informed. Since 1925 he had been working on the physiology of blood, specifically the protein haemoglobin that gives blood its red colour.

Blood: for anyone who wanted to learn the secrets of life, here surely was the place to start. Blood and life have been intimately connected in literature from the earliest times: when Virgil's sybil tells Aeneas that she sees the Tiber 'foaming with much blood', he would not be in any doubt that her prophecy entailed the loss of many lives. As one of the four humours of Hippocrates, blood was a focus of attention for practitioners of the healing arts long before the advent of modern medical science, and we still talk of people having 'sanguine' temperaments. Another even stronger metaphorical tradition linked blood with heredity: 'Am I not of her blood?' demands Sir Toby Belch, desperate to establish his relationship to his aristocratic niece.

By the early twentieth century the true physiological role of blood was clear: blood went hand in hand with breath, that other powerful metaphor for life. Surgeons and scholars since classical times had observed that blood changes colour, from the bright scarlet of arterial blood to the purplish-red of blood in the veins. German chemists in the nineteenth century established that blood carried both oxygen and carbon dioxide, and linked the change in colour to oxygen's uptake and release during respiration. In 1862 Felix Hoppe-Seyler in Tübingen discovered that blood owed its red hue to a single constituent, for which he coined the term haemoglobin in 1864. The name reflected the protein's composition: a combination of the red, iron-bearing haem group with a much larger protein component known as a globin. Soon afterwards George Stokes, Lucasian Professor of Mathematics at Cambridge, completed a series of experiments showing definitively that respiration depended on haemoglobin, and that the iron-containing haem group was critical to its role as an oxygen carrier. Numerous intriguing details quickly followed. Ordinarily iron mixed with oxygen irreversibly changes its state, which is why it oxidises – or rusts – if left unprotected. But each atom of iron in blood carries a molecule (two atoms) of oxygen without undergoing this change: in the body, almost all blood becomes oxygenated without being oxidised.

Max listened eagerly as Felix told him about his own work. Using spectroscopy, he had studied the physical properties of various forms of haemoglobin, and for the first time had isolated and crystallised the protein from fetal blood – which grabs oxygen more avidly than adult blood so that it can extract what it needs from the blood of the mother across the placenta. Felix dismissed Max's proposal that he might set out to solve the molecular structure of haemin, a red pigment that can be extracted from blood once the haemoglobin has broken down. He gently pointed out that Professor Hans Fischer at the Technische Hochschule in Munich had already synthesised haemin (for which he won a Nobel prize in 1930). The chemical formula was therefore known: a crystallographic study would reveal nothing more of interest. Instead Felix suggested a much more challenging project: to find the crystal structure of haemoglobin itself. Max was immediately excited at the prospect. Protein crystallography was in its infancy, and Bernal was its chief prophet. He would surely approve of adding haemoglobin, a molecule of supreme physiological importance, to the very short list of proteins under crystallographic examination.

Haemoglobin had exactly the right combination of interest and mystery to make it seem a perfect choice for a research topic. No one had any idea how it worked: a crystallographic study of the molecule might for the first time reveal how its structure adapted it so perfectly to carry and release oxygen. A combination of youthful optimism and inexperience blinded Max to the difficulties that would stand in the way of success. All he wanted to know was where to get crystals for his experiments. Haemoglobin was relatively easy to crystallise – indeed, in the middle of the nineteenth century it had been the first protein seen in crystalline form. Two American chemists, Edward T. Reichert and Amos P. Brown, published an entire atlas of haemoglobin crystals, 600 photomicrographs from 109 animal species, in 1909. They were trying to make a point about evolutionary relationships: they chose haemoglobin because it was common, in similar but not identical form, to all vertebrates. Haurowitz did not have haemoglobin crystals to hand at the time of Max's visit, but suggested he look up the biochemist Gilbert Adair as soon as he got back to Cambridge.

Max was enthused by his meeting with Haurowitz, but didn't hurry back to Cambridge: after the disappointment of the previous month, there was still some serious climbing to do. He and Lotte squeezed into an ancient Peugeot (*e pur si muove*, he joked in a letter to Evelyn) belonging to Lotte's friend Edi Schaar and rattled from Vienna to Salzburg and up into the mountains where more friends gathered for two weeks of mountain air. He was relaxed and happy: 'We were refreshingly vulgar, a very agreeable contrast to my usual life in England.'

Returning to Vienna he made a point of looking up all the friends he had not seen for a year, including his former lab partner Gretl Schloegl who was about to get married.

We had only 4 hours to spend together & had great difficulties in telling each other everything we wanted to – I had to increase the speed of my speech by a factor of three & we talked all at once most of the time. We were so frightfully pleased to see each other again & to realise that we had not become strangers during all the long time, which had seemed a century to me, after we had spent three years almost continuously together & I nearly fell in love with her all over again ... & I met the three boys, Weissel, Granichstaedten & Hoefft & compared them with the many people I had met during the past year & I thought they were nicer than most of them ... And in general I remembered the words of a Dutchman who had returned to Holland ... after having seen half the world: 'And then I realised that Holland was the most beautiful country in the world.'

He had to leave Vienna sooner than he wished, as he had found on his return from the mountains a summons from Seligman requiring his presence on the Jungfraujoch on 30 August for a short planning trip. 'I left Vienna on the evening of the 29th', he wrote to Evelyn, 'and I felt more homesick for the next 24 hours than I had ever been before in my life.' It is a letter of extraordinary poignancy – Max would not return to Austria until 1949, by which time his friends and family would be scattered and Vienna devastated by war.

Leaving his heart in Austria, he took his head back to Cambridge and shyly announced that he now had a subject for his thesis, which he later remembered produced a discouraging reaction.

My fellow students regarded me with a pitying smile. The most complex organic substance whose structure had yet been determined by X-ray analysis was the molecule of the dye phthalocyanin, which contains 58 atoms. How could I hope to locate the thousands of atoms in a molecule of haemoglobin?

With Bernal's enthusiastic backing he felt able to ignore the doubters. He believed it would not be proper to approach Gilbert Adair for crystals without a formal introduction. Adair was 'a terribly modest man, a Quaker, an endearing character but very diffident' in Max's words. So, with Bernal as an intermediary, he arranged for Gowland Hopkins's daughter Barbara Holmes to invite both him and Adair to a lunch party. Introductions satisfactorily effected, Max lost no time in putting his request. Whether Adair ever knew of his ulterior motive is not recorded, but a few weeks later, he appeared unannounced in the crystallography lab with some tubes of haemoglobin crystals. Apologising for their small size – about half a millimetre across – he said they'd probably be no use: but Max looked at them under a microscope and saw that they were extremely promising.

They were crystals of methaemoglobin, a form of the protein that develops if blood is left exposed to the air. In this form the iron oxidises irreversibly, but such brownish-red crystals are more stable than oxyhaemoglobin and sufficiently similar to be useful. Adair had extracted his haemoglobin from horse blood, then available 'by the gallon' from horse butchers in Cambridge. Max set about learning the technique of crystal preparation, but found it frustrating working alongside Adair in his laboratory because of his 'eccentric habits': 'He never washed any of his glassware nor let anyone else touch it until he actually needed it so that all his benches were as cluttered as in an alchemist's shop.'

Fortunately, the following year the Quick Professor of Biology David Keilin offered him bench space in his 'clean, tidy and warm' lab in the Molteno Institute of Biology and Parasitology, a small Cambridge University research institute. Max worked here gratefully for more than a decade until his growing research group finally acquired space of its own for biochemistry. Keilin, born in Moscow and brought up in Warsaw, was a modest and kindly biochemist and a natural teacher who became something of a father figure to Max. While Bernal drew admiration, Keilin inspired warm and genuine affection. His tidy and methodical approach, rather than Bernal's slapdash brilliance, ultimately provided the model for Max's development as a scientist.

Max's summer in Vienna and Prague had kept him away from the lab for more than two months, and on his return he found he had a lot of catching up to do. It took a while to get back up to speed. 'The damn crystals are not growing, the usual misery of the crystallographer,' he complained in a letter to Evelyn Machin in September. Haemoglobin, it seemed, would have to wait. In November he told Evelyn he was doing 'a strictly crystallographic work' with Peter Wooster, in order to learn 'all the theory I should have learned last year'. There was a large body of theory that underpinned the analysis of crystal structures: to many people the term 'crystallographer' means someone who works on developing this theory, rather than just using the technique. It requires a solid background in maths and physics, which Max well knew were his weak points. Wooster was doing him a favour in forcing him to do the theoretical work, but he complained that it was keeping him from his new project: 'I am supposed to do another work on haemoglobin with Bernal, but I have not got any time to do it.'

Apart from work in the lab, he had several other calls on his time. Planning for the Jungfraujoch expedition the following year entailed making high-level presentations on glaciology:

> Just imagine me standing in front of 4 university professors ... and talking about a subject which I scarcely understand myself. I put in as many jokes as possible and they did even condescend to laugh at them.

At the same time he had been drawn into the activities of the Cambridge Scientists Anti-War Group to the extent of translating German documents for them, in support of their campaign on the inadequacies of government advice on civil defence. He told Evelyn that two of his translations had been given to the Labour leader Clement Attlee the week before, adding 'I earned 7/3' [about 36p today, but the sum represented about half of his daily budget at the time]. With blatantly false modesty he told her, 'In spite of all the compliments which people make in my presence ... my English is nothing like perfect,' and added that he had recommended her to Wooster for further translation work as he had no time to do any more. It seems clear that he did the translations to help his friends and perhaps because of the financial incentive, but remained detached from the passionate political commitment of most of his fellow scientists.

Girls were another preoccupation, and Evelyn evidently encouraged him to tell all:

> I started a love affair with the lady who prints my photographs and who is most amusing and intelligent and pleasant. When I found out that my predecessors included Bernal and all the most brilliant scientists I was rather bucked at first about the good company I had joined but later I thought the company was rather too numerous and decided to withdraw gracefully. However, it was highly entertaining and I learnt more about the soul of Cambridge in an evening than otherwise in a year . . .

Soon afterwards he had a visit from Meta Steinschneider, one of his circle of Vienna friends, who stayed in Cambridge for a few weeks. Meta was tall, vivacious and intelligent and 'her presence helped me to dozens of invitations', but there was no more to their relationship. It was around this time that he met Anne Hartridge, daughter of Hamilton Hartridge, Professor of Physiology at St Bartholomew's Hospital in London. Anne worked as a technician for the professor of medicine at the same hospital, who was conducting research into diseases of the blood. She was friendly with Evelyn and her mother, and often visited the family home in Guildford; she had also spent several summer holidays travelling in Austria with her father.

> I suppose Max and I had a basis for friendship – I already loved his country and feared for its future, and he could understand the research into blood diseases at which I worked. Soon I grew to admire Max's exceptional qualities, intelligence, courage, kindness and humour.

Max found a welcome at the Hartridge home in Northwood, and he and Anne saw each other regularly for outings in London. Over the next two years he began to entertain the hope that they might one day marry.

In the autumn of 1937 the Cavendish laboratory was in a state of uncertainty. Its illustrious head Lord Rutherford had died in office in mid-October: most of the crystallographers, including Max, turned out for his funeral in Westminster Abbey. Max had worked in the Cavendish for a whole year without ever meeting him, something he told Evelyn he now regretted:

> It would have been quite easy to get an invitation to one of his house parties for his research students, but I thought to postpone it until I became 'somebody' . . . And somehow you never realise that someone

is in the same league with Newton if you see him passing your window every other day and hear him telling little stories in his lectures.

Becoming 'somebody' was clearly an object that Max was not embarrassed to admit to his oldest and closest friend.

> There is always plenty of excitement on the road to success in science . . . The profound wish to achieve something spectacular, to remain with science for the rest of my life and to remain in England provide plenty of thrill. Of course the atmosphere in the Crystal Lab is more friendly than in most of the labs in the world. There, at least, I have not got to fight my way through a hostile camp, but just got to make use of all the opportunities offered to me. Bernal is as friendly and encouraging as ever . . .

For all his homesickness when he left Vienna in August, this letter to Evelyn written at the beginning of November 1937 indicates that Max had already turned his back on his native country. His commitment to science was absolute. Science provided him with a set of ideals, an ambition to succeed, and, most of all, a circle of friends and colleagues who shared his passion. And science and Cambridge, in Max's mind, had become inextricably linked: he would never, in his long scientific career, entertain the possibility of voluntarily moving anywhere else.

Bernal, however, had formed no such association. His experience of Cambridge had taught him that, whatever his brilliance, his unconventional lifestyle and passionate political beliefs would keep the doors to higher office there firmly closed. A much more congenial opportunity had arisen: the left-wing physicist Patrick Blackett was moving from Birkbeck College in London to succeed Lawrence Bragg as professor of physics at the University of Manchester, and Bernal had accepted the offer to become *his* successor as head of physics at Birkbeck. Birkbeck is a college of London University that specialises in degree-level evening courses for adults; its egalitarian and aspirational philosophy suited Bernal perfectly, but also, more to the point, under Blackett the physics department had gained considerable standing in research. Completing the circle, Lawrence Bragg (who had previously succeeded Rutherford to the Manchester chair) was coming to Cambridge as Cavendish Professor in the autumn of 1938. Fankuchen was to go to London with Bernal, but Max decided to stay in Cambridge to complete his PhD. How he would manage after his father's £500 ran out, he had no idea.

Bernal was already spending a lot of time in London, where he stayed in Hampstead with Margaret Gardiner, who had given birth to his son Martin in March 1937. Towards the end of that year, he fell ill with jaundice and stayed in London for several weeks. It was during this period that Max finally got down to work on his new haemoglobin project, and he reported his progress in a series of letters to Bernal. However 'horrible' he found the minerals he had worked on in his first year, they had given him a good basic training in the practical side of crystallography. He had always been happiest with hands-on bench work rather than wrestling with the theoretical parts of the subject. The first haemoglobin crystal he tried, a flake teased from a mass grown painstakingly over several weeks, was disappointing, but soon he was able to report better news: 'I found three good haemoglobin crystals of perfect shape, not twinned, on Saturday . . . The crystals are not flakes but thick plates and so far are behaving very well.'

As Bernal had taught him, he had mounted one of his tiny haemoglobin crystals, kept wet in a fine glass capillary, placed it in the X-ray beam and taken a series of photographs along each axis of the crystal. Perhaps surprisingly, given his later delight in telling the story of his career, he never recorded how he felt the moment he saw his first haemoglobin pictures. After Dorothy Crowfoot had developed her first insulin photograph in early 1935, she wandered the streets of Oxford in a daze in the small hours of the morning, until advised to go home by a kindly policeman. Dorothy Hodgkin (as she became) later described the moment as the most exciting of her life – which had included winning the Nobel prize. Max must have been similarly euphoric. Good protein X-ray diffraction pictures were still extremely rare commodities. His were superb: Hodgkin later wrote of them as 'the most beautiful protein X-ray photographs yet seen'.

Beauty, in the eye of a crystallographer, lies in an image that consists of a regular array of sharp spots, which confirms that the atoms in the crystal are arranged in a regular, repeating pattern. Most beautiful of all are the images that have spots right out to the edges, far from the central blank where the X-ray beam is shielded to prevent it from obscuring the much weaker diffracted beams. These outlying reflections indicate that not only are the molecules arranged according to a regular pattern, but that internally each molecule is identical to the others. It means that the photograph has detected repeating layers of atoms down to 2 Å (2 angstroms, 0.2 nanometres, or two ten-millionths of a millimetre) apart, a separation characteristic of neighbouring atoms in a molecule. In principle, therefore, the photograph holds the key to the three-dimensional arrangement of the atoms.

In practice there is a complication, known as the 'phase problem'.

Electromagnetic waves, such as light or X-rays, have peaks and troughs, just like waves in the sea. In order to calculate the position of an atom, you need to know the position in this cycle, known as the phase, of all the diffracted rays that are contributing to each spot in the diffraction pattern. With simple molecules it is possible to make an educated guess at the phases, and try various alternatives until you hit the one that matches the actual pattern (the 'trial-and-error' method). But with large, complex molecules such as haemoglobin, there are far too many alternatives. To solve the phase problem was the cherished dream of protein crystallographers, still very far from realisation when Max took his first haemoglobin photographs.

Max did admit to showing them proudly to anyone who would stand still long enough, only to retreat in confusion when asked what they meant. Over the next few weeks, reporting regularly to Bernal and with further tutelage from Fankuchen, he painstakingly made the measurements that would begin to answer his questions about the structure of the haemoglobin molecule. There are three main types of data in an X-ray photograph: the positions of the spots, their intensity, and the symmetry of the overall pattern. From these data, Max was able to calculate the size of the unit cell – the smallest repeating element of the crystal structure – and to infer that it contained two molecules of haemoglobin. The main new piece of information he gleaned was from the symmetry. The diffraction data allowed him to conclude that the molecule had a two-fold axis of symmetry: in other words, it was made of two identical halves. 'This was the first structural information that X-ray crystallography had given about a protein, and I was proud of it,' wrote Max later.

In what must have been a month of intense activity in the lab, he and Fankuchen worked simultaneously on another protein, the digestive enzyme chymotrypsin. Enzymes are of enormous interest to biochemists, as they are the body's catalysts: without them none of the chemical reactions that keep us alive would take place. Only in the previous decade had American chemists, including John Northrop at Princeton, succeeded in purifying enzymes sufficiently for them to crystallise. Northrop sent the chymotrypsin to Bernal, whose triumph with pepsin in 1934 had made him world famous among protein chemists. Max found that wet chymotrypsin also produced 'perfectly definite reflections at spacings as low as 2 Å'. For comparison he and Fankuchen also photographed the chymotrypsin and haemoglobin crystals in the dry state. Drying altered the size and shape of the unit cells, but the most obvious difference was that the elusive reflections from low spacings disappeared: removing water did not just mean that the molecules packed more closely together, it destroyed the perfect regularity of their internal structure.

The paper describing their results came out in *Nature* in March 1938, paired with a paper from Dorothy Crowfoot and Dennis Riley in Oxford reporting successful X-ray photographs of wet crystals of the milk protein lactoglobulin. It was a collective triumph for Bernal and his extended circle, establishing that the pepsin photographs of four years previously had been no fluke, and giving the subject of protein crystallography a new momentum.

Max had little heart for celebration. A week earlier a colleague had passed his window while he worked in the crystallography lab, and told him that Hitler had annexed Austria. He later wrote:

> There are certain events, like the death of a loved person, which you dread so much that you cannot contemplate their happening, even if all the evidence tells you that they will. The threat to Austria, which I loved, had not been uppermost in my mind when I decided to come to Britain; I came for scientific reasons, but I should have taken the threat more seriously.

He was no longer a visiting foreign national: he was a refugee, like so many others. He ran into his friend Fritz Eirich in London soon afterwards, looking for jobs for himself and Hermann Mark. Mark was sacked for his Jewish heritage, but had also been a close associate of Dollfuss and his successor Schuschnigg; he was arrested and briefly imprisoned until he managed to bribe an official in the new regime to release him. Eirich escaped to England on the pretext of going to a scientific meeting (taking his wife and most of his belongings); Mark and his wife and children also arrived in London via a 'skiing holiday' in Switzerland, having converted his wealth into platinum wire which he disguised as coat-hangers. When Eirich saw Max he told him he was lucky – he was already settled in England. 'Probably I am,' Max wrote to Evelyn, but continued: 'I feel as if something had died within me and I find it almost difficult [sic] to write a moderately cheerful or coherent letter.'

The *Anschluss* placed his parents, brother and sister, plus countless other more distant relatives and friends, in grave danger. Because the British newspapers had been predicting it for a week, Max had written home urging his parents to leave Vienna, but not until 12 March, when the German army crossed into Austria – greeted by ecstatically cheering crowds – did they make a move. Fortunately Max's brother Franz, who was based at the Prague office of the family firm, happened to be visiting and had a car with Czech number plates. They immediately activated the plan devised in 1933: all four of them piled into the car, taking little more than they stood up in, and drove straight for Prague without being stopped. In the days that followed tens of thousands of Austrians were arrested and imprisoned.

Max himself was in no such physical danger, but his future had become highly uncertain. He had only enough money to last a few more months, and it now seemed unlikely that he would get more from his father. As Bernal had already left Cambridge, it would be up to Lawrence Bragg to decree whether or not he was able to stay in the Cavendish, and Max entreated Bernal to intercede on his behalf. Wooster had suggested that the University might create a job for him by making him an assistant 'who acts as a connecting officer between the biological and crystallographic departments', and thought the heads of all the relevant departments might be able to propose the idea. By the middle of April it became obvious that if Bernal had talked to Bragg about this, he had been spectacularly unsuccessful.

I must apologise [Max wrote to Bernal] for disturbing your 'peaceful' existence in London again with my troubles, but I think I am in rather bad need for your help and advice. Wooster asked Bragg on Sunday whether he had given any thought to the problem Perutz in the meantime. The answer was not only negative, but Bragg definitely declined to be interested at all in the matter as in all personal staff matters and told Wooster to talk to Appleton [the acting head of the Cavendish] about it. Appleton, of course, knows nothing about me nor about the structure of proteins and said that the University does not, as a rule, appoint officers who have not yet reached the degree of a PhD . . .

Appleton suggested that Max should apply for help to the Society for the Protection of Science and Learning (until 1936 known as the Academic Assistance Council), a body that had been set up by the British academic aristocracy in 1933 to provide short-term grants and find posts in Britain or the US for refugee scholars from Germany. Max at first impatiently dismissed this idea, and instead urged Bernal to keep trying with Bragg.

Firstly my chances of getting a grant from them seems [sic] very slight in view of the recent rush of refugees to this country. Secondly, grants are not given in order to finance important research, but in order to relieve scientists of distinction from immediate distress. I am not a scientist of distinction and my distress is not immediate but only imminent.

Do you think you could have another talk with Bragg if there is an opportunity and tell him all the fame his laboratory will achieve, if biological work is continued there?

Bernal replied soothingly that Bragg had given him an undertaking that Max would be 'looked after'. Bragg did write a letter of recommendation to SPSL on Max's behalf, as did Bernal, in the expectation that he would swallow his doubts and apply for a grant. Fortunately, it was not until after this that Bernal sent Bragg Max's paper on the radioactive nodules. The disdain it provoked can have done nothing to help Max's case. Max applied unsuccessfully for a Royal Society fellowship, but Peterhouse gave him a grant of £30. What saved him financially in those uncertain months was Seligman's glaciological expedition (see Chapter 5): all Max's costs were more than covered from the beginning of June until the middle of September.

His parents remained a worry. They stayed in Prague for six months. Max met Franz in Paris on his way to Switzerland, a highly unsatisfactory meeting from his point of view.

That my own brother should have become such a miserable hypocrite, dripping of vanity, more tactless than any American and a criminal fool altogether, that was rather a blow to me. He told me all the unpleasant facts he could possibly think of . . .

Max did not elaborate on the nature of these 'unpleasant facts', but they probably related to the family fortunes, which had been in decline for some time. Hugo's firm was virtually bankrupt before the *Anschluss* took place, and Dely had infuriated her sisters by trying to cover up the problems with under-the-counter exchanges between their factory and his. Max had never liked his brother, but his outburst may have been a reaction to being forced to look at his adored parents in this unwelcome light. More happily, he and Lotte were briefly able to forget their predicament when she joined him in the mountains for a fortnight's skiing and climbing. They talked about her future: he urged her to join him in Cambridge, but they decided that, if she hadn't managed to find a job within three months, she would go on to New York. His mother also visited, taking a plane from Prague to Zurich to avoid passing through Nazi territory and meeting Max at a hotel in Wengen, where she described the Anschluss and their flight to Czechoslovakia. She seemed 'relaxed and happy'; it was the last time he was to see her in such a buoyant mood.

Max was appalled to discover that his father and Lotte had actually returned to Vienna during this period to collect more of their belongings and to place their furniture in storage. Dely evidently regarded their difficulties as purely temporary and expected to return to her former lifestyle in due course. She was firmly disabused of this belief in September 1938, when Neville Chamberlain's Munich agreement with Hitler ceded the Sudetenland to Germany, and with it the factories belonging to Hugo and to Dely and her sisters. They had now lost not only their home but also their business. Franz at once left Europe for the US: he had spent a year there in 1928 and had a visa and contacts in New York. Lotte decided against joining Max in Cambridge; she too considered that her best hope lay in America. She was initially to work as a care assistant in a children's home in Brooklyn (later she qualified as a social worker in New Orleans before pursuing a career at Yale). She boarded Holland America Lines' luxurious *Statendam III* to make the transatlantic crossing, alerting Max that it would make a stop at Southampton. He went there to bid her farewell, an occasion he later described as the saddest of his life. Max had not been taken in by the British media's euphoria over the Munich agreement, and thought it highly likely that Britain would soon be at war. He faced the real possibility that, at the time he most needed a friend and confidante, he was losing his beloved sister for good.

Hugo and Dely went to Zurich, where they had a little money saved on which they could survive for a while, but their position was precarious. They had only a temporary residence permit, and there was no guarantee that, were war to break out, the Swiss authorities would not return them to Austria. Max immediately tried to persuade them to come to England, but the problem of how all three of them were to live initially seemed insurmountable. Never had they been so much in need of *Protektion*.

By the time he returned to Cambridge as Cavendish Professor in 1938, though still under fifty, Lawrence Bragg had acquired the status of a grand old man of British science. His research interests were now mainly in the structure of solids, particularly metals. It was his father Sir William Bragg who had first inspired Bernal and Astbury to extend the field of X-ray crystallography into biology. Nevertheless, Max kept hoping that his new professor would visit the crystallography laboratory and look at what he had been doing. Six weeks went by without a sign of the great man. Eventually, Max could stand it no longer, and overcame his reluctance to approach a senior figure without prior introduction. He collected up some of his best X-ray photographs, left the mineralogy lab and went up to Bragg's imposing office in the main laboratory building in Free School

Lane. Encouraged by the moustached and balding Bragg's benign air, he took out the photographs and showed them to him. Immediately, he recalled, Bragg's face lit up with enthusiasm, and he began to ask about his future plans. This was Max's cue: he could not make plans, he was quite without funds, his parents were stranded in Switzerland . . . But Bragg was already hooked. 'Our fancy was fired,' he told the author Horace Judson, by Max's photographs. 'He wanted to go on, and I was all for his going on.'

The Cavendish had a reputation for aloofness from other departments, but this was not Bragg's style. No biologist himself, he sounded out the local leaders in the biological field as to the best means of keeping the haemoglobin work going. David Keilin at the Molteno Institute was a key ally. His research into the biochemistry of respiratory enzymes received research funding from the Rockefeller Foundation in New York, benefiting from Warren Weaver's enthusiasm for the application of physics and chemistry to biology. Bragg decided to apply to the Rockefeller for funds for a post for Max. He had a preliminary meeting in London on 23 November with Walter Tisdale, the Rockefeller's representative in Europe. Tisdale was very impressed with Bragg's willingness to collaborate with other departments, but told him 'pointedly' that 'a proposal to aid Perutz would not be warmly received'. Bragg immediately assured him that his motivation was not to help Max but to provide himself with a 'capable technician' in order that he might himself develop the field of protein structure. After the meeting Tisdale noted in his diary that a 'smallish' grant to bring X-ray work and biology together could be the start of something bigger. He felt that '. . . eventually this scientific center of the British Empire will follow in at least some respects those cooperations which in other centers are proving of value.'

Less than a week later, Bragg sent off an application to the Rockefeller's Paris office for a research project on the structure of haemoglobin, to be headed jointly by himself and David Keilin, the only costs of which were to be £275 per year to pay for an assistant, £100 for a new Phillips sealed X-ray tube and a small sum for incidental research expenses. The Foundation, whose files reveal an organisation dedicated to giving the maximum encouragement to scientists with the minimum of bureaucratic fuss, agreed almost immediately: Max was to start his new job on 1 January 1939. The Rockefeller documents confirm that the funds were to help Bragg get started in studies of biological molecules, not to support Max:

In assuming the professorship at Cambridge [Bragg] proposes to continue studies with X-rays and feels that it is a logical time for him to apply his efforts to the analysis of crystals of biological interest. He has already arranged for a cooperation with Prof Keilin, who will prepare crystals of various haemoglobin compounds and also certain enzymes . . . There is in Cambridge an assistant who has been working in the field of crystal analysis . . . it is Prof B's request that the Foundation permit him in this period of transition to retain the services of this well-trained technician who will be of great value in helping Prof B in his orientation towards his new field of interest.

Max always acknowledged this development as one of the key turning points in his career. It gave him financial security at a time of great uncertainty, so that he could guarantee to support his parents. From a scientific point of view, it represented Bragg's personal engagement with the haemoglobin project, which not only gave Max some desperately-needed theoretical back-up, but also secured the medium-term future of the research within the Cavendish Laboratory, where no one could say it properly belonged. Although his PhD thesis was still incomplete, he had taken the vital step from research student to research scientist. How far he would be able to pursue his research would now depend on his own endeavours.

3

'The most dangerous characters of all'

My lord, wise men ne'er sit and wail their woes,
But presently prevent the ways to wail.

William Shakespeare, *Richard II*

Professionally, Max had every reason to be pleased with life at the beginning of 1939. He had a secure job for two years at least, and was already recognised as one of the pioneers of a new branch of science that was making waves throughout the scientific community. He had a clear idea of the experiments he needed to do, and the means to do them, an extremely supportive boss in Lawrence Bragg, and wonderfully inspiring mentors in David Keilin and Desmond Bernal.

In contrast, his personal life was in turmoil. He was effectively stateless, his Austrian travel documents worthless since the *Anschluss*. His immediate family were safe, but their long-term future was as uncertain as his own; uncles, aunts and cousins who were still in Austria or Czechoslovakia were potentially in great danger. He had no family close by with whom to share his anxieties. Strict censorship meant he was unable to communicate freely with his Austrian friends, and none of his Cambridge acquaintances provided the same bond of instant understanding. Outside the ferment of anti-fascism in Bernal's lab, Hitler's action was by no means universally seen as a bad thing. In his daily visits to Peterhouse, Max found himself among those with a very different perspective.

In College . . . they admired what Hitler had done to revive the virility of Germany, and regarded the Jews' stories about concentration camps as atrocity propaganda designed to draw England into the war against Hitler . . . In some ways one couldn't help sympathising with that point of view. While my scientific colleagues detested Chamberlain and regarded Munich as a betrayal, my other acquaintances breathed a huge sigh of relief at Munich, believed Chamberlain, and thought that one should concede to the Nazis their just aspirations and avoid a war at any cost.

As a moderate, tolerant democrat, Max felt himself isolated. The one person who shared his view and provided the sympathetic, supportive refuge he so desperately needed was, to Max's own initial surprise, a German. The only Germans he had come across previously had been the loud and arrogant young Nazis he found in the huts on his alpine skiing and mountaineering expeditions, whom he detested without reservation. Paul Ewald, in contrast, was a tall, handsome, good-humoured and highly cultured physicist aged about fifty, who had lost his chair at the Technical University in Stuttgart because of his opposition to anti-Semitism. It had been Ewald's PhD thesis that prompted Max von Laue to think of the first X-ray diffraction experiment. He arrived in Cambridge in 1937 and immediately became part of the community in the Cavendish, organising seminars (the first time this had happened, according to Max) and acting as a friendly adviser to younger scientists.

His mother, wife and children all came with him and, having lost all their belongings, they set up home in what Max described as a 'dismal house in a dilapidated street' which they furnished with 'oddments picked up at the weekly auctions on the Corn Exchange'. When Max visited, he found them living happily as if they had never known anything better, and their cheerful acceptance of their lot made a great impression on him. He began to spend rather more time there than he did in his own digs, and they treated him as one of the family – the Ewalds' daughters were not much younger than he. From Paul Ewald he learned a valuable lesson that was to colour his approach to people he met for ever after, as he recalled in an appreciation: 'Ewald's upright and loveable character cured me of my blinkered prejudices and made me realise, like Pasternak's Zhivago, that there are no nations, only people.'

Max's immediate problem was how to bring his own parents to Cambridge from Zurich. The immigration regulations stated that older foreign nationals wishing to reside in the UK had to raise the sum of £1,000 and find someone in the country to guarantee that their expenses would be covered. With his Rockefeller salary, Max could provide the guarantee, but the £1,000 – equivalent to almost £40,000 in purchasing power today – was a different matter. In a memoir dictated shortly before his death, Max told a story of how he raised the money. Perhaps emboldened by his success in asking Bragg for help, he approached the richest man he knew for a loan. Gerald Seligman, the gentleman amateur glaciologist who had employed him all summer on the Jungfraujoch, agreed to lend him £500 on the condition that Max raised the matching £500 himself. Max's father wrote to him giving the impression that he could raise that sum by selling one of Dely's rings, and with that information Seligman made the loan. Max discovered afterwards that Hugo had never sold the ring after all; he had just looked at the prices of rings in a shop window and noted the value of one that resembled Dely's. It shook him that his father, whom he always admired, could be so devious, though he rationalised his behaviour as 'a misjudged desire to appear to acquiesce to my demand and simultaneously to spare my mother the distress of selling her jewellery'.

There are some inconsistencies in this story, but, whatever the details, the British authorities were persuaded of the Perutzes' solvency and Hugo and Dely finally arrived in Cambridge in March 1939. Any joy at their reunion was short-lived. Unlike the Ewalds, his parents proved quite unequal to the task of accepting their changed circumstances. Since his arrival, Max had lived happily in typical 'digs' of the time: a furnished room in a house on the outskirts of the city with a shared bathroom, breakfast provided, other tenants occupying the other rooms. He thought that rather than booking his parents into a hotel, he would find them something similar, but when they arrived and saw what was on offer, they were simply appalled.

Instead, they found a furnished house to rent at 15 Emmanuel Road, right in the heart of Cambridge, and Max was forced to move in with them to help spread the cost. It was less than five minutes' walk from the Cavendish Laboratory; but Max did not think it a good long-term solution. He established that if they put together all their remaining financial resources they could buy a small, semi-detached house in one of the newer estates around the town. Dely dug her heels in. She would not consider buying a house 'until her furniture arrived from Austria'. One has to smile at the thought of her grand pieces being shoehorned into a 1930s semi, but of course it never happened: after war broke out, her furniture was destroyed in the allied bombing of Vienna. The couple never again had enough money to buy a house (Max may have been over-optimistic about the possibility anyway), and spent the rest of their lives in a series of 'hideous lodgings and furnished flats'.

Max's relationship with his parents became increasingly difficult from the time they arrived. He was still the beloved youngest son; yet whereas in Vienna this had meant freedom and indulgence, now it entailed a burden of emotional and financial responsibility. By virtue of his financial independence and more or less established position in England, he had become de facto head of the family, but it was always going to be awkward to fulfil this role in relation to a father he greatly respected and an imperious mother. Hugo and Dely depended on him to save them from destitution, but were often unwilling to take his advice. He later complained that Dely was 'ruthless' in demanding his help, and never gave him any thanks. He saw his parents as 'traumatised' by their unaccustomed poverty: the mother he remembered as a sparkling socialite and elegant hostess now spent her days bemoaning their loss and complaining incessantly about their situation.

She seemed helpless in the face of the smallest tasks, such as planning what to buy for the next day's meals, though she could still dress like a Paris fashion plate if the occasion demanded. There were vague hopes that she and Hugo would be able to earn some money by acting as agents for Brüder Perutz, which Max's second cousin Felix Perutz was valiantly continuing to manage from Budapest, but nothing much ever came of this idea. Although his parents frequently upset Max by turning down his direct offers of financial help, he knew only too well that when their meagre funds ran out it would be he who kept them from starvation. Sorting out some financial arrangement that was acceptable to them, affordable for himself, Lotte and Franz, and adequate for their needs would preoccupy him for years to come.

The stress told on his health. Max had been sickly as a child, but since his teens he had enjoyed a tougher constitution, honed on the mountain slopes. Now, he seemed to revert to the state in which ill-health was an ever-present threat. In May 1939, for example, he wrote to Bragg to excuse himself from the lab on the grounds that he had 'caught sun stroke' the previous Sunday (in May? in Cambridge?), and expressing the hope that he would be well enough by Friday for a pre-arranged meeting with Dorothy Hodgkin and her assistant Dennis Riley. Bragg replied rather brusquely that if he had not recovered in time, he should wire to cancel the meeting – as the whole point of it was to discuss collaboration on the haemoglobin work, it was hardly worth Hodgkin and Riley making the tedious trip from Oxford to Cambridge if he wasn't going to be there.

As the months wore on, it became impossible to deny the inevitability of war with Germany. Once Neville Chamberlain had made his sombre broadcast on 3 September 1939, Max and his family were no longer merely refugees but 'enemy aliens'. By this time 55,000 people had fled to Britain from Germany, Austria and Czechoslovakia. What to do about them was a question that increasingly exercised the government of the day, and what ensued has been exhaustively researched by Peter and Leni Gillman for their book *Collar the Lot!*. As recently as the spring of 1938, the Committee of Imperial Defence had received a recommendation that, should war break out, aliens should be 'encouraged' to return to their countries of origin. Fortunately, someone in the Home Office realised that, given the circumstances under which most of them had left, such a policy would lead to uproar and it was never adopted. Mindful of the outcry during the First World War, when 29,000 aliens had been interned in conditions of great hardship, ministers initially agreed that there would be no mass internment. Therefore, on 4 September, the Home Secretary, Sir John Anderson, set out a filtering policy designed to incarcerate only those who were a threat to national security. He agreed to recognise the status of those who were refugees from Nazi oppression.

By the end of October 1939, 120 tribunals had been set up around the country to examine aliens and decide whether they should be put into category 'A' for immediate internment, 'B' for restricted freedom or 'C' for unrestricted freedom. Max and his parents duly appeared before the Cambridge tribunal when summoned by the local police in November. With letters of support from their British friends, they had no difficulty in convincing the local worthies facing them that they were no threat, that they were of good character and that they were refugees from Nazi oppression. They were all given a 'C' rating and allowed home. By March 1940, tribunals across the country had heard the cases of 71,600 aliens, of whom only 600 had been interned. More than 60,000 had been given a 'C' rating.

As a chemist, Max fervently hoped to be given work in support of the war effort, but in response to his request he was simply told to carry on with his research. So throughout the 'phoney war' from September 1939 to March 1940, he continued to work quietly in the Cavendish, taking his turn at fire-watching in the evenings. His interest in his subject received added impetus when he spent the book token Evelyn had given him for Christmas on Linus Pauling's newly-published text, *The Nature of the Chemical Bond*. Pauling (1901–94) was one of the most innovative chemists of the twentieth century, and, at the California Institute of Technology (Caltech) between the two world wars, had laid much of the theoretical foundation for an understanding of molecular structure. Max found his book the most exciting he had ever read.

His book transformed the chemical flatland of my earlier textbooks into a world of three-dimensional structures. It stated that 'the properties of a substance depend in part upon the type of bonds between its atoms and in part on the atomic arrangement and the distribution of bonds', and it proceeded to illustrate this theme with many striking examples. Pauling's imaginative approach, his synthesis of structural, theoretical and practical chemistry, his capacity of drawing on a wide variety of observations to prove his generalisations, and his vivid writing drew the dry facts of chemistry together into a coherent intellectual fabric for me and thousands of other students for the first time.

In March 1940 Max successfully submitted his thesis on X-ray crystallographic studies of haemoglobin and could now proudly call himself Dr Perutz. Perhaps mindful that it would become difficult to continue research in the UK, he prudently sent a copy of the thesis to Isidore Fankuchen, who had fled back to the US on the outbreak of war and now held a junior research post at the Massachusetts Institute of Technology (MIT).

His parents remained a worry: Hugo's health was poor; how he and Dely were to survive financially was still unresolved. Hugo tried to pick up the threads of the textile business by selling fabric samples to clients such as Marks & Spencer. It was extremely difficult to stay in touch with Felix Perutz in Budapest, however, and in practice Hugo and Dely were still dependent on Max. His salary was comfortable for a bachelor who had learned to be frugal in his habits, but it was scarcely adequate to support three adults. Though he had no wish to leave England himself, Max believed that his parents would be safer and more comfortable in America. Having left continental Europe so late, they were rather far down the waiting list for the all-important quota numbers, but all felt hopeful that in due course their time would come, and that they would join their older children.

The position of refugees in Britain changed radically in April. Hitler had sent his armies into neutral Norway, and stories began to circulate in the British press that a 'Fifth Column' of sympathisers within that country had aided the Nazis. As fear of a German invasion grew, enemy aliens became the target of press scaremongering about the existence of a similar Fifth Column within Britain. Although there was no evidence that the public at large believed these stories, the Home Office architects of the existing policy on internment began to lose influence to those in the War Office and secret services who favoured a harsher approach. The day Chamberlain resigned as Prime Minister and Winston Churchill took over, 10 May, Germany attacked the Netherlands, Belgium and Luxembourg. On 11 May Churchill's first Cabinet meeting took the decision to intern all male aliens between the ages of sixteen and sixty in the south-east and east, the regions that would be in the front line of an invasion from Europe.

So it was that Dely opened the front door of 15 Emmanuel Road on the morning of Sunday 12 May – a day of cloudless blue skies, the trees in new leaf – to find a policeman on the step with a warrant for Max's arrest. The policeman said it would be 'just for a few days', but Max knew better. He packed for a long journey, including food and books among his other essentials, and bade farewell to his distraught parents. He had no time to tell his girlfriend Anne Hartridge what had happened.

Anne was not the only one who had to wait for information. Weeks passed before he was able to send a letter to his parents, or to receive news from them. He and a hundred or so others from Cambridge were first held in a former school at Bury St Edmunds, where they were 'herded into a huge empty shed' – actually the school gym. There Max noticed a fellow internee engrossed in studying a spot of sunlight that had forced its way through the blacked-out skylights, and which he had caught on a piece of paper. The man told him he was studying sunspots. He was Hermann Brück, a Cambridge lecturer, who in 1957 would become Astronomer Royal for Scotland. The mathematician Hermann Bondi, then an undergraduate at Cambridge, arrived the same day. A fellow internee recalled that they all slept on the concrete floor with one blanket between three. Yet in the first letter that he was allowed to write, on 20 May, Max reported to his parents that conditions in this first camp 'were quite pleasant, we slept in comfort and got plenty to eat'. It could be that conditions improved after the first night: he writes that the camp 'was made pleasant by an extremely nice officer (20-year-old undergraduate from Cambridge) who got for us all the comforts on his own initiative.' But he may have been deliberately econom-ical with the truth: it is clear that the main purpose of this letter is to allay his parents' worries:

> The duration of our internment cannot be foreseen at the moment but our position is by no means hopeless! . . . There is assembled among us the most distinguished crowd of people I have ever seen together, all the Refugees from Cambridge University, undergraduates, research students, lecturers, fellows of Colleges. We have already started a kind of University consisting of lectures and courses. I am secretary of the Natural History Faculty. The people are really most pleasant and I have acquired some very nice new friends . . . There is, I think, nothing you can do for me at the moment . . . And there is a good chance that things are not as bad as you may think. Keep well and don't worry too much.

This letter was written on the day that Max and his fellow internees arrived at Huyton, a half-built council housing estate near Liverpool. Here conditions were hardly pleasant or comfortable. The men slept several to a room in unfurnished semi-detached houses. As the system for censoring mail had yet to be established, there were still no letters from home, and the huge backlog of outgoing mail meant that neither internees nor the families they had left outside had any idea of each other's welfare – or, indeed, their whereabouts. While they were at Bury St Edmunds, they had been allowed books and newspapers, and could follow the progress of the war on the wireless. By the time they reached Huyton, such sources of outside information had been banned and they were left to rely on rumours of doubtful reliability. However, in his next long letter home, Max mentioned little of this.

If the internment were not such a wretched waste of time as far as my research is concerned (plus a few other ifs) one might regard it as a good school of human companionship . . . We are not suffering any great hardships, except that we are not allowed to receive any mail, but the fate of each of us is greatly afflicted by the loss of time the internment involves . . . At the beginning the attitude of our guards in this camp was rather horrible, because they thought that we were all spies. But by now they have convinced themselves that we are harmless refugees and are quite friendly.

The same day he wrote to Evelyn, who, married with two young children, remained one of his closest friends. To her he was much franker about the treatment the internees had received:

Yesterday . . . we arrived at a new camp where people regarded us as a bunch of spies [and] committed shocking cruelties against some of our old and sick people. The officers apologised . . . The plight of many of the refugees here is ghastly, I have never seen so much tragic misery among capable and intelligent men.

Max later remembered the bafflement of his camp commander, an ageing officer who had fought in the First World War.

Then a German had been a German, but now the subtle new distinctions between friend and foe bewildered him. Watching a group of internees with skull-caps and curly side-whiskers arrive at his camp, he mused 'I had no idea there were so many Jews among the Nazis.' He pronounced it 'Nasis'.

For his first week or two, Max seems to have been content to let matters take their course. Fritz Eirich had by this time joined him in the camp, as had a number of other scientific acquaintances. Despite the cramped conditions, which made it difficult to study in peace and quiet, the lectures and courses continued. Max was pleased to be able to occupy himself with organising them, although he wryly observed that 'the titles of the lectures are often more ambitious and interesting than the lectures themselves'.

By late May his tone began to change. The emphasis was still on reassurance: 'I am very well, sleep well, eat enough and there is sufficient opportunity to wash (shower-baths).' But at the same time he expressed a growing realisation that as he was now entirely unable to alter his own situation, his parents might be mobilised to do so. He sent a list of brisk requests, bordering on the peremptory: to send four shirts as 'there is no possibility yet of sending things to the laundry'; to go to the Cavendish at the end of the month and collect his salary; and to send 10 dollars to Lotte for her birthday if possible ('Do it on the day you receive this letter, please.'). If they know of any means by which his release might be effected, he asks that they 'please follow it up with all energy and determination'. He also asks that they 'try every possible means' of getting him an American quota number, the essential prerequisite for emigration to the US.

Fear that the internees might escape from Huyton led to another move in mid-June, this time to the seaside resort of Douglas on the Isle of Man. In wartime it was unlikely that the boarding houses on the promenade would have been filled with holidaymakers, so the internees made a convenient substitute. Central Camp consisted of a row of Victorian terraced villas, hastily surrounded with barbed wire. Max lived in House 7, sharing a room with 'two bright German medical researchers', who 'opened [his] eyes to the hidden world of living cells'. Food was still short, but at last the internees began to receive mail, although they themselves were allowed to write only to relatives, only in English (the same applied to letters they received) and only on standard-issue 24-line letter forms. They were not ill-treated, but some of the officers and men guarding them still harboured suspicions. The camp university continued, but Max found the enforced inaction intolerable. He applied for his release on the grounds of the national importance of the work he was doing with Bragg, and of securing the Rockefeller grant without which he could not support his parents.

Meanwhile his parents threw themselves into helping Max out of his predicament. They inferred from his letters that his immediate priority after his release was to leave the country. Accordingly, they applied for an American quota number for him, and Dely began to pack up his belongings with the intention of sending them to Franz to await Max's arrival in the US. They also made enquiries about other options that now seem extraordinary until one appreciates the very real fear that Britain would fall to a Nazi invasion. Hugo met a friend who was just setting off for Shanghai, and told Max that going there would be feasible and not too expensive. Twenty thousand European Jewish refugees entered Shanghai between 1939 and 1941, the city being one of few in the world that required no visa; since the late nineteenth century it had had a flourishing Jewish community known as 'Little Vienna'. Dely undertook to plead with Lawrence Bragg to apply for Max's release; Max too wrote repeatedly, but Bragg seemed initially reluctant to act.

Max's increasing frustration is clear from the letter he wrote to his parents on 17 June:

> For Heaven's sake shake my professors (Bragg, Keilin, Bernal) and friends out of their complacency and impress upon them that I don't want futile safety and degrading idleness, but active work in our war effort, and that unless they make a real effort for my release I shall in all probability remain here for years, that you will starve and my carreer [sic] will be ruined without having rendered the country any service by my sacrifice ... Wire any decisive developments.

Decisive developments there were, but not in the direction he hoped. Bragg, by his own later admission, thought he 'ought to abide by the decision of the Government, and have faith in the wisdom of the course it was thought necessary to take'. Accordingly he did nothing to obtain Max's release throughout May and June. Meanwhile, attitudes to aliens continued to harden. As the retreat from Dunkirk began in late May, a secretive Cabinet committee, dominated by War Office and secret service representatives and called the Home Defence (Security) Executive, took over responsibility for policy on aliens.

In June Mussolini took Italy into the war on the side of Germany, prompting Churchill to utter the infamous line 'Collar the lot!' in response to the question of what to do about the thousands of Italians working in restaurants and elsewhere in Britain. All Italian men who had arrived within the previous twenty years – and even a few who had been in Britain for longer and were naturalised – were accordingly rounded up and added to the population of internees. Meanwhile the Home Defence (Security) Executive had begun to negotiate with Canada to take those internees thought to present the greatest threat to Britain's security – 'the most dangerous characters of all'. They persuaded the Canadians that there was an urgent need to deport 7,000 aliens, consisting of German merchant seamen who had been captured on the high seas by the British navy, Italians, and prisoners of war.

On 21 June the Home Office agreed to mass internment of male aliens from Germany and Austria. This time when the knock came at the door of the Perutz household, it was for Hugo – sick and elderly as he was. On 25 June he was taken to the Isle of Man and interned in a camp just a few miles away from Max, but with no means of communication between them. Dely was left alone in Cambridge with her family scattered across the world. The next day, she wrote despairingly to Max: 'Never in my life was I so helpless as I am just now' – but it was to be months before he received the news of his father's fate.

One day towards the end of June, army doctors arrived and vaccinated all the unmarried men under the age of thirty in Max's camp. They were told they would be leaving within a few days, but not where they were going (though most guessed either Canada or Australia). Max just had time to dash off a letter to his mother, in which he made a testament that granted power of attorney to his friend from the Jungfraujoch expedition, Tom Hughes. If she misunderstood the significance of this act, she was not left in doubt for long: on 6 July she received a telegram from the camp commander informing her that Max had left for Canada the previous Tuesday.

Of the 6,750 men who eventually arrived in Canada, 2,700 were refugees: hardly 'the most dangerous characters'. The second ship to leave, the SS *Arandora Star*, sailed from Liverpool on 1 July with more than 1,500 men on board, almost half of them Italians and another 400 or so German or Austrian prisoners of war and refugees. Nearly half of them drowned when she was torpedoed and sunk the next morning at dawn. Ignorant of this disaster, Max and many Cambridge students and young researchers gathered in Liverpool to board the HMT *Ettrick*, an 11,000-ton troopship. On 3 July they were hustled aboard and packed into the holds, about 1,200 alien internees in one and several hundred German prisoners of war in another. As evening fell, the ship steamed out of the port, her destination still a mystery to most of her 'passengers'.

Even the officer in command reported that there were far too many prisoners and aliens for the quarters. Sleeping arrangements took the form of a hierarchy of squalor: an upper tier of refugees in hammocks swung above a second tier lying on tables, beneath which a third tier slept on the floor. Barbed-wire doors were locked at night. Only after news of the sinking of the *Arandora Star* reached the *Ettrick* did the crew issue lifebelts but, as in the case of the doomed liner, the ship's officers held no lifeboat drill. The internees were fed – inadequately – morning and evening; they were well aware that the prisoners of war received twice as much, because they alone enjoyed the privileges of the Geneva Convention. No one was allowed to visit the toilets at night, the motion of the ship led to the inevitable consequences and the resulting 'sickening stench' only compounded the problem. They protested vigorously, but to no effect: the commanding officer refused to meet their representatives, and called them 'scum of the earth'.

A leader emerged in the unexpected form of the 29-year-old Prince Friedrich of Prussia, who had been living in England and whom Max later recalled with admiration.

To this revolting scene [the Prince] ... restored hygiene and order by recruiting a gang of fellow students with mops and buckets – a public-spirited action that earned him everyone's respect so that he, grandson of the Kaiser and a cousin of King George VI, became king of the Jews.

Despite the Prince's efforts, the appalling conditions led to outbreaks of dysentery and other illnesses. Max was among those who succumbed: he ran such a high fever that he passed out. The authorities had provided no medical arrangements, but some of the German doctors among the internees had established a sick bay, where Max found himself when he recovered consciousness. By that time the *Ettrick* was steaming into the mouth of the St Lawrence River.

On balance, Max's parents were relieved to hear of his deportation. 'We must be glad if Max is safe,' wrote Dely to Hugo, 'and not think how terrible it is for us, to be separated from him, without having had an opportunity to say good by [sic] to him.' She immediately focused on getting him a job in North America, entirely justified in believing that this was the strategy Max would favour. He had already made the difficult Atlantic crossing. He had asked her to get him a US quota number. Many believed that a Nazi invasion of Britain was likely, and that any Jewish refugees they found would be shipped back to the death camps. So she undertook to mobilise Max's influential contacts – skilfully deploying her Viennese instinct for *Protektion* – to set him up at an American or Canadian university. She went to see Bragg and Keilin, and wrote to Bernal, all of whom said they would help, and provided some comfort in her distress by telling her how highly they all thought of Max's work. By this time, Bragg was beginning to feel some remorse for his former inaction, and he wrote to the Rockefeller Foundation proposing that they find Max a job in an American university. 'We cannot carry on the work here without him, he is a key man,' he added.

Bernal wrote to Linus Pauling at Caltech in California, arguing that an invitation to Pasadena, even to give a few lectures, might win Max's release from internment. He suggested that Pauling would find Max 'a very useful addition to your lab', and extolled his skill in preparing, mounting and photographing protein crystals. Bernal also saw Max's transfer to the Pauling lab as a way of getting experimental protein crystallography to 'take root' in the US, at a time when the prospects for continued pure research in the UK looked extremely bleak. His own group at Birkbeck had been dispersed at the beginning of the war, and his lab later received a direct hit in the Blitz. He himself was entirely occupied with advising the government on scientific and engineering aspects of warfare: 'Ce n'est pas magnifique, mais c'est la guerre,' he quipped grimly to Pauling.

At the same time there was a dramatic change in the attitude of the government to the refugees. The trigger was a remarkable letter written by a young officer called Merlin Scott. Scott was on duty when the fifth (and as it turned out, final) consignment of deportees was boarding HMS *Dunera* at Liverpool on 11 July. This shipment included 450 survivors from the *Arandora Star*. Scott was appalled to see soldiers and policemen searching the internees' battered suitcases on the quayside and pocketing any valuables. They herded the men up the gangway at bayonet point, to jeers from the watching crowd. As soon as the ship had steamed away, Scott wrote to his father Sir David Scott, who was assistant under-secretary in the Foreign Office, saying how abominably he thought the *Arandora Star* survivors had been treated. The letter rapidly made its way round the Foreign Office and then the Home Office, neither of whose ministers had any idea that the men had been deported again.

Soon afterwards the House of Commons debated the issue of refugees' treatment, and MPs roundly criticised the military authorities. As a result, the Home Office won back responsibility for internment policy. From 5 August onwards, it began to send home batches of Category C internees from British camps. It was a remarkable example of a single compassionate individual turning the tide against apparently more powerful interests. If Merlin Scott was aware of what he had achieved, he did not have long to reflect on it: he died during the British advance into Libya early in 1941, fighting Italian troops. He was twenty-one years old.

In the middle of September Hugo's release finally came, and he and Dely were reunited at a London hotel. For those who had been deported, however, it was much harder to undo the injustice of internment. The *Dunera* continued her two-month journey to Australia, one of those on board being Fritz Eirich, Max's friend and fellow chemist from Vienna. It was three years before he was released in Australia and returned to Britain.

By deporting the internees, the British authorities had surrendered responsibility for them to their Canadian and Australian counterparts. The Canadians had been expecting thousands of dangerous and desperate fascists, and acted accordingly. When Max and the rest of the *Ettrick* consignment disembarked at Quebec on 13 July they were driven through the town in police vans with sirens wailing, the streets lined with jeering crowds. They were first housed in an army camp high on the citadel overlooking the St Lawrence River, known as Camp L. It was well equipped, the wooden huts heated, plentiful food supplied and clothing provided – in the form of thick navy blue jackets with a large red bullseye on the back.

Their status, however, had undergone a radical change. The Canadian government classified them as 'Prisoners of War, Class 2, Civilians'. There were some advantages in this, in that the Geneva Convention ensured that they were decently fed, clothed and housed; but it was a further appalling blow to men who were unswervingly loyal to the allied cause and hated the fascist regimes of Germany and Italy. It also meant that were Britain to lose the war, they would be sent back to their 'home' countries to face certain extermination. Max later wrote of his anguish at his situation:

> To have been arrested, interned and deported as an enemy alien by the English, whom I regarded as my friends, made me more bitter than to have lost freedom itself. Having first been rejected as a Jew by my native Austria, which I loved, I now found myself rejected as a German by my adopted country.

As prisoners of war, the inhabitants of Camp L were surrounded by barbed wire and moved about under the baleful gaze of guards armed with machine guns. On arrival, they were photographed, fingerprinted and stripsearched, and any remaining valuables were taken away. Max dropped the contents of his wallet out of the window before the search began, and easily managed to recover his few valuables afterwards. To their horror the refugees found that there were Nazis in the same camp – the Canadians initially recognised no distinctions among the internees – and it was all they could do to avoid being allocated to the same huts. The men were locked in at night, 100 to a room in bunk beds, and the weather was stiflingly hot. For several weeks they were permitted no communication with their families, nor did they receive any of the letters and telegrams sent enquiring after their welfare. Worse than any other deprivation was the loss of freedom, especially when the spectacular view beyond the barbed wire was bounded by the mountains of the northern United States.

In that country lived Max's brother and sister, Franz and Lotte, able to go about their life and work without hindrance. How could he reach them? He thought about means of escape – cutting through the barbed wire and dashing for the border? Jumping on a goods train? – but soon recognised the impossibility of such a reckless act. Reason would be his salvation – somehow the Canadian and British authorities, or possibly the Americans, must be persuaded of his innocence and of his value to the war effort as a scientist – but, without the means of communicating even with his family, it was difficult to see how he could argue his case.

There were compensations that made life bearable. To his delight, Max found that his friend from his student days, Heini Granichstaedten, was in the same hut and they chose bunks one above the other. Together they could indulge their Viennese sense of humour, reminisce about their trip to the Arctic in 1933 – how long ago that seemed – and speculate wryly on the 'regulation-ridden minds' of their captors.

Several members of the 'camp university' that had continued from Bury St Edmunds to Huyton and Douglas also fetched up in Camp L. They lost no time in re-establishing lectures and classes. Max took the opportunity to learn more of the subjects that had been most lacking in his own scientific education: mathematics and theoretical physics. Hermann Bondi taught vector analysis, a branch of mathematics used in applied fields such as engineering and astronomy as well as physics. Max, who was never much of a mathematician, was lost in admiration for his ability to deliver his lectures without notes, chalking up complex equations on the blackboard entirely from memory.

Klaus Fuchs, a tall, spare German Protestant and former member of the German Communist Party who was now a researcher from Edinburgh University, taught the classes in theoretical physics. Max found his lectures so good that for the first time in his life he thought he understood the subject. Like his fellow scholars, Max knew nothing of Fuchs's political beliefs, and found him 'very distant, not a person with whom you made friends'. After his release, Fuchs was employed on the atomic bomb project in Britain and the US, all the while passing on scientific information to the Russians. He was eventually unmasked in 1950, receiving the comparatively lenient sentence of fourteen years in jail, of which he served nine.

Max himself taught the principles of X-ray analysis of crystals, though without the means to carry out practical experiments his courses must have been somewhat dry. So the time passed in an unreality of high-level academic discussion, physical confinement and complete detachment from the outside world. The first communication the internees were allowed to send home – more than a month after their arrival – was a standard issue postcard giving the bald information that 'Prisoner of War X' had arrived at Camp L and was 'safe and well'. Max sent one to Lotte in New Orleans. Within the following week, they were allowed to write letters for the first time, and Max wrote first to Anne and second to Lotte. He was distressed beyond measure that the stationery they were obliged to use had 'Prisoner of War Mail' printed in thick black capitals on the front. Having had no news from the UK, he could only imagine that hostility to 'enemy aliens' had continued to increase, and that for his mother to receive such a letter would put her under immediate suspicion. He told Anne: 'It is with great reluctance that I write to you on this [paper], but after 7 weeks I feel I cannot wait any longer.' He went on in a hopeful vein:

> Now things do not look quite as black. In optimistic moments I often believe that I shall see you before X-mas. Only my body is in Canada. When I don't work my thoughts are in England and I think of you and my people and friends and I dream of all the marvellous things I shall do when I come back. As camps go this one isn't bad ... I am still together with several very nice friends; I don't know what I should do without them.

It was from Lotte, on 23 August, that he received his first letter since leaving England. For the first time he heard of all the efforts his parents and friends had been making to secure his release, which cheered him immensely, but he was still concerned about his parents, especially his father: 'Most important is to get father released and both parents over the ocean.'

Max was now determined to return to England. From what he had been able to discover, it seemed that even were he released in Canada, it could take years to satisfy US immigration regulations. (Events proved him correct: none of the internees succeeded in entering the US from Canada before 1945.) As soon as he could, he fired off letters to Bragg and Keilin, urging them to get him back at all costs. At the end of August, Bragg wrote to Dely to say that Max had asked him to let her know that he was safe and well, and commented that: 'He seems keen to return to England if it can be arranged.' This was almost certainly the first Dely knew of Max's wanting to come back to the UK. The last letter she had received from him, before he left for Canada, urged her to progress his application for a US quota number. She had spent two months doing everything she could to achieve not only his release, but his employment in an American university, including badgering Bragg to persuade the Rockefeller Foundation to transfer his grant. She had all Max's belongings packed up and stored in the college room of his friend Tom Hughes, and was on the point of despatching them to Franz. As she corresponded with Hugo during the months of his internment, they constantly reassured each other that emigration was the best hope for Max.

In the middle of August, Dely wrote to Hugo in great excitement: it seemed that Max's foresight in sending his thesis to Fankuchen had paid off in the form of access to a potentially powerful source of *Protektion*. Fankuchen himself, who was not a diligent correspondent, had never bothered to acknowledge receipt of the thesis, but he had shown it to Professor M.J. Buerger, a highly respected crystallographer at MIT. Buerger had written to Max, not knowing he was interned. Dely breathlessly told Hugo that he had said

> that the xray photos were amazing and unic [sic], and he ask for the permission of Max to publish them in his new book. I was told that this Professor Buerger was a famous man ... I did not loose [sic] a day in sending a copy of the letter to Max, and the letter itself to Lotte, asking her to answer it ... Such a connexion may at this moment be of great importance for him.

The Rockefeller Foundation had acted as soon as it heard from Bragg, his warm support for Max backed up by letters from Bernal, Keilin, Tom Hughes, Gerald Seligman and others. Testimonials from researchers in the US including Fankuchen established to the officers' satisfaction that he was 'quite first class' and not 'just a pair of hands for Bragg'. Both Linus Pauling at Caltech and Buerger at MIT said they would be happy to have Max working with them, though neither had the funds to support him. Accordingly, in mid-October the Foundation approved a grant of $3,200 to the New School of Social Research in New York (which was active in finding academic posts for refugee scholars in the US), to fund an associate professorship in biophysics that Max would hold at either Caltech or MIT.

Max heard nothing from England, other than via Lotte, until the beginning of September. The first missive, which he must have greeted like Noah seeing the dove arrive bearing an olive twig, was a small parcel containing two Penguin books, a satire and a detective thriller, sent by Anne. On 7 September letters arrived from Bragg, Wooster and Bernal as well as his mother, and another came from Hermann Mark in New York. At last he could grasp the full details of Dely's energetic efforts to get him to the US. But what he seized on was the news that Bragg wanted him back. Immediately, he asked David Keilin to countermand all her arrangements.

> There has been a serious misunderstanding. Apparently last July my mother asked you and Professor Bragg to get me a job in the US ...

There is no question of being released in Canada, let alone be allowed to work. My only hope is that I can return home to England . . . Do you think you could be so kind and get in touch with my mother, Professors Bragg and Bernal and try to make things move . . . I am sure that justice will prevail and it will be realised that we aren't Nazis.

Max explained to Lotte the dilemma that made the release of those interned abroad so much more complicated than that of those at home.

Competent for our release is solely the Home Office, who has no authority to release us here but can only order our return. We cannot communicate with foreign consulates . . . but only with the Swiss consul . . . My passport has been impounded . . . Please tell mother to do all in her power to get me back; never mind the dangers. In the old climbing days we used to take greater risks for smaller gains. If there were a 70% risk of death in my return I should take it at once to regain my freedom.

Max had come to understand that, despite the Statue of Liberty's supposed welcome to 'huddled masses yearning to breathe free', the US certainly did not want to shoulder the burden of Britain's alien refugees. The treatment he had received since his arrival in Canada gave him no sense that North America would take a liberal attitude to those in his position. Everything he cared about was in England, and it was England that had given him a home and friends when Austria rejected him. Returning to Cambridge became his only object. Writing to Anne, who was very much part of the world he yearned to see again, he declared:

I should like to swim across the ocean if there were as much as a 1 per cent chance to reach England alive and I should be glad to have seen the last of this continent where, as far as I am concerned, the only unlimited possibility is the duration of my internment. It will be 5 months this week.

During October, the Canadian military authorities decided to organise the occupants of the camps by religion. By this time, more or less all the Nazis had been removed from Camp L, and the remaining prisoners were designated 'refugee-type' – but still classified as prisoners of war. Now, Roman Catholics were shipped off to a well-equipped army camp, Camp A; Jews and Protestants were dumped at Sherbrooke, south-east of Montreal and near the US border. Known as Camp N, it consisted of two sooty, leaking railway sheds. In the words of an official report, 'Five cold water taps, six latrines, 2 pissoirs, were the sole sanitary provisions for 720 men.' The prisoners threatened to go on hunger strike and to refuse to cooperate. Their captors merely counter-threatened that they would be locked up in the dark, starved and, if necessary, transferred to a Nazi camp.

Max was a Catholic, but chose to go into this 'purgatory' with the 'heretics and Jews'. His friend Heini was a Protestant; many of the scientists who had formed the core of the camp university were Jews. Wishing to stay in their company, Max swapped identities with another inmate. If he had known what was in store, he might have thought twice. While the commandant at Camp L had encouraged its academic activities by providing a room for quiet study, at Sherbrooke the men could only mill in their hundreds. Building work to bring sanitary facilities up to some minimum standard continued for seven weeks after Max's arrival, and the angry inmates spent their time 'locked in futile arguments over trivial issues'. Under these circumstances his 'assaults on differential equations petered out in confusion'. When the commandant discovered Max had given a false identity, he was locked up for a few days in an unfurnished steel cage in the local City jail. He smuggled books into the jail inside his baggy plus fours to pass the time; but these ran out before the term of his sentence, and the sergeant-major on guard refused him any more. Sleeping on the bare wooden floor, he became infested with scabies, which caused an unbearably itchy rash.

In November the camp commandant gave Max a cable from his father saying that the Home Office had ordered his release. Another cable brought the offer of the associate professorship from the New School of Social Research. Max wrote by return, saying he would be glad to accept, but that he had no means of obtaining a US visa as long as he remained in Canada. Privately he was sceptical that the job offer would be enough to cut through the red tape of the US immigration authorities on any reasonable timescale. At the same time he told the commandant that despite the risks of the sea crossing he would like to return to the UK, a choice that 'drew the admiring comment that I would make a fine soldier'.

Eventually he was put on a train going east with about two hundred others. After a long journey through snow-covered forests, they arrived at another camp near Fredericton in New Brunswick. It was warm, efficient and well kept; but Max, by now something of a connoisseur of prison camps, was past caring: freedom was all that interested him, as he told Lotte:

My release order arrived and I was transferred to another camp . . . It is not known when we shall leave . . . This is a purely Jewish camp, in a way the best I have seen. However, I am sick of waiting and fed up to the brim. I just can't understand how a man with a 10 years imprisonment term does not commit suicide ... I am so tired of it all and so homesick that I nearly cry when I think of it.

He spent almost the whole of the dark, freezing month of December in the camp, but Christmas 1940 finally found him on the move once more – to Halifax, Nova Scotia, where he met the Home Office civil servant, Alexander Paterson, who had been sent to arrange the release and repatriation of those he considered to be no security risk. Paterson was a sympathetic, efficient and energetic prison commissioner, whose drive and integrity did much to unlock the diplomatic log jam. The Canadian authorities would not release the prisoners in the country, nor would they allow them to be moved without a military escort – which they would not provide, as '[their] internment was Britain's affair'. The British government responded by sending a single British captain to escort the prisoners back to the UK. On 26 December Max and 280 others – including both Granichstaedten and his father, who had been in a different camp – embarked on the Belgian liner *Thysville* for the return to the UK. They set sail as part of a motley convoy of cargo ships and other vessels, dodging the U-boats while guarded only by a lightly-armed cruiser and a single submarine. The journey was a marked contrast to Max's passage in the other direction: 'I slept in a warm cabin between clean sheets, took a hot bath, brimful, each morning, ate my meals from white table linen in my friends' company, walked in the bracing air on deck or retired to read in a quiet saloon.' The journey also gave him a chance to extend his appreciation of music: the ship had a collection of 1,200 records and there were concerts every day: 'In the camps, concerts were often the only diversions and I learned to appreciate them and on the way back I had leisure to listen to Symphonies and operas and greatly enjoyed them.'

After almost three weeks – they could travel only at the speed of the slowest cargo boat in the convoy – they docked at Liverpool one grey January morning. For many of those on board, including Granichstaedten, their internment was still not at an end – they were taken back to the Isle of Man until the Home Office had confirmed their release. Max was luckier – he was briefly interviewed, but within three hours was handed a train ticket to Cambridge and told to register with the police as an enemy alien when he arrived.

To be back on British territory, able to take a train, to buy a cup of tea, to talk to anyone he liked! The dinginess, deprivation and danger of wartime

Britain could not dampen Max's jubilation at being a free man again. On reaching London he telephoned his parents – it was the first they knew of his return, and they were overjoyed. As it was too late to get to Cambridge he went to Anne Hartridge's house in Northwood, turning up on her doorstep unannounced with his battered brown suitcase in his hand.

He was smiling and sunburnt and I was so *very* happy to see him. But somehow we were both a little shy, and I found myself saying something very ordinary about him looking as if he had just been on a luxury cruise.

He stayed overnight and continued his journey to Cambridge the following morning, where his beaming parents met him at the station. At the end of the month he wrote to Evelyn, a letter full of the euphoria of his return.

I had spent days and nights imagining what coming home would be like, but the real thing surpassed all expectations. People seem to have forgotten all the grudges they ever had against me and to remember just nothing but what a jolly good fellow Perutz had always been . . . I believe the past fortnight has certainly been one of my happiest and perhaps the most successful I ever had.

4

Home and homeland

I know not whether the conviction of being loved, be more delightful or the corresponding one of loving in return.

Charles Darwin to his sister

On his release, Max was immediately caught up again in the excitement of research and the joy of solving problems. Bragg seemed genuinely delighted to have him back and treated him with such kindness that Max inferred (probably correctly) that he had a slightly bad conscience about not having acted earlier to obtain his release. Max wrote to tell the Rockefeller Foundation that he would not now be going to Caltech after all. With admirable patience, Warren Weaver cancelled the grant to the New School of Social Research and approved another to pay Max's stipend for a further year in Cambridge. With teaching staff in short supply, Bragg invited Max to mount practical demonstrations for the students in Peter Wooster's classes, his first experience of teaching.

Virtually no one else in the Cavendish was pursuing civilian research – Bragg was now occupied with advising naval researchers on sound ranging and underwater sonar for the detection of enemy submarines. For Max there was little on offer in the way of war-work other than fire-watching, and Bragg encouraged him to take up his protein research where he had left off. He immediately picked up a suggestion Bernal had made in their joint 1938 paper on haemoglobin and chymotrypsin. Wet crystals gave a beautiful diffraction pattern: once they dried in air they shrank by almost half their volume and produced only a very limited array of spots. Bernal surmised that much of the water in the wet crystals filled the spaces between the molecules, and that if you could catch the crystal at two or three stages between wet and dry, you might be able to distinguish between diffraction spots due to the arrangement of the molecules, and spots due to their internal structure. If there were intensity differences between the related but not identical sets of reflections from each stage of hydration, he suggested, it might provide the means to solve the elusive phase problem.

Max set to work. Gilbert Adair had given him the crystals he used in his pre-war work, but for the new experiments he grew his own, working in the Molteno Institute under the fatherly eye of David Keilin. Horse blood from the local slaughterhouse was his raw material. To separate the red cells from the other blood components, mainly white cells and liquid plasma, he spun test tubes of blood in a centrifuge so that the heavier red cells separated at the bottom and the rest could be removed. To crystallise the haemoglobin, he mixed the red cells with a salt solution that broke open the cell membranes; the haemoglobin would then fall to the bottom of the tube and, over a period of weeks, form crystals. A good crystal for X-ray analysis needed to be about one millimetre across; it took a combination of luck and skill to get good crystals, the skill being in getting the concentration of salt just right. It did not always go smoothly, and under Keilin's patient guidance Max learned to make up several tubes at once, each with a different salt concentration, to be sure that at least one of them would do the trick. By the end of April he was able to report to the Rockefeller Foundation that experiments were 'in full swing again'.

Having picked out the most favourable crystals under a microscope, Max would pack them carefully in his bicycle basket and pedal back to the Cavendish to take his X-ray photographs. At first he used the same technique as before, photographing the crystals sealed in a fine glass capillary with some of the solution from which they had crystallised. The new step was to let the crystals dry, and take more photographs before the drying was complete. By trial and error Max found that he had to dry the crystals extremely slowly, over a period of weeks, in order to catch the intermediate stages. When he finally succeeded, he saw that rather than changing continuously, they jumped from one stage to the next, giving him four stages that he could use as a basis for comparison; he also found that he could arrest the drying at each stage to give himself time to take photographs. His studies of the crystals showed that almost all the shrinkage happened in one dimension: layers of haemoglobin molecules in the crystal packed more closely together as the water between them disappeared, but the thickness of these layers themselves did not change, nor apparently did the structure of the molecules.

It was a thorough and careful study, and Bragg was delighted, writing at the end of the year to tell his friend and fellow crystallographer R.W. James in Cape Town that Max had done 'a very pretty piece of work on proteins' from which they hoped 'to get some direct evidence of the form of the molecule'. The hope turned out to be premature, but the work was the first such study of its kind on proteins and Max's paper, published in *Nature*

the following year, was the first example of the approach that would ultimately be successful: the comparison of data collected from related but not
identical forms of protein molecule. He wrote, somewhat ruefully, to Lotte:
'It's quite an important [paper] and constitutes the first step in the solution of the great problem. In peacetime it would have made a certain
amount of stir, but now very few people remain who still devote themselves to academic problems.' Nevertheless, the work was enough to renew
Bragg's confidence in him, and to ensure that the Rockefeller continued
to pay his salary.

While his first year back in Cambridge was successful professionally, Max's
personal life remained a struggle from which the lab was a welcome haven.
As soon as they could find a suitable house he moved back in with his
parents. Dely, who had been so resourceful on her own when both her son
and her husband were interned, at once resumed the routine emotional
demands that made his life so difficult. At the age of twenty-seven, he had
to account to her for his movements and submit to her reproaches if he
spent more than one or two evenings out. He had to listen to her complaints
about other members of the family: her brother-in-law Arthur who was in
Budapest and had sent several telegrams asking if they were all right, but
no money; and Hugo's cousin Josef and his wife Valerie, who were living
in Cambridge but about whom she had written to Hugo: 'How could you
imagine Josefs being a help to me? He is too ill and she too selfish and too
stupid.' Josef, whom Max described as 'by far the nicest relative we had . . .
a great man, and literally loved by all who knew him', was dying, and his
wife hardly in a position to take on Dely's concerns. He died in the autumn
of 1941. Max was with 'Tante' Vally in the hospital when he died, and
wrote to Lotte that she behaved 'very nicely and with much dignity'.

It was not just living with his mother that was difficult. With his strong
accent and foreign name, Max's alien status was all too obvious. He had
learned since his arrival in Cambridge in 1936 that it was almost impossible to predict the response he might receive from a stranger, and in
wartime it was even more uncertain.

> I had a good idea that the attitude, say, among the English upper class
> went in various shades from the supporters of Mosley to others who were
> great liberals and abhorred the fascist regime in Germany. I also realised
> that among the working class there was much less sympathy for Germany
> than there was among the upper class. I felt very much when I came
> back from internment how nice the lab mechanics were to me; they
> had no antagonism whatsoever.

Nowhere did he feel hostility against him more acutely than in his college, Peterhouse, where he described the atmosphere as 'icy'. The wartime Master, Paul Vellacott, was said to have been gassed by the Germans during the First World War and to be vehemently anti-German; Max assumed that the Master perceived him as 'one of them', despite the obvious evidence to the contrary, and that he communicated the same attitude to the Fellows (with the exception of Max's former tutor Charles Burkill, with whom he remained on friendly terms). Officially, too, his status entailed certain restrictions: he was not supposed to be out later than 10 or 11 p.m., and was required to ask permission from the local police if he wanted to leave Cambridge, even for an overnight visit to London. With occasional all-night experiments at the Cavendish, and his girlfriend in London, one imagines that Max, with his Austrian disdain for rules and regulations, routinely flouted these conditions.

One day in July 1941, he received two letters. The first reported that a young German doctor called Heinz Meyerhof, whom Max described as 'the only internee with whom I formed a close, affectionate bond', had died in a Canadian internment camp. Later, he recalled that Meyerhof had told him he would be a great scientist one day, 'which had never occurred to me'. The letter asked that Max, together with the distinguished German-born neurologist Wilhelm Feldberg, should break the news of her only son's death to Meyerhof's mother. They did as they were asked: 'It was really terrible. The poor woman collapsed on hearing that awful news and we didn't know how to respond.'

Max's own grief at Meyerhof's death was not the only reason he found it difficult to cope with this distressing scene. He had suffered another loss of his own. The same post that brought the news of his friend's passing had also brought a letter from Anne Hartridge, saying that she could not marry him. Since his return from Canada he had found it difficult to gauge her true feelings, though he believed she loved him in return. In fact, though she was touched by his attentions – he once bicycled from Cambridge to London and back to see her – she never thought of him as more than a good friend. Realising that the deprivations he had suffered might make him yearn even more strongly for a home and a wife, though they had never previously discussed marriage, she wrote to make it clear that he should not ask her.

Max was 'doubly shattered' by the loss of his marriage hopes and the realisation that Anne had not loved him as he thought. For many weeks he could not bring himself to tell his parents – Dely had doted on Anne – or even Lotte what had happened. Anne, however, had very sensibly realised that she and Max were not well suited and, as things turned out, she did him an enormous favour. Within a few months he was able to write coolly that 'We were too different in temperament to be a good match.'

This calm acceptance, recorded in a long letter to Lotte in November 1941, came about because he now had a basis for comparison. During the summer Max had an appointment to see Esther Simpson, the administrator at the Society for the Protection of Science and Learning, which had moved to Cambridge to avoid the London bombs. He went to see if they could do anything about his friend Granichstaedten, who had been languishing in the Isle of Man ever since they returned from Canada. The SPSL was now interceding on behalf of internees as well as finding academic jobs for refugees, with some success. Max had previously written an eloquent letter pleading for the organisation's help, and was now following it up in person. When he arrived, however, Miss Simpson was not in the office. Instead, he was 'received by a gorgeous girl who was combing her lovely fair hair'.

The girl was Gisela Peiser, a German refugee one year his junior who had recently begun work in the office as a bookkeeper, translator and general assistant. Born in Berlin and raised as a Protestant, she had moved with her family to Zurich on Hitler's rise to power in 1933, from where her Jewish-born father was able to continue his work as a manager in a civil engineering firm. However, the Swiss authorities allowed him to enter only on condition that members of his family would not seek work in Switzerland. Herbert Peiser accordingly sent his son Steffen to St Paul's School in London, from which he won a place at Cambridge: by the time Max and Gisela met, he had done some research in crystallography and already knew Max slightly. Gisela would have liked to study medicine, but her father blocked her ambitions. She left school before taking her examinations, trained as a medical photographer at the technical university in Zurich, and moved to London in 1938 to work at St Bartholomew's Hospital. After the hospital was evacuated to St Albans in 1940, research work more or less dried up, and she lost her job. However, she had also acquired basic office skills in Switzerland, was fluent in English, French and German, and so was ideally qualified for SPSL. In the course of her work she had seen Max's letter about Granichstaedten.

I was very impressed by the letter Max wrote – the thought that had gone into it. When he appeared at the office, I was very interested to meet the person who had written such an impressive letter. As I wasn't busy, we had a chance to talk for a few minutes.

It was enough for Max to declare in his later account, 'I fell in love with her before I left.' In the ensuing weeks they saw each other as often as possible, and in October they became engaged. However, they decided to keep the engagement secret at first, anticipating – with some justification – that neither set of parents would greet the news with unalloyed joy. Max was fairly sure his mother would want him to marry an English girl, and Gisela's parents would have no opportunity to meet Max until after the end of the war. In his November letter Max finally revealed the emotional upheavals of the previous five months – 'I got engaged, and not to Anne' – to Lotte. He devoted a closely-written page and a half to an almost forensic description of Gisela's physical appearance and personality, penned with an impartiality rather surprising in one newly in love.

She has lovely features and a great deal of pleasantly brown hair, beautiful teeth and a fascinating smile and kindly eyes coloured like a rainbow. She is a bit smaller than I, and of rather large build. She is 26 and born in Berlin which is rather a blow ... She has none of the evil characteristics of her Nationality ... She has a sense of humour, though a bit different from ours in some things, and she has a sense of proportion which I greatly admire and an extraordinary amount of common sense. She is very intelligent and also intellectual, but in a nice, unpretentious and unobtrusive way ... She seems to have a genuine interest for science ... She has a social conscience to which is added a great deal of experience as a social worker at that Society. Finally I think she is efficient and industrious in her work ... One rather serious snag is that as soon as we talk German her northern accent makes me freeze up; there are so many unpleasant associations connected with it and I couldn't get over it yet. So we gave it up and stick to English, which Gisela talks very correctly ... but with an accent that sounds quite pleasant. A less serious snag is that her life in Zurich seems to have had a decisive influence on her taste as regards clothes. She also has Swiss legs.

At about the same time, Gisela wrote to her parents to tell them the news. She was less preoccupied with appearance, but tried to get to the heart of Max's personality:

He [is] ... not very tall, dark, very young-looking ... He has an amazing amount of common sense, of cool and sober vision of any person or situation coupled with great sensitiveness for other people's feelings and reactions. His own reactions seem to be dictated by a marvellous mixture of intelligence, feeling and warmth. I should not think that his feelings will run away with him easily, but they run deep ... He is Catholic, but non-practising. He is a non-Aryan and would be very cross if I used this word a lot ... he hates all this racial terminology and way of thinking.

It is difficult to overestimate how lucky Max was to meet Gisela when he did. This was not just the first girl he met on the rebound from Anne – this really was his ideal match, someone who shared his taste in books, music, people and landscapes, with a sharp mind allied to a steady temperament. She would not play the emotional games he endured with his mother, nor would she fret about the time he would have to devote to research. The history of Max's relationships with women since his teens had mostly been one of rejection and failure, and it seems likely that he would have married anyone who appeared to return his affection. He was perhaps more fortunate than he ever realised that the first woman to do so with conviction should also be so perfectly suited to be his wife.

As Max had feared, Dely didn't see Gisela in the same light. When the couple finally got round to breaking the news to Max's parents, she reacted badly. Although, as Max told Lotte, 'Father likes Gisela a lot and I think Mother would if she wouldn't want to marry me', Dely had a list of objections. She thought Gisela 'too old' at twenty-six, but above all too German, which Dely thought would hamper Max's career. Fundamentally, she was hurt that another woman had a hold on Max's affections, especially as it was obvious that 'Gisela ... is much nearer to me already than Mother ever was.' She turned the emotional screws once more by fretting that if Max moved out and ceased to pay his third of the rent, she and Hugo would not be able to manage financially. Around this time Max had a serious attack of the 'gastric trouble' that was to plague him for the rest of his life. He spent almost the whole of the Christmas vacation in bed on a 'strict milk diet', and made a partial recovery in early January only when a different doctor pronounced that he was suffering from malnutrition as a result of his previous treatment. He confessed to Lotte that he was suffering from a 'low level of temper', which, given his habitually optimistic outlook, could only have meant that he was profoundly depressed. The weather in Cambridge didn't help.

I always belittled the effects which people alleged the Cambridge climate had on you & now I see that suddenly, from one day to another the cold and dampness become insupportable evils, particularly if you sit here all year without interruption.

Hugo was doing his best to make a contribution. At the age of sixty-three, the formerly wealthy industrialist and manager had gone to a college in Letchworth and trained as a lathe operator so that he could get a job in a factory. His courage and determination contrasted vividly with Dely's incessant bemoaning of her lot, redoubled when Hugo was no longer available to help with housework, and Max admired him for it. At the same time he was surprisingly sympathetic to his mother, whose tasks can hardly have been onerous in a household of three adults in a few rooms. From his sickbed he wrote to Evelyn: 'It is frightful to watch her being overworked always every day and hardly to be able to do anything to relieve her.' He assured his mother that he would continue to pay his share of the rent even after moving out. With Gisela's wages of £4 a week and his Rockefeller salary he thought they would be 'quite well off', especially as he now discovered that Gisela's parents were wealthy, something he had never suspected. Lotte chipped in with an offer of $20 per month for the parents, and Max asked Franz to contribute the same, so Dely's economic arguments were gradually undermined. She nevertheless continued to exercise her 'marvellous instinct for "poisoning" the atmosphere between us', Max grumbled to Lotte.

Meanwhile the Peiser parents sent a letter giving their blessing. While they had not managed to meet Max in person, they had received detailed reports on him both from Gisela's brother Steffen and from her aunt Annemarie Wendriner, who lived in London with her family and was very close to her niece. These had been uniformly positive about Max (his 'delicate tommy [sic]' being the only concern), moderately reassuring about his financial situation, and clear-sighted about his parents. Annemarie dismissed Dely as a 'silly woman', 'one of those poor creatures who can't forget the past', but spoke admiringly of Hugo: 'He seems to be at peace with himself and does with quiet dignity what present conditions require.'

Hugo finally landed a job at a workshop in Cambridge in early March 1942 – initially for six nights, mercifully soon changed to a day shift, for three pounds and twelve shillings (£3.60) a week. This was what Max had been waiting for, so that he could fix the date of his wedding: he and Gisela settled on 28 March. Making the arrangements was a further source of anxiety, but the decision to opt for the Cambridge registry office (against strong opposition particularly from Gisela's parents, who wanted to give her a full-scale traditional wedding in London) was perhaps the easiest, as Max explained to Evelyn:

It is really impossible for us to do all the arranging that would be necessary for us to marry in church and Gisela doesn't like the idea of two
people of different religion being married in a church belonging to a
third and neither of us believing in either of them.

The day was a great success: 'Everybody, including the parents, was
as happy as could be.' Hugo and Dely, Fritz Eirich's wife Wopp, and
Gisela's brother Steffen acted as witnesses to the brief ceremony at the
Shire Hall. Gisela wore navy blue trimmed with white; Dely was so well
turned out in a red two-piece, hat, gloves and pearls that Annemarie
Wendriner wrote to Gisela's parents that 'for the sake of her decent
looks I gladly put up with some of the deficiencies of her character'.
Afterwards everyone went to London for a party at the Wendriners'
house in Holland Park, which was 'not only gay but *gemütlich*'. The
guests ('only very close friends and no relatives') included Evelyn and
Max Machin, Anne Hartridge and Tom Hughes; Heini Granichstaedten
was invited but was unable to make it from Edinburgh, where he had
gone to live on his eventual release from Douglas. Steffen read out an
address written (in English) by Gisela's father, a moving and heartfelt
speech that began with a quotation from *Henry V* in tribute to England's
stand against the Nazis and ended with a reference to the couple's shared
experience of oppression:

> [Y]ou have witnessed changes so radical and revolutionizing as, in the
> ordinary course of history, would suffice to fill up centuries. These moving
> events have, for both of you, been – to a large extent – cruel but they
> have simultaneously given you the immense benefit of an experience to
> make you wise far beyond your age. You have seen wealth and splen-
> dour melt away, power and might crumble into nothingness . . . But you
> have, at the same time, learned the wonderful lesson that the world will
> not submit to material power and violence and that in the end the
> subtle forces of mind and heart will always prevail and triumph.

The bride and groom spent three days in London before setting off for
a week at The White Lion, a modest inn in the village of Hartley Wintney
in Hampshire. They made excursions to Salisbury and Winchester, which
Max described as 'the prettiest town I ever saw in this country'.

Max and Gisela set up their first home by renting a room in a house
belonging to a woman whose husband was away in the army. It was cheap
and clean, and Max pronounced himself well satisfied.

Gis and I are very happy, in fact I cannot remember being as contented as I am now ... I am terribly glad to have got away from home. The views of Mother and me are diametrically opposed on most subjects and her increasing despondency drove me nearly crazy.

The happiness clearly had an effect on Max's health. By June Annemarie Wendriner reported that he had put on a lot of weight and was eating 'practically everything', much to his mother's surprise. (One can't help thinking that getting out of her orbit had a great deal to do with it.) He and Gisela settled into a routine, which he set out in detail in a four-page letter to her parents. Although Gisela was still working at the SPSL office, it is clear that shopping, cleaning and cooking were all her sole responsibility: soon after the marriage she gave up working on Wednesday afternoons so that she had more time for domestic tasks. Even so, the young couple employed a 'char' to come in twice a week and help her. Max would leave work at 7 p.m., expecting to find his supper on the table when he arrived home. After dinner, if they did not have guests for coffee, Max would 'read or write or work' while Gisela 'mends or does her accounts or irons – anyway she always finds something to do'. These early days set the pattern for the rest of their married life. Gisela's priority was always to manage the household with such efficiency that Max never needed to trouble about anything other than the family finances, and could devote all his attention to his work. He always acknowledged that the happy home she gave him 'was the ground on which my work grew and prospered'.

A conference on the application of X-rays in industry provided the first opportunity for Max and Gisela to entertain as a couple: Dorothy Hodgkin and her husband Thomas came to dinner, and the following day the newlyweds threw a coffee party for ten people. Trivial events, perhaps, but for Max, after two years of dislocation, disappointment and family discord, they were a confirmation that he had finally come home. 'Gis and I are happy together and I think it will last,' he told Lotte. 'We haven't even had a single row so far.'

By their first wedding anniversary they had moved into their own flat, a tiny attic conversion close to the centre of Cambridge known as the Green Door (a euphemism for the back door of the Old Vicarage in Thompson's Lane). The house, with a church on one side and a synagogue on the other, dated from the sixteenth century and was owned by a tobacconist and his wife, Mr and Mrs Kitteridge, who lived on the ground and first floors. A lecturer in the Cavendish had been renting the top-floor flat, and when he moved away Max and Gisela took over the lease (at a rent of 25 shillings, now £1.25, a week). Both found it enchanting, as Max enthusiastically told Evelyn.

> All the lines are crooked and all the angles oblique . . . with some nice pieces of furniture, flowers in the window boxes, pleasant pictures and a spot of redecorating . . . it looks most *gemütlich* and attractive. We both love it . . . The flat is selfcontained, with all modern conveniences including a bath in the kitchen. The parents regard it as too bohemian and were a bit shocked at first . . .

The flat had no telephone, central heating, washing machine or refrigerator, leaving one rather to wonder what Max counted as a 'modern convenience'; the answer probably is that it had electricity and an indoor lavatory, by no means universal in modest homes of those days. It was boiling hot in the summer and freezing cold in the winter, and the roof leaked if it rained heavily, but for the first time they were in their own home rather than rented rooms, and Max described the seven years they spent there as very happy. The Green Door was to feature in the history of molecular biology for years to come as one researcher after another succeeded the Perutzes as tenants.

Throughout the summer and autumn of 1942, Max and his assistants continued to work on their haemoglobin study and, as he reported to his paymasters at the Rockefeller, 'new and exciting results kept cropping up every few weeks'. At the same time, he began to acquire some teaching experience. In the autumn term of 1942 David Keilin asked him to give a series of eight lectures to the biochemistry students in the Molteno Institute on X-ray crystallography as applied to biology. The first attempt did not go down as well as he would have liked, but Max quickly got the hang of what he was doing:

> In my first lecture I introduced lattice theory, trigonometric functions, and Fourier series very lucidly, I thought, but half the students failed to turn up for my second lecture. That sobering experience made me look around for other ways of explaining X-ray diffraction. The following year I replaced my forbidding lecture with a non-mathematical, largely pictorial introduction called 'Diffraction Without Tears'. This time my audience stayed . . .

The University was sufficiently impressed with the first lecture series to ask Max to make it an annual feature. The number of students remained small in wartime, however, and Max was delighted when, at the end of December 1942, he was finally recruited to war research. The detail of the project, codenamed Habbakuk [sic], is covered in the next chapter, but it meant that for the first eight months of 1943 he lived and worked in London during the week, and from September to November he was in Canada and the US. In order to make this trip, he needed British travel documents: the US Embassy refused to issue a visa to an enemy alien. The Chief of Combined Operations, who had taken a personal interest in Habbakuk, was Louis Mountbatten, a cousin of King George VI and a distinguished naval officer. He requested the Home Office to have Max naturalised as a British subject on the grounds that it was essential to the national interest. A Special Branch policeman, 'Sergeant Smith of Scotland Yard', appeared one evening to interview him at his London digs. To Max's amusement he asked a series of perfunctory questions as to his loyalty, previous convictions (Max admitted to cycling without lights in Cambridge) and contacts with aliens overseas, adding that normally all the answers would be checked 'but in your case we won't bother'.

> I was excited and in good form and told him the story of my life as best and short as I could. He seemed gratified with all he heard . . . Told me all about the bombs that nearly hit him, the plane that came down in his back garden, the 15,000 egg deal on the black market which he discovered. Of course he had lots of stories that would stagger me if only he could tell them . . . He only departed after two hours.

The only other formality was for Max to swear an oath of allegiance to the King before a solicitor. Four days later, on 30 August 1943, Max's naturalisation certificate came through. The head of the Nationality Division told him it was the quickest job the Home Office had ever done, and that fewer than fifty naturalisations had been granted since the war began. It was a moment of the profoundest joy to Max, shared by all who knew him well. He stood drinks for all his colleagues on the project, all Englishmen and women who probably didn't fully appreciate the significance of what had happened, but were quietly amused that their serious colleague was capable of such unbridled delight. Max reassured himself in a private diary that he belonged among them: 'Everybody was very friendly. I felt to-night that I must be popular amongst these people who share my happiness.'

Gisela was of course thrilled – her own naturalisation papers, automatically granted by virtue of her marriage, would arrive a few days later. Max tried to put his feelings into words when he wrote to Lotte with the momentous news the day after his arrival in New York on 17 September:

> Any success I may have had in my war work has been rewarded in the most overwhelming manner conceivable, quite out of proportion with

the importance of my achievements . . . I was transformed into a British subject on Aug. 30th. You will imagine my feelings . . . The knowledge to have a home once more and a country to belong to and to have achieved one of the most cherished ambitions of my life was quite overwhelming.

All that was needed to complete the picture of domestic bliss was children, and to the couple's great joy Gisela found herself pregnant in the spring of 1944. All the same, the prospect threw up a new concern. Max wrote to his lawyer that he and Gisela were 'very anxious our child should not have the handicap of a foreign name'. He does not elaborate, but the implication is that he believed his name had been a barrier to his own acceptance in some parts of British society. The lawyer advised that Max could not change his child's name without changing his own. Gisela had raised the possibility that he might do this, as very many other refugees had done, but Max decided after some thought that it would 'do more harm than good' – presumably in relation to the scientific reputation he had begun to acquire as Perutz. Instead, they hit on the compromise of giving the child an additional English-sounding surname that could be used instead of Perutz 'at school etc', and which the child could then adopt by deed poll on reaching the age of twenty-one. 'Michael Perth Perutz' was the name suggested, but this was clearly inappropriate for the little girl who arrived on Boxing Day 1944. She was named Vivien Angela Perth Perutz. When her brother Robin Noel was born five years later, he also had the extra surname: but after the war Max and Gisela lost their anxiety about playground chauvinism and both children have always been known by their father's surname.

Shortly before Vivien's birth, Gisela gave up her job at SPSL (much to the Committee's regret), and until the children went to school she devoted herself to them full time. She then took up counselling as a trained volunteer with the Marriage Guidance Council – more interesting than bookkeeping, though not the medical career she had once dreamed of. Max adored his children: as far as his work permitted, he played with them, wrote them stories and always took an interest in their activities. However, it clearly never occurred to him that jobs such as cooking, cleaning, shopping and washing would be undertaken by anyone other than Gisela or the au pair girls they began to employ after Robin's birth. He had also got into the habit of using his wife as a secretary, for instance in helping him to compile a set of index cards on all his reading. Over the years there are hints in the letters that Gisela found her responsibilities – which she took extremely seriously – something of a strain, and, given the limitations of their first proper home, which Max detailed in a letter to Franz, this is hardly surprising:

It is not easy in our minute little flat, where we have no room for the baby who sleeps in our bedroom in the day time and our sitting room at night. Besides you must remember that we possess no washing machine and no bathroom – everything has to be done in the tiny little kitchen. Of course there is also no refrigerator, vacuum cleaner or telephone and none of the shops deliver any goods to your house. There is no proper garden – the baby [in her pram] is put into a sort of no man's land between the churchyard and the synagogue, and the people from the synagogue complained once that Vivien interrupted their service by her crying. Nevertheless we have much to be thankful for . . .

Despite their religious doubts both Max and Gisela were in favour of Christian upbringing, and Vivien was duly christened to the delight of all.

Mother put all her jewels on and looked extremely elegant. She still always manages to look as though she had just emerged from a shop in Paris, and is by far the smartest dressed woman in Cambridge . . . Vivien looked very sweet in a christening frock that had already been worn by Gisela and her mother . . . Father admired me because I looked after the baby, a thing which he could never do . . .

The christening took place on 15 April 1945, just after news reached Britain that Vienna had been liberated from the Germans without the long siege that Max had feared. 'Vienna is free,' he wrote, though his joy was premature: the eastern part of Austria, including Vienna, remained occupied by the Soviet army for another ten years.

With the end of the war in Europe, news began to arrive of the fate of friends and relatives. Most shocking were the deaths of Max's uncle Arthur (Hugo's elder brother) and his wife Mitzka. They had been living in the Hotel Hungaria in Budapest when the Nazis arrived in March 1944. Although they were not picked up in the first wave of arrests, Arthur Perutz refused to listen to those who begged him to flee, saying he was too old and ill (he had had a foot amputated because of a circulatory disorder) to go into hiding. Three days later he and his wife were arrested and jailed in Budapest. Felix Perutz, who had sent his wife and children to safety in the countryside and astonishingly kept the family business running from Budapest throughout the war, bribed everyone he could think of to let them out, but to no avail. Later in the year, Arthur was transported to

Mauthausen and Mitzka to Auschwitz: both either died on the journey or were gassed.

Felix's mother Ida had thrown herself out of a window to her death in 1941 as the German troops approached Prague. Max's cousin Leo, brother of Gina Haurowitz, also died in Auschwitz; Leo and Gina's mother Marie died of 'exhaustion' – probably starvation – in the Lodz ghetto in Poland. Many other more distant relatives who had stayed in Prague died in the camps: the Jewish memorial at the Pinkas Synagogue in Prague lists many Perutzes among the 77,297 Czech victims of the Nazis.

Max compared notes with Gina's husband Felix Haurowitz, who had accepted a professorship in Istanbul and taken his family there in April 1939. Through Haurowitz and others he learned of the terrible toll the war had taken on their family. He sent his commiserations to Gina on the loss of her brother Leo and her mother, and added that they were 'all very upset' about Arthur, especially Hugo who 'broods over it when he lies awake at night'. But his letters passed briskly on to talk of other things, principally Max's work on haemoglobin which he described to Haurowitz in detail. It was not a time for looking back. Max's mood comes over clearly in a letter of 21 May 1945, just after VE day: 'I am very happy here and extremely fond of the country and its people. I am particularly happy in Cambridge and look forward to staying here after war [sic] when I am supposed to get some definite job at the Cavendish Laboratory.'

Later in the year, by which time he had secured a research fellowship and the prospect of new collaborators, he reflected further in a letter to Lotte on the happiness he now enjoyed. His thoughts were prompted by the first Cavendish Dinner to be held since before the war, a glittering black tie occasion for all the Cambridge physicists, that took place in December 1945:

> As I walked across Trinity Great Court . . . with all the cheerful lights where all had been in dismal darkness for so many years, I felt pleased for once with fate. To have survived the war, to be happily settled with a family as a British subject, to have a job at the Cavendish Laboratory and a room in the grand new building which I saw being planned and where I dreamed to work one day, to have at last a first-rate team to work with, all this good fortune crossed my mind as I walked up to the Old Combination Room where seventy physicists, mostly young, cele-brated the end of the first term's work in peace time.

5

Mountains and Mahomet

There is a real world independent of our senses: the laws of nature were not invented by man, but forced upon him by the natural world.

Max Planck, *Philosophy of Physics*

Max had known from the moment he arrived in Cambridge that he had found an ideal setting in which to pursue a career in scientific research: had he stayed in Vienna, he would never have achieved as much. However, in other respects Cambridge fell far below his ideal – literally, in the case of its geographical location.

Cambridge rises fifteen metres at most above the surrounding fens; Max was by nature and upbringing a man of the mountains. By the time the war in Europe ended in June 1945, he had climbed nothing taller than a few damp and rocky peaks in Snowdonia and the Lake District since his Jungfraujoch trip in 1938. It was an appalling deprivation. Max blamed the poor health and spirits he had suffered on the gloomy Cambridge winters. As soon as such a thing became possible, he began to plan a trip to Switzerland. He had not yet met Gisela's parents – nor they their new granddaughter – and it was only natural to want to make a family visit. But Max had another motive: 'My own prime purpose was to become once more an active mountaineer instead of having to live on the glory of the past, and of course to get fit.'

The Peisers pressed them to come as soon as possible, and eventually they decided to leave for Switzerland at the beginning of May 1946 and not return until the end of June. This evidently caused a certain amount of consternation in the Cavendish. Max was almost the only member of staff who had remained in post throughout the war; Bragg had just obtained him an ICI research fellowship, which secured his future for a further three years, but Max seemed to feel not the slightest pang of guilt.

Bragg was a little staggered when I asked for 2 months' leave, but I don't think he will raise any serious objections, seeing that I am one of the more hard-working members of the department. I am already enjoying myself studying maps and planning trips!

The young couple and their baby daughter arrived in Zurich after a long train journey in early May. The tickets had entirely exhausted their funds, but Gisela's father generously supported them throughout their visit. The Peisers lived in an elegant flat on one of the hills overlooking the lake. Fresh from the deprivations of wartime Britain, Max and Gisela were astonished at the apparent affluence:

> Everyone goes about in new clothes, on new bicycles, wheels his children in new prams, lives in houses which are newly redecorated and sometimes also new ... All Gisela's friends live in flats such as we haven't even started to dream about, and some of them aren't really richer than we.

Max found his father-in-law Herbert Peiser to be 'a perfectly charming old man, with a subtle wit and humour, an immense knowledge, a fabulous memory and many other enviable gifts'. His wife, Nelly, was a very different proposition, however. An artist and potter, she had developed paranoid tendencies that led to her suspecting even her nearest relatives of conspiring against her. Max did his best not to become involved, but eventually he found the injustice of her attacks insupportable.

> Gis had a terrible time in Zurich, partly as a result of the shock of finding her mother in that state, partly because her mother is full of imaginary grudges and jealousies against her and singled her out for attack. I ... got so annoyed when she complained to me about Gis that I told her that in my view she is just imagining things, that Gis is an angel of patience and that she should stop her scenes. This only made things worse, of course.

Max was able to escape from this rather fraught situation for part of the time as he had scientific engagements in the city, giving a talk on haemoglobin at the prestigious technical university ETH, and discussing both protein research and glaciology. He also re-equipped himself for skiing, having lost the Norwegian hickory-wood skis that Gerald Seligman had bought him in 1938. With some relief, he and Gisela left the Peiser household after two weeks for a planned holiday in the mountains. Leaving Vivien in the care of her doting grandparents, they borrowed bicycles and travelled by bus and bike to the alpine resort of Lenzerheide, where Gisela had spent many of her childhood holidays. They stayed on a local farm and were spoiled to death by the proprietors, who lavished on them quantities of eggs, butter and cream – rare delicacies in Britain where rationing continued to be at least as severe as it had been during the war. The scenery was at its picture-postcard best: the mountain-tops covered with snow, while the meadows round the village were filled with alpine flowers. The sun shone every day: Max revelled in the sensation of it burning his skin, and experienced an exhilaration he had not known for almost a decade.

But this was not enough. After a week they packed their skis and took the little train to Pontresina, a village surrounded by larch and pine 1,800 metres above sea level in the Engadine valley. Having settled into their lodgings, they looked for a guide to take them ski mountaineering. Max had been a member of the Swiss Alpine Club before the war, and he asked to be introduced to the local representative Herr Golay, the village pharmacist. Golay's intimate knowledge of unorthodox routes in the Alps was to delight Max (and later his son Robin) for many years to come. On this occasion he did not accompany Max and Gisela, but gave them the keys to the Coaz Hut, up at 2,385 metres, at the foot of a summit called 'The Capuchin' and which could be reached only via a long walk up the Val Roseg.

For all her Swiss holidays, Gisela had never attempted anything more strenuous than gentle meadow walks. They were now above the snowline with not a soul in sight. Fear and loneliness reduced her to tears, but Max was not for turning. They eventually reached the Hut, a solid two-storey stone chalet with space for thirty to sleep, and found it quite deserted. In the morning they awoke to the breathtaking view of the surrounding mountain tops, and breakfasted surrounded by families of marmots who played and tumbled nearby. They stayed a week, Gisela content to spend her days basking in the sun while Max went off skiing on his own:

> I am not good enough really either on foot or on skis, and I prefer looking round and making no effort. It certainly is lovelier than we could have dreamt in the seven long, grey years in England . . .

Then came the climax of the trip, the whole reason Max had been so keen to 'buy gravity' by climbing so high: putting on their skis once more, they sped through virgin snow all the way back to Pontresina. For Max, it was the ultimate reassertion of the freedom that had been so cruelly denied him in 1940. However, it was to be the only time Gisela would accompany him on such a taxing expedition. He reported to Evelyn that: 'Gisela likes the views, but can do without the 4–6 hour ascents which most of these high mountains require.'

Gisela returned to Zurich, where she found Vivien in great form, having made good progress towards walking. Max remained alone, though he did feel it necessary to excuse himself to his in-laws for his long stay in the mountains:

It will help me to consolidate my recovered physical strength, and to stand the stress of work at home without the tiredness, colds and tummy upsets which have hindered me much during the past four years. I wish Gisela could have the same length of stay, as she was in very poor condition too when we left Cambridge, but she does not seem to flourish up here as I do.

At the end of June the sun came out, and Max set off, alone, on a journey he had long dreamed of making, cycling to the Italian lakes and then returning to Pontresina from where he finally managed to climb one of his favourite peaks, Piz Palü. Soon afterwards he returned to Cambridge ahead of Gisela. Discovering that Evelyn had had a difficult time with the birth of her third child Blaise, he lost no time in persuading her of the recuperative merits of a Swiss holiday.

There is something about the mountains and the flowers, the pine-woods and the tidy little villages, the hot sunshine and the delicious food that turns a nervous war-weary middle-aged-feeling 'flu-ridden lowlander into a vigorous youth full of pep and ideas. In fact I felt like 40 when I arrived and like 20 when I left.

The story of this visit illustrates a facet of Max's character that many who knew him only bent over a laboratory bench would scarcely have recognised. His passion for climbing and skiing could at times outweigh all his other interests, including his scientific research and his family. It was more than just an emotional attachment – it was physically necessary to him to spend part of each year at high altitude if he were not to succumb to ill-health and low spirits. When he had left Austria in 1936, it was not obvious that he would be able to meet this need: skiing, then as now, was an expensive hobby for anyone based in Britain. His lucky break, that had not only given him regular fixes of mountain air, but had also led indirectly to his acceptance as a British subject, was Gerald Seligman's Jungfraujoch expedition of 1938. A wealthy amateur glaciologist, Seligman was interested in studying how the small, fragile snowflakes that fall on the top of an actively forming glacier in winter are transformed into huge ice crystals at the glacier's snout in the valley – a process that takes hundreds of years. Max, knowing nothing of the subject at this stage, thought the problem 'humdrum', but could not resist the temptation of a summer in the high Alps.

Max's credentials as a mountain man were already impeccable. A mountain diary started in his teens first lists a summer expedition on a glacier near St Moritz in 1926, when he was twelve years old, and over the following decade records an impressive tally of peaks in the Austrian Tyrol and Swiss Alps that he had ascended, either on winter ski tours or summer climbing expeditions with family and friends. Piz Palü, the 3,905-metre peak he climbed again in 1946, is recorded matter-of-factly ('Verhältnisse wunderbar' [conditions marvellous]) during a family trip to St Moritz when Max was sixteen. Almost every Christmas, Easter and summer holiday produced entries for the diary.

His first preparatory trip to the Jungfraujoch with Seligman in August 1937 he recorded as an almost entirely wasted climbing opportunity: '31.VIII–8.IX 1937. Fast die ganze Zeit mit Gletschforschung beschäftigt [almost the whole time busy with glacier research]'. He did ascend the two nearest peaks: the Jungfrau (4,158 m) with Dr Philip Bowden of Cambridge and other companions; and a few days later the Mönch (4,105 m) on his own. Clearly he didn't think this was enough:

> We . . . had four days of creeping about glaciers, climbing into crevasses and endless discussion on snow and ice . . . Dr Bowden . . . and I observed with much regret that it might look rather funny if we went away on our climbs after Seligman had paid all our expenses. Consequently we had to pretend a fierce interest in glaciology and to make sweeping and final statements on the mechanism of glacial movement, which were substituted by other final theories half an hour later.

The research party set off in earnest in early June 1938. The Jungfraujoch Research Station had been opened in 1930 as a centre for high altitude research projects in all disciplines. It hugs the flank of the Jungfrau in the Swiss Alps near Grindelwald, 3,300 metres above sea level, but requires no particular skill in alpinism to reach it: a wonderful mountain railway, completed in 1912, drops you more or less at the front door. Max reported to Evelyn that life there was:

> luxurious . . . I have got a pleasant room to myself, magnificent food, a fairly intelligent occupation, a good wireless set, plenty of books, grand surroundings and considerable peace of mind. I never read any newspapers and try to forget about my numerous worries.

It was all much more fun than he expected, thanks largely to the presence of the young physicist Tom Hughes: Max credited his 'inexhaustible fund of dirty stories' with the party's general good humour. His only grumble was that Seligman would not let him climb by himself, and no one else in the party was interested. He had been there nearly two months before Lotte arrived for her visit, and he was able to bag some more peaks for his diary.

In the meantime, Max's doubts about the interest of the research had gradually begun to evaporate. His chief role was to examine ice crystals collected at various depths near the top of the Aletsch glacier, and to compare their size and orientation with the crystals at the glacier snout. Swiss scientists had already established that glaciers move at a rate of a few centimetres per day, and that the centre of a glacier moves faster than the edges, but the internal mechanism that accounted for this movement was unknown, and the subject of much debate. Did layers of ice slide over one another? Or did the whole glacier deform in some continuous fashion, for example by individual crystals moving in relation to one another? No one had looked at the structure of the ice crystals in a sufficiently rigorous way to answer this question.

Max began by organising the rather special laboratory he would need to study the crystals. It consisted of a cave in the stationary ice that clung to the rock below the research station, reached via an ice grotto that had been dug for tourists decades earlier, with a constant 'blister-ingly cold' temperature of $-4°C$ – 'It looks very romantic and blue with a dark low passage through the glacier leading up to it, amongst tourists only the pretty girls are admitted – in consequence we have had no visi-tors so far.' There he set up a polarising microscope, a camera, and cutting equipment designed to shave off sections of ice half a millimetre or less thick that he could mount on glass microscope slides. In order to work in this room, he wore a large, padded airman's boiler suit with a sponge over his nose and mouth to prevent his breath fogging the microscope's eye-piece, and bags of hay on his feet to keep them warm. If the hilarious photograph of this outfit is to be believed, he left his head bare.

To collect samples, he and Hughes would ski down the glacier to the deepest crevasse they could find (the deepest sample they collected was at 28 metres). Max would then climb into the crevasse via a rope ladder, dig away the first 40 cm or so of the wall to reach ice that had not been weath-ered, and then saw out a small cube of ice which he dropped into a thermos flask cooled with solid carbon dioxide. Then he had to climb back up the ladder, plod back up the glacier to the laboratory, and carefully cut the ice into sections. They also hammered vertical rows of steel pins into the crevasse walls, returning a couple of weeks later to see how far each pin had moved out of line, and therefore how much faster the top end of the line was moving than the bottom. 'Notwithstanding the destruction of our apparatus by an avalanche soon after the experiment had come into operation', laconically records the scientific report of this study, the first of these experiments found that for each metre of depth, the ice travelled one millimetre less in a day than the ice at the surface.

Max plotted the size, shape and orientation of the ice crystals from each sample, and saw clear differences from samples taken from different levels in the glacier. Near the surface, where in summer the ice melts during the day and freezes again at night, he found that the crystals sat at right angles to the glacier surface – in other words, aligned with the temperature gradient from warm to cold. Here, there are distinct air spaces around the ice crystals, allowing the warmer meltwater to percolate around them. At greater depths, the ice is denser, with few, isolated air spaces so that water no longer flows. Here, where the temperature stays much the same all the year round, the ice crystals no longer line up neatly, but point in all directions. Max and his colleagues concluded that as the glacier flowed and settled, the ice crystals slowly tumbled against one another until they lay at every orientation. Further down the glacier the picture changed again: the crystals increased in size, from a few millimetres to as much as 10 cm, and once again appeared to adopt a common orientation. Max's studies suggested that the shearing action of the glacier moving at different rates at different depths forced them to take the line of least resistance, with their flat bases parallel to the direction of shear. Huge layers of ice slipped over one another, and at the same time individual crystals slipped on their bases and merged with their neighbours under the huge pressures.

It wasn't the secret of life, but Max and his colleagues had collected important new data and added to the sum of human knowledge on glaciers. Seligman was extremely pleased with the outcome, and Bernal pronounced himself 'very excited'. Max himself felt ashamed of his earlier disdain for the project: 'This taught me to be humble in my approach to Nature; wherever you look closely she will reveal to you some beautiful, unsuspected new facet.'

The results were written up for a short paper in *Nature* in early 1939, and in a much longer and more detailed one later in the year in the *Proceedings of the Royal Society*, co-authored by Max and Seligman and communicated by Lawrence Bragg. (This was the very journal that had rejected Max's paper on the radioactive nodules with such hauteur only a year previously.) 'A crytallographic investigation of glacier structure and the mechanism of glacier flow' still stands as the authoritative voice on the subject.

Three years later, Max's knowledge of ice had a bigger impact on his prospects for building a life and career in Britain than he could possibly have predicted. His chance came as one of the more bizarre outcomes of the fascination with boffins that characterised Winston Churchill's conduct of the Second World War.

Max had worked quietly in Cambridge after his return from internment in January 1941, steadily making progress in his work on the swelling and shrinking of haemoglobin crystals. Then, around the time of his marriage to Gisela, in the spring of 1942, there was a surprising interruption to his daily routine. A mysterious telephone call demanded that he present himself at an apartment in the Albany, the elegant building in Piccadilly that provided a London base for a number of Members of Parliament, distinguished writers and others. The door was opened by:

> A gaunt figure with a long, sallow face, sunken cheeks, fiery eyes, and a greying goatee, who was camped out amid piles of books, journals and papers, and cigarette butts lying scattered on oddments of furniture. He looked like a secret agent in a spy film and welcomed me with an air of mystery and importance . . .

This bizarre individual was Geoffrey Pyke, who had somehow contrived to become attached as a scientific adviser to Combined Operations Headquarters, the War Office department recently set up under Lord Louis Mountbatten to coordinate the armed services in their projected invasion of Europe. Pyke – long-haired, unwashed, unshaven, usually without socks – sharply divided those who came across him: while Mountbatten regarded him as eccentric but brilliant, many others thought him dangerously unhinged. Pyke had worked as a journalist in Berlin during the First World War, been imprisoned there as a spy, escaped, made – and lost – a fortune on the metal exchanges, founded the experimental Malting House School in Cambridge where Susan Isaacs laid the foundations of educational psychology in Britain, and organised a secret survey of public opinion on war in Hitler's Germany.

When the Second World War broke out he was determined to contribute to the Allied strategy. Through the Cabinet Minister Leopold Amery he obtained an introduction to Mountbatten, who liked unorthodox thinkers and took Pyke under his wing with surprising indulgence. The same open-mindedness led Mountbatten, a great-grandson of Queen Victoria, and distinguished naval officer, to appoint the known Communist J.D. Bernal as Scientific Adviser to Combined Operations early in 1942. (There he came nominally under the direction of Captain Tom Hussey RN, who was Director of Experiments and Staff Requirements – and my great-uncle.)

The reason for Pyke's mysterious summons to Max in the spring of 1942 was to ask him about tunnelling in glaciers, in support of his scheme to use units of highly-trained special mountain forces to defeat the Germans. He had been sent to the US to liaise with the American military, but they soon got fed up with him. While he was sidelined, and living in a hotel room in New York, his fertile mind began to develop another idea. The Allies desperately needed forward bases from which their aircraft could conduct raids on occupied Europe. The idea of floating airfields had knocked around Combined Ops for a while, but the expense in materials and production seemed prohibitive. Pyke's brainwave was that they should use ice. The more he thought about this, the more he believed it could offer a decisive strategic advantage. He began to write a document setting out his thinking for Mountbatten. No scientist himself, he needed evidence that the idea was technically feasible. The collaborator he found (probably on Bernal's advice) was Professor Hermann Mark, then at the Polytechnic Institute of Brooklyn, who had recommended Max to Bernal in 1935. Mark began a series of studies of the strength and freezing rates of ice, and on 24 September 1942 Pyke's proposal, more than 250 pages long, finally arrived on Mountbatten's desk.

Anyone reading this document today must be baffled that Mountbatten didn't quietly have Pyke confined to an institution. It was headed 'Mammoth Unsinkable Vessel with functions of a Floating Airfield'. Pyke explained in his covering letter: 'The cover name for this second project, because of its very nature and partly because of you, is HABBAKUK, "par ce qu'il était capable de tout."' He misspelled the name of the Old Testament prophet Habakkuk whom Voltaire had described as being 'capable of everything'. No one seems to have corrected him, and the dozens of thick files in the National Archives referring to the project all retain the misspelling. He argued that the Allies needed 'such a number of aircraft carriers that you need have no more reason to hesitate for fear of losing one than you need to hesitate in risking, say, a hundred tanks'. As icebergs were unsinkable and melted slowly, and as ice could be made for a fraction of the cost of steel, he proposed the construction of a large fleet of enormous 'bergships'. They should be heavily armed and used to attack and destroy occupied harbours, after which they would spray 'supercooled water' to immobilise personnel and vehicles. The rest of the document suggested strategies for the remainder of the war, involving bergships in every conceivable situation including the invasion of France.

Mountbatten was no fool. Knowing that he did not have the scientific background to judge the technical basis of Pyke's scheme, he handed it to Bernal. Bernal realised at once that much of the document was science fiction – supercooled water had been made only in tiny quantities in highly artificial conditions. However, his nature was always to encourage rather than discourage – Max would say later that he 'lacked critical judgement' – and he wrote a summary saying he thought there was enough in the bergship idea to be worth pursuing. Mountbatten was delighted that Bernal had condensed the original '250 page novel' into a couple of pages 'without really losing any of its essentials'.

Lord Cherwell, Churchill's chief scientific adviser, an Oxford physicist known to all as The Prof, saw that building an untried vessel out of an untried material was very risky and opposed the plan from the start, but Mountbatten had Churchill's ear, and personally put the idea to the Prime Minister. On 7 December 1942 Churchill sent a minute to the Chiefs of Staff Committee saying:

> I attach the greatest importance to the prompt examination of these ideas ... the advantages of a floating island or islands, even if only used as refuelling depots for aircraft, are so dazzling that they do not at the moment need to be discussed ...

He saw the need to control the resources devoted to the project, suggesting that instead of building the vessels from scratch, they should cut them out of the Arctic ice sheet and tow them south.

> The scheme is only possible if we make Nature do nearly all the work for us ... The scheme will be destroyed if it involves the movement of very large numbers of men and a heavy tonnage of steel or concrete to the remote recesses of the Arctic night.

A few days later the Chiefs of Staff Committee gave Habbakuk the go-ahead. By this time Bernal and Pyke had suggested that the Canadians should also be recruited because of the need to carry out any large-scale construction in a cold climate. Cherwell seized on this suggestion with alacrity, proposing that Pyke and Bernal should be sent to Canada as soon as possible.

A minute of 27 November mentions the need to take on a 'scientific assistant'; the following month a further document (clearly typed by some hapless clerk at Bernal's dictation) lists research work to be carried out in

Cambridge by 'Dr Max Pellets'. On 23 December Bernal visited Cambridge and obtained Bragg's permission for Max to work for Combined Ops, undertaking a series of small-scale experiments on the strength of ice in the University's Low Temperature Research Station. Max himself remembered that this visit was preceded by another summons to see Pyke:

> [He] told me, with the air of one great man confiding in another, that he needed my help for the most important project of the war – a project that only he, Mountbatten and . . . Bernal knew about. When I asked him what it was, he assured me . . . that he had promised to keep it to himself, lest the enemy or, worse, that collection of fools on whom Churchill had to rely for the conduct of the war, should get to hear about it.

Throughout January Max dutifully made small beams and blocks of ice and subjected them to a variety of stresses and strains. In deference to the secrecy of the project, none of his written reports refer to ice. Instead, he used the term 'piccolite', a codeword dreamed up by Mark and Pyke the year before. He found that as a construction material, 'piccolite' was rather unpromising. On the one hand it was brittle, cracking rather easily – and worse still, unpredictably – when under tension. This was disastrous for a long vessel expected to navigate through the high seas, where it might at any moment teeter on the peaks of hundred-foot high waves. While it was less likely to crack under pressure, ice gradually flowed like toffee under its own weight.

Everything changed at the beginning of February when Pyke sent Max a file of Mark's experimental results. Pyke had failed to mention up to that point that since July the previous year, Mark had been working on the 'microreinforcement' of ice by mixing fibres such as cotton wool or wood pulp with water before freezing it. Mark was an expert on plastics, and knew that some of the early plastics, such as Bakelite, had been strengthened in this way. The criss-crossing of the fibres within the ice seemed to stop cracks spreading – the same principle as using steel wires to reinforce concrete. As soon as Max saw the results, he advised all those involved with the project to cease work immediately on 'pure piccolite' and carry out the same tests on 'filled piccolite' instead. Lieutenant Douglas Grant, a naval architect who had the highly delicate job of acting as secretary to the various Habbakuk committees, first suggested that 'filled piccolite' should be known as 'pykrete' to avoid confusion. Making up codenames seemed to be one of the ways the staff of Combined Ops kept themselves amused and fostered their group identity: Churchill was

'God Almighty', Mountbatten 'Mahomet', and Pyke (more rarely) 'John the Baptist'.

Grant set Max up with a new, larger lab in London, which meant that he would have to live there during the week. Excitement about finally being asked to do something worthwhile overrode all other feelings, as he wrote triumphantly to Evelyn: 'I am engaged on war-work now ... it's something that will really help to win it ... I meet many interesting people and am enjoying myself on the whole ... only drawback is that it is in London and I only see Gis at weekends.'

Max still didn't know what the pykrete was for and, as he was spending some time in the COHQ offices while the new lab was organised, this was making things difficult. The pragmatic Grant saw that further secrecy was counter-productive.

> Accordingly I took the bull by the horns and went to see [Brigadier Sir Harold] Wernher, explained the position and pointed out that we would get far better cooperation from Perutz if he were not handicapped by not being fully aware of the object of his experiments. I also pointed out he was, so to speak, our brightest pupil. Brigadier Wernher eventually agreed, when I told him Perutz would not be shown the relevant letters from God Almighty and Mahomet, etc, nor those referring to the possible strategic uses of the object. This has cleared the air considerably and Perutz is going about his work with even greater zest.

Max's new lab was in a cold store five storeys underground below Smithfield meat market in the City of London, with an assistant from Cambridge – a physics student called Kenneth Pascoe – and four naval ratings to do the heavy lifting. During one of his lunch breaks he felt a tap on his shoulder and heard a familiar voice say, 'Hello Max!' It was his old friend from Vienna Fritz Eirich, just back from Australia after three weary years of internment. 'Speaking of never meeting acquaintances in London!' Max wrote to Gisela. 'I shouldn't like to work out the chance on a statistical basis.'

Pyke and Bernal had left for Canada towards the end of February to liaise with a team of Canadian scientists and engineers under the direction of Dean Mackenzie, Deputy Director of the Canadian National Research Council. They confidently expected that they would need Max to join them sooner rather than later. It was at this point that the question of his naturalisation first came up. Despite representations from Mountbatten, Wernher and Bragg, the Home Office was extremely reluctant to process his application, naturalisations having been more or less put on hold since the aliens scare of 1940. The best Wernher could achieve was an undertaking that, if it should prove to be absolutely necessary for Max to travel to the US, the Home Office could issue the necessary papers within a week.

Lieutenant Grant sent Bernal and Pyke a series of long and chatty 'News Flashes' throughout March and April. On 8 March he visited Max's underground lab, which had been in operation only a week, and was struck by the 'rather petrifying spectacle' of Max and his team going about their work in electrically-heated flying suits, bathed in blue light from fluorescent bulbs. A concrete mixer kept the wood pulp mixture in constant motion to stop it settling, while Pascoe and the sailors laid the sheets of pykrete out to freeze in a wind-tunnel powered by buzzing electricity generators. They had just made their first block of pykrete; presenting Grant with a hammer, Max invited him to do his worst. Grant aimed a blow with all the force of his service training, only to see the hammer bounce off without making the slightest impression.

A week later, Max had to lay on a demonstration for Mountbatten himself. Grant had noticed that his own arrival in naval uniform had excited gossip among the Smithfield meat porters and warehousemen. He dreaded to think what might ensue if the Chief of Combined Operations and other assorted brass-hats turned up in full fig, so asked them to come in 'mufti'. On the appointed day he and Mountbatten, Brigadier Sir Harold Wernher, the Canadian High Commissioner Vincent Massey and the US Liaison Officer Captain Tollemache left COHQ by the back entrance, dressed 'like insurance brokers, undertakers and what have you', and sped to the City in a plain van.

On arrival, the whole party got into insulated flying suits in preparation for the 'very effective little show' that Max had arranged. First, he took them through the construction of pykrete, from the raw pulp, through its combination with water in the concrete mixer, to its spreading in successive sheets in front of the wind-tunnel. Then, he invited them to attack a block of pykrete: both Mountbatten and Massey had a go with hammers and chisels, comparing pykrete with 'pure piccolite'. Finally, Brigadier Wernher produced his .45 pistol and fired it at blocks of each substance: a bullet penetrated the ice, making a large crater, while one fired at the pykrete merely made a small dent and fell out again. Massey was all for recommending that the Canadian scientists should immediately move from research to construction.

The Canadians had already built a 1,100-ton model section of a Habbakuk craft (roofed in tin to disguise it as a boathouse) on Patricia

Lake in Alberta. They had subjected it to numerous tests, including weapons tests, and pykrete had come through with flying colours. However, little of this had been reported back to COHQ. While Grant fired off regular reports to Pyke and Bernal, traffic in the other direction was woefully thin. A brief cable had appeared from Pyke in mid-March to say that the Canadians had still to solve the problem of creep, the gradual flow of ice under its own weight that would lead any large vessel slowly to subside into a shapeless blob. Grant became increasingly impatient with his disorganised scientific advisers. Max added his own reproachful pleas:

> It would greatly assist my work if I had more news from you. For all I know people over on your side may be planning on completely [different] lines and my research here may be futile except that by some carefully designed stunts I helped to persuade the big noises of the feasibility of Habbakuk.

For want of further guidance, Max decided to devote himself to the creep problem. He designed a device that would put steady pressure on a block of pykrete and enable him to measure the degree of creep over days or weeks. He planned to set this experiment up in the Low Temperature Station at Cambridge rather than the Smithfield cold store, and got the research assistant he had recently been able to recruit for the haemoglobin work, Edna Davidson, to agree to record the readings each day. But the experiment had an inauspicious start. Max travelled to Cambridge on the train, his precious samples of pykrete in a thermos flask that had been borrowed from the Prime Minister himself.

> About 5 miles outside Cambridge a sudden bang in the thermos startled the passengers in the compartment and they began to make anxious enquiries as to the thermos's contents . . . I spent the rest of the journey in terrible doubt as to the fate of my specimens . . . On arrival in Cambridge at 11 pm I had to climb into the premises of the Low Temperature station and get the thermos into one of their cold stores, where I unpacked it and much to my relief found the specimens still alive. The collapse of the thermos had apparently been due to the pressure of the solid CO_2 which I had had to put in for cooling purposes . . .

While work on the properties of the material went well, all those who had been engaged in the practicalities of making it into a ship were growing increasingly pessimistic, apart from Bernal and Pyke. In the middle of May, Pyke, Bernal and Dean Mackenzie returned from Canada and everyone met in London to try to resolve their differences. Mackenzie told Mountbatten that they still did not have enough evidence that the project was feasible. The vessel's designer had given up on pykrete and begun developing an alternative design made of timber and supported on floating pontoons. Bernal refused to accept that the project was in trouble. It was left to Max, together with Jon Rivett, an architect whose previous experience had been in designing air raid shelters, and E.S. Green, a refrigeration engineer, to estimate the costs in men, time and money of building one bergship in Canada by June 1944.

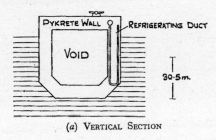

(*a*) VERTICAL SECTION

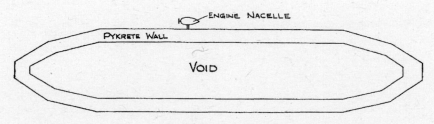

(*b*) HORIZONTAL SECTION

Schematic diagram of the construction of the 'bergship', Habbakuk II

The favoured design was known as Habbakuk II. It was a self-propelled aircraft carrier that, at 2,000 feet, was twice the length of the *Queen Mary*: its displacement of 2.2 million tons made it twenty-seven times heavier. It took the form of a hollow, square beam, tapered at either end; the internal space would provide two decks of aircraft hangars, while the upper flight deck served as a runway large enough for bombers. Its pykrete walls were to be up to forty feet thick, to withstand torpedoes, threaded with refrigeration ducts conveying compressed air at −30°C from the vast onboard refrigeration plant. For propulsion, the Admiralty designers came up with the idea of attaching twenty or more electric motors distributed around the hull: the ship had to be able to make at least 7 knots to avoid being blown off course. No one ever satisfactorily solved the problem of how it should be steered. To make this monster, Max, Rivett and Green calculated, would take 12,000 men fifty weeks at a cost of £2 million. None of them seemed to think this unreasonable, and indeed wartime construction did engage huge quantities of manpower at high cost. But would it be worth it?

In August there was a dramatic development. Mountbatten got Habbakuk onto the agenda for the Quadrant Conference between Churchill and Roosevelt in Quebec at the end of the month. He took Bernal with him – decidedly not normal procedure – ordering him to get a haircut and join the *Queen Mary* for the transatlantic crossing at a few hours' notice. At the Château Frontenac Hotel in Quebec, Mountbatten laid on one of his tried and tested demonstrations for the chiefs of staff of both the US and British services. He fired his revolver into blocks of ice and pykrete at some danger to the assembled dignitaries: Grant reported that he himself had been hit in the thigh by a ricocheting bullet, and Mountbatten in the stomach, while another account says that the bullet clipped the trouser leg of the US Admiral King. The sound of gunfire caused uproar among the aides waiting outside, who burst in to find nothing but a pile of shattered ice and a crowd of genially laughing generals. When it came to Churchill and Roosevelt, Bernal went to the Citadel to make the less risky but equally convincing demonstration – held on this occasion in a silver punchbowl – of pykrete's ability to survive in boiling water as its outer layers turned to an insulating mush.

Roosevelt set up a joint Anglo-American committee, based in Washington, with a view to constructing the vessel on the US West Coast. On hearing via cable of the success of the demonstrations, Max began a daily diary, in a ministry-issue cardboard-covered notebook, to chart the progress of the project to its triumphant conclusion:

24 August 1943, Tuesday

Yesterday the famous cable had come from Mahomet [Mountbatten] that P [pykrete] had won in a canter, against everybody's expectations that an assembly of ancient Generals and other brass at Q [Quadrant] would prefer steel or wood.

The following day a message arrived to say that he, along with Rivett and Green but not Pyke, were to 'proceed to America forthwith'. There was no alternative but for the Home Office to deliver on its promise to have him naturalised within a week, with the joyful consequences recorded in the previous chapter.

They were due to sail for the US on Saturday 4 September. Max's parents were as excited as he was, and Dely plied him with lists of nylon stockings and other goods unobtainable in Britain that she wanted him to send. Gisela came up to London on the Thursday so that they could be together for his last twenty-four hours (much of which she spent typing letters at his dictation, sorting out his affairs in Cambridge). But just as they steeled themselves for the moment of parting, all ended in anticlimax: despite various high-level cables to the British Embassy in Washington and the US State Department, no US visa had been issued for Max. He would have to follow the others by air. While he waited, the progress of the war began to shift decidedly in favour of the Allies, with Italy's forces surrendering and the Russians pushing the German forces back. 'It is quite obvious that [Hitler] is finished,' Max noted in his diary on 10 September. The visa finally arrived the following day, stamped in his brand-new British passport. Max could not resist taking it out over and over again and surreptitiously admiring it. His flight was booked for the following Wednesday.

Before leaving he went to say goodbye to Pyke, whom he found 'rather pathetic': 'He has done his bit, of course, and is not wanted any more. With Mahomet leaving without even saying goodbye to him, poor Pyke is left in a vacuum.' Another outcome of the Quebec conference was that Mountbatten had been appointed Supreme Allied Commander in South East Asia, and so Pyke's days as Habbakuk's champion were over. After the war Pyke found no further outlets for his invention: increasingly isolated, he died of an overdose in 1948 at the age of fifty-six.

The journey to the US exposed Max to innumerable new experiences. The first was the flight: a short hop from Poole to Shannon on the four-engined 'Golden Hind', before transferring to the much larger 'Yankee Clipper' flying boat for the Atlantic crossing. The fare for the trip was $525, leading Max to reflect that his importance to the war effort had been much overrated. Everything aboard the Clipper was the height of luxury for its twenty-two passengers. Rationing did not apply, and Max enjoyed a dinner of roast beef three hours into his flight before curling up to sleep in his comfortable seat. 'Dawn found us travelling over a sea of fleecy clouds while the horizon turned first yellow, then crimson, with gorgeous clouds above us ... Most of the elderly Americans seem to have done the trip before and are indifferent to these sights, probably indifferent to Nature anyhow.'

The plane refuelled in Newfoundland and New Brunswick before smoothly touching down on the water at La Guardia Air Port in New York: the journey from London had taken thirty-four hours. Max came ashore tired and jet-lagged – it was just after midnight British time – only to be questioned for almost an hour by an immigration officer who noticed that his shiny blue British passport was only five days old. Max told him his life story (leaving out the bit about his 'involuntary sojourn' in Canada), produced letters from officials, and trusted to his own good conscience, but all seemed to no avail. After months building up his confidence as a trusted member of the British establishment, he was once again reduced to pleading his case before one who saw him as a potential enemy. The officer began to ask about his family. Yes, he had a brother and sister in the US. Where did they live? Max remembered with a jolt that his brother Franz's house had been searched by the FBI because he had corresponded with a prisoner of war in Canada. Finally he mentioned that Lotte lived on Prytania Street in New Orleans. 'Prytania Street? But that's where I was born!' said the official – and Max was through.

Max was amazed by the cleanliness of New York compared with London, the volume and speed of the traffic, the abundance and quality of the food, and the 'large population of negroes'. He was fascinated to see that Navy submarines came in and out of the harbour at New London fully in the public eye – a thing unheard of in wartime Britain – and mystified by the absence of bicycles, even in the quiet, suburban district where Franz and Senta lived in a colonial-style wooden house with a large garden. (Later he was told that no one cycled for fear of being run over.)

He quickly got the hang of the subway ('you can get anywhere with a nickel') as he visited the Rockefeller Center to discuss his grant with Frank Hanson, and Brooklyn Polytechnic for a joyful reunion with Hermann Mark and 'Fan' Fankuchen. Bernal turned up there too, on his way back to the UK from Washington, looking 'younger and fresher than for a long time'. After much argument about the inclusion of civilians as well as naval officers on the Anglo-American Habbakuk Committee, he and Rivett had been accepted, but as Bernal was now returning to London, Max would represent him in further discussions. Nothing had yet been decided, and indeed Bernal feared that the rest of the committee would resign in his absence. Max began to get the first vague inklings that his extremely expensive trip, so vital to the national security, might turn out to be of no value to anyone.

He headed down to Washington where the rest of the team greeted him like a long-lost relative and desperately pumped him for news from Britain. There was nothing for them to do except wait for the US-led committee to make a decision. Max spent his days sight-seeing, working his way down Dely's shopping list, and trying to keep up with the scientific literature in the beautifully air-conditioned Library of Congress. He liked Washington's spacious streets and elegant buildings, with the exception of the Capitol which he found 'dignified but hideous', and was pleasantly surprised by the quality of the collection in the National Gallery. The American lifestyle, with its focus on comfort, convenience and efficiency, attracted him enormously, and he told Gisela that the Americans he met confounded his prejudices.

> The unpleasant side of American life – if it exists in the way we pictured it at home – is not apparent to the visitor. People are open, helpful and friendly wherever you go . . . It never happens to you here that you wander helplessly about the deserted streets of a town, longing for food and shelter. There is always a drugstore handy. . . . There are few eating places that are as dreary as Lyons or ABC.

At the beginning of October Max's colleagues decided to send him back to New York to consult further with Mark. This visit to Brooklyn, however, was much more significant for Max's future standing in the protein science community. On his last day there, Fankuchen introduced him to John Edsall, the professor of chemistry from Harvard Medical School who was to become one of Max's most trusted mentors. Max found him 'a quiet and very pleasant man of about 45 or 50 who had actually read my thesis and nearly all my papers – my thesis seems to have had a wide circulation in this country thanks to Fan.'

Soon afterwards Max left for Canada. He told Gisela with studied understatement that his second entry to the country, after a long but comfortable train journey, 'was an improvement on the first': throughout the trip he carefully evaded his hosts' conventional question as to whether it was his first visit. Ottawa was decked in its best autumn colours, but, as he told Gisela, that was not the only natural beauty that struck him: 'The people here are far more good-looking than in New York or Washington and it was altogether a delightful sight. It's rare in Washington to see a really pretty girl, still rarer in New York, while this place seems to be thick with them.'

Over the next few days he held discussions with Dean Mackenzie and his colleagues that clarified all points of disagreement and misunderstanding between the two research groups that had arisen months before – and might have been sorted out then if Pyke and Bernal had bothered to send more reports.

In his absence, the Anglo-American Habbakuk Committee had finally met. Advised that building the vessel would take 40,000 tons of equipment and need ten refrigeration plants covering a total area of fifty to seventy-five acres, the Chairman Admiral Moreell concluded that: 'It is strongly recommended that the Combined Chiefs of Staff be informed of the seriousness of the problems involved in constructing Habbakuk II of pykrete . . .'

The whole Habbakuk project was now on the back burner, the flame dying out altogether by the end of 1943. Other technical and strategic advances had made floating airfields unnecessary. Bombers and fast fighters had increased their range; the U-boat threat in the Atlantic had been neutralised; in the Pacific – the only remaining theatre where 'mastery of the oceans' was still an issue – the Americans were being highly successful in the 'island-hopping' strategy that would enable them to advance from the Philippines to the Asian mainland. Habbakuk had proved to be, in Max's words, a false prophet.

Today it reads as an astonishing story – of technical ingenuity and serious research in the service of a science-fiction fantasy, of charismatic visionaries and hard-headed managers battling for control of a misguided project. At least, Max mused in retrospect, his work on the project never led to anyone's death. He remained in the US until mid-November, taking up John Edsall's invitation to lecture on haemoglobin at Harvard. He was disappointed at the response, or lack of it: only thirty people came, and he had the impression they found it hard to follow 'though I left out . . . vector diagrams and reciprocal lattices'. However, it was a further opportunity to talk to Edsall, and he also called at MIT to meet Professor Buerger, the crystallographer who had asked to reproduce illustrations from his thesis. For a young scientist from wartime Britain, these were rare opportunities to expand his circle of scientific acquaintance and further establish himself among the select few engaged in protein crystallography.

After waiting weeks for a berth he finally found space on the *Queen Elizabeth*, crammed with 14,000 US troops en route for the European invasion.

> After six days we steamed up the Firth of Clyde, where a large Allied battle fleet lay assembled in the gloomy winter morning, the sinister grey shapes anchored between the dark, cloud-covered mountainsides lending drama to a scene that looked like a Turner painting of a Scottish loch.

No one at Combined Ops HQ was particularly surprised to hear that Habbakuk was no more. Reflecting on his own experience some months later, Max told Evelyn:

I had a very interesting time, met people like Louis Mountbatten and others whom you normally only see from a distance, and saw some of the higher Government machinery at work. Perhaps my most valuable experience was in administration and committee work . . . I would ten times rather do research.

The happiness of Max's reunion with Gisela was marred by his poor state of health. Before he left the US he had had another bout of his gastric problems, and he spent part of December in the Middlesex Hospital in London, undergoing an investigation that came to no definite conclusion. After Christmas 1944, he got 'steadily worse'. Always a believer in going to the top, he consulted Sir Arthur Hurst, the country's leading gastro-enterology specialist, who had moved to Oxford from his private clinic at Windsor on the outbreak of war. Hurst, who was himself descended from an earlier generation of German Jewish immigrants, treated him kindly and prescribed a mixture of codeine, a painkiller, and atropine, a muscle relaxant. As Max told Evelyn, the treatment

got me back on my feet within 48 hours and very nearly cured me. I still have a hangover as these things take a very long time to get well, but I can work normal hours and do any amount of physical exercise. I had 2 weeks climbing in North Wales at Easter and that transformed me from a weak convalescent into my old self again . . . I started learning tennis so as to keep it up when mountains are out of reach.

For the remainder of the war, Max returned to his haemoglobin research. Although it was going well, to his disappointment neither a college fellow-ship nor a university lectureship seemed to be coming his way with the advent of peace. With no certainty that the work would be supported after his ICI fellowship ran out, it seemed sensible to keep his hand in with research on ice. It was also good for his morale to have another problem to work on that stood some chance of getting results in a relatively short time.

Under Gerald Seligman's energetic leadership, glaciology in Britain enjoyed a resurgence in the post-war years. The Association for the Study of Snow and Ice, which Seligman had founded in 1936, reconsti-tuted itself as the British Glaciological Society; to fill the gap left by the demise of the *Zeitschrift für Gletscherkunde*, it launched the *Journal of Glaciology*, which remains the pre-eminent international journal. Max was a co-opted committee member and regularly attended meetings and discussions. Remarkably quickly (perhaps because it was of no conceivable value to national security), the Habbakuk project was declassified at the war's end, and Max wrote up his own investigations on the mechanical properties of ice and pykrete to present at a meeting of the Society in November 1946.

His work on creep had started him thinking again about the processes at work in glaciers, in which the ice is subject to the sustained stress of its own weight, just as the walls of a pykrete Habbakuk would have been. Under the auspices of the British Glaciological Society, he convened a Glacier Physics Committee largely consisting of Cambridge physicists ('in order to facilitate formal and informal meetings') with expertise in friction, geophysics, plasticity and even polar exploration. It was a high-powered body including two senior physicists, Sir Geoffrey Taylor and Professor Egon Orowan. The committee drew up a plan of laboratory research and field-work to collect data on the plastic properties of ice and the flow mechanisms in glaciers. The lab work Max largely delegated to a PhD student, John Glen, but the field study he kept under his own direction.

The question that was still unanswered was whether ice flows like a very viscous liquid, or like a plastic substance such as a metal. Liquids flow faster in direct proportion to the stress applied to them. Metals remain rigid up to a certain level of stress, at which point they become malleable and can be rolled into sheets. Their malleability also depends on temperature. Exactly how either process would predict what occurs in glaciers was still unknown. Some suggested that a process of 'extrusion' would cause a layer in the middle of the glacier to flow faster than either the surface or the lowest layers. The only way to solve the problem was to find out how fast a glacier flows at every level. To do this, suggested the geophysicist on Max's committee Dr Edward Bullard, they should bore a hole vertically through the glacier from top to bottom, line it with a flexible steel tube and then come back each year to measure how far the tube had moved from the vertical as the layers of the glacier moved at different speeds. To make the measurements, the plan was to lower an instrument called an inclinometer (which incorporated a simple pendulum that would swing to one side or another as the instrument tilted) down the tube and take readings at intervals along the way.

Max later spoofed the discussions of the committee, which combined practically-minded experimentalists and abstract theorists:

Gerrard: What happens if the pipe goes down in the form of a spiral? We shall never be able to interpret the inclinometer readings.

Bullard: I don't see why the pipe should not go down straight, do you Orowan?

Orowan: As a matter of fact this can be calculated quite simply. You just assume a glacier of infinite extent flowing with the velocity of light through which you push an infinitely elastic tube infinitely slowly. Then the curvature of the tube can be calculated from a simple Bessel function.

Seligman: I'm sorry, Orowan, I am afraid I don't quite follow your argument.

The plan was completed, the Royal Society put up most of the funds, and the date was fixed for the summer of 1948. Max chose a spot just below the Jungfraujoch Research Station as most likely to answer the remaining questions about glacier flow, as well as being within easy reach of sources of power and transport. As his assistants he hired two graduate students, John Gerrard and Geoffrey Hattersley-Smith. Neither had skied before, but he assumed they'd soon get the hang of it. He had two vital pieces of equipment made: the electric heating element that would be screwed on to the first piece of steel tube and used to bore the hole in the ice; and the inclinometer crafted in the Cavendish workshops and encased in a shiny brass tube.

From the moment Max set off on the expedition a series of disasters befell him whose cumulative impact was so farcical (in retrospect) that he wrote it all up as a comedy drama. First of all there was a dock strike in London, so that he had to transport all the heavy equipment to Folkestone by lorry and negotiate with customs officers against the clock to get it shipped to Switzerland. The heating element had only just been ready in time for him to collect it from the left luggage office at Victoria Station before he set off for Folkestone. To be on the safe side, he decided to carry it himself, rather than sending it as freight with the rest of the equipment. Gerrard and Hattersley-Smith had gone on ahead to start learning to ski when Max, accompanied by Gisela and three-year-old Vivien, began their journey by train and boat – Gisela and Vivien were to visit the Peisers in Zurich. Shortly after leaving Calais in a hot and crowded second-class couchette, Max got up to take a suitcase down from the overloaded luggage rack. It dislodged a package, which promptly fell out of the open window. Checking his possessions, Max realised with horror that the package contained the heating element without which they could not even begin the experiment.

In fluent French, Gisela pleaded with the conductor to organise a search as soon as the train reached Boulogne, offering 10,000 francs reward for the safe recovery of the parcel. He agreed to arrange for a railway employee to search the embankment on his bicycle, and to forward the package that evening. The family continued their journey to Switzerland, Max in a state of acute misery and agitation as he confessed to the rest of his research party what had happened. Two days later, a telegram arrived from the station master at Boulogne to say that the parcel had not been found. Against her parents' objections, Gisela then returned and organised her own search. Despite valiant efforts by the local police, boy scouts and schoolchildren, the heating element never turned up.

What to do? Max found a local firm in Bern who made soldering irons and other heating devices, including one designed to burn a neat, tapered hole in beer barrels. As there was no time to design a new device from scratch, the firm modified some of their beer tappers for Max's expedition, assuring him that as long as he was using them in water rather than in air they would not burn out with the extra load needed to bore such a deep hole. Greatly relieved, Max returned to the Research Station – only to find that Gerrard and Hattersley-Smith had both broken their legs while practising the elementary skiing manoeuvres he had taught them before going down to Bern.

It was now nearly two weeks since Max had left Cambridge, and the project seemed about to collapse beneath him. In desperation he appealed to the Avalanche Research Station in Davos. This produced the mountaineer André Roch, who was later to participate in the Swiss 1952 Everest expedition that provided much valuable information for Edmund Hillary's successful ascent the following year. Roch immediately took the research party under his wing. He was immensely able and good humoured: it was he who devised the method of sledging the fifty lengths of three-inch diameter steel tube from the research station down to the bore-hole, and taught Max how to make an igloo to store the equipment safely overnight. His joviality kept the whole team in good spirits throughout an expedition that had by no means seen the last of its troubles.

After two weeks' work in immensely difficult conditions – record blizzards occurred throughout early July – the tubes were ready, and a tripod constructed to support them as they were lowered vertically into the hole. Drilling finally began on 22 July, at the painfully slow rate of 1.5 metres per hour at best – only for two out of the three Swiss heating elements to burn out on the first day. Max sent them away for repair, but each time he replaced an element all the tubing that had been inserted so far had to be hauled out again and then put back. Desperate to finish, they worked on through a thunderstorm: lightning struck the drilling rig and fused the electricity supply, but miraculously no one was hurt. In the confusion, however, a large pipe wrench fell into the borehole and jammed the pipe. Somehow, all these problems were surmounted – on 3 August they hit rock at 137 metres.

Joy was shortlived: the inclinometer gave inconsistent results, a week of 100 kilometre per hour blizzards made work impossible, and one of the remaining student volunteers went down with pneumonia and had to be nursed back to health. After five days of intensive testing, Max got the inclinometer working, and at last, in mid-August, the weather cleared. On a day of breathtaking beauty such as Max had not seen since his last trip to Austria ten years before, he and two of the students completed the inclinometer readings.

The original plan had been to repeat the readings every six months. In the event, André Roch took the next set a month later, and then he and Max met again at the Jungfraujoch in 1949 and 1950 to collect further data. Each time, Max took an improved inclinometer, the first designed by Charles Jason, a physicist from London who had been on the 1948 trip, and the second a modification of Jason's design by Tony Broad, the instrument maker who had joined Max's research team in the Cavendish Laboratory. Comparing the measurements across the three readings showed clearly that the surface of the glacier did indeed move fastest, and there was no fast-moving layer further down.

Max's alpine adventure provided the first experimental means of testing the theoretical work on the way ice behaves under stress that John Glen had conducted in the Cavendish cold laboratory. Glen had come up with a law relating creep rate, load and temperature that confirmed that ice did not flow like a viscous liquid. Applying it to Max's data showed that movement within the ice itself accounted for about half of the rate of 35 metres per year that the surface of the glacier moved: the rest was due to the whole glacier slipping down the mountain on its base. Another physicist, John Nye, showed, on the basis of the Jungfraujoch data and readings from a subsequent expedition in Norway, that the law generally held good in glaciers. Since then the study of glaciers has continued apace, taking on additional urgency as the massive glaciers of Greenland and Alaska as well as those in the Alps retreat each year with the rise in global temperatures.

Max's own career as a glaciologist effectively ended with this experiment. There was no doubt of his standing in the field: in September 1947, while the Jungfraujoch work was still in the planning stages, he was invited to join an Anglo-Norwegian-Swedish expedition to the Antarctic, planned for 1948–1950, as the senior glaciologist. 'This is of course quite out of the question,' he told Lotte, though he was pleased with the recognition that the invitation implied. He did accept an invitation to make a short trip to the Rockies in 1953; but when a sudden breakthrough in the protein research put all thought of glaciers out of his mind, he bowed out of the party and never returned to research on ice.

As with the Habbakuk project, the research on flow in glaciers was anything but a waste of time for Max personally. Apart from the satisfaction of having his name attached to some of the fundamental papers in the field, there was much valuable experience to be gained. As an exercise in collaboration – between theorists and experimentalists, between British and Swiss researchers – it was a model of its kind. Max personally undertook the organisation of the expedition, and while he proved characteristically accident-prone, he showed great resourcefulness and determination in surmounting all the problems. After his return, he was much in demand to speak about the research, often to lay audiences; he gave a talk on glaciers on BBC radio that was subsequently published in *The Listener* magazine. He was delighted when the Antarctic Place-Names Committee named a glacier in Crystal Sound on the Antarctic Peninsula after him.

Mountains, of course, continued to play an important part in Max's life: barely a year passed without a visit to Switzerland or Austria to climb or ski. The expense was a problem in the early days. In 1947 and 1949 Max solved this by leading ski-mountaineering courses, first from Pontresina in the Engadine and second from Kühtai in the Austrian Tyrol.

For the Pontresina trip at Easter 1947, Max gathered a group of nine enthusiastic young men: Roger Hanna and John Gregory were physicists from the Cavendish, while several others were medical students from London. Accompanied by Gisela (who did not, however, join the mountaineering expeditions), they travelled first to the pretty village of Bivio where they were to acclimatise for a week and practise their skiing. The party then trekked over the Julier Pass to Silvaplana in the Engadine Valley, wading through six feet of wet snow. Roger Hanna remembered that, like racing cyclists, they took turns to lead the party before falling to the back exhausted. However, once they arrived in Pontresina, further up the valley, things suddenly improved, and they had ten days of perfect weather.

Max was in his element. He took his party, sometimes delegating the leadership to his friend of 1946 Herr Golay, up eight peaks over 3,000

Max's mother Dely Goldschmidt in 1904, the year of her marriage to Hugo Perutz

Hugo Perutz, 1906

Max aged about two, with his Bavarian nanny Cilly Jetzfellner

Max aged seven (seated), with his brother
Franz and sister Lotte, in Reichenau

Pupil at the Theresianum, aged fifteen

Relaxing with friends Fritz Eirich and Pussy Gatzenburg in a mountain hut, 1931

Students in the organic chemistry laboratory at the University of Vienna, 1935. Max (with glasses) is seventh from the left in the back row

The Villa Perutz, the family's holiday home in Reichenau

One of Max's many stunning mountain photographs, showing Gerald Seligman with Dr Philip Bowden and a guide on the Jungfraujoch in 1937 or 1938

Max with his mother around
the time of his engagement
to Gisela in 1941

The photograph Max
had taken to send to his prospective
parents-in-law in Switzerland

(*Above left*) Gisela Peiser before she
left for England, on the terrace of
her parents' flat in Zurich

Max and Gisela were married at the
Shire Hall in Cambridge in March
1942, with Wopp Eirich, Dely and
Hugo Perutz and Steffen Peiser (who
took the picture) as witnesses

Gisela by Max, taken during their first post-war Swiss holiday

Max with his three-year-old daughter Vivien during the 1948 Jungfraujoch expedition

Max with his son Robin on a visit to Austria in 1952

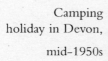

Camping holiday in Devon, mid–1950s

Max in the ice cave on the Jungfraujoch where he examined ice crystals under the microscope in 1938

Max tests the inclinometer, Jungfraujoch 1948

Return visit to the Jungfraujoch igloo, 1950 or 1951

Max sets off skiing during the 1968 Hirschegg protein workshop

Max's unmarried sister Lotte retired to Vermont, where he visited her whenever he could

Family holiday at Feder in the Dolomites, 1994: Gisela, Max, their grandchildren Timothy and Marion, children Robin and Vivien and daughter-in-law Sue

Max's mentors: (*above*) Felix Haurowitz,
(*right*) John Desmond Bernal, (*below right*)
David Keilin and (*below*) Lawrence Bragg

Sketch of Max
by Lawrence Bragg

metres high, demonstrating all his mountain skills: the ascent with skis, ropes, crampons or ice-axes depending on the terrain and gradient, the exhilarating descents through powder snow, the sure choice of routes in a sometimes confusing landscape. On one occasion Hanna's ski came off and disappeared down the mountainside. 'Roger – you're in a mess!' remarked Max cheerfully. Then he gave Hanna his own pair of skis, put on Hanna's one remaining ski and accompanied him one-legged down the several hundred metres of descent it took before they found the missing one. After that everyone in the party was instructed to use an extra pair of bootlaces as safety straps.

Roger Hanna later remembered that everyone in the party was good enough to tackle whatever Max proposed, and Max wrote in his diary on the day before the final ascent of Piz Palü: 'Whole party in tremendously good trim and keen as mustard'. However, there is no doubt that he ran the whole trip at the extreme limit of their capability, which he regarded (probably with justification) as greatly inferior to his own. Given what they had just achieved, the members of his party would probably have been surprised at the account of them he gave to Lotte and Franz soon after his return:

> My 'Course in Alpine skiing' was a roaring success in the end . . . I led my party without a single mistake, incident or accident of any kind and I got a great kick out of the feeling of their confidence in my leadership.
>
> One of my difficulties was the very low skiing standard of my pupils. It was impossible to think of skiing on a rope with them, and I had to adjust my programme so as to avoid all skiing over dangerously crevassed glaciers . . . My own skiing was very satisfactory all the time on any kind of snow.

Later in his career Max found ways to mix work and mountains when, in the late 1960s, he organised extremely popular workshops on protein structure at Hirschegg, a ski resort in Austria's Klein Walsertal, just south of Munich. When Vivien and Robin were old enough he took them skiing as well. Trips to Switzerland came up every couple of years for the young family, because of the necessity to visit Gisela's parents (who often paid the bills), and always included a summer or winter holiday in Lenzerheide or some other resort. In August 1958, when Max was in Austria to oversee the sale of the Villa Perutz in Reichenau, the Perutzes took a joint holiday in Alt Aussee with Dorothy and Thomas Hodgkin and their children, who were then aged 12, 16 and 19. Robin in particular proved an apt pupil at both skiing and climbing, and Max was delighted to revisit all his old favourites with his son at his side. Piz Palü, the mountain that held so many happy memories for Max, was not just the highest mountain Robin ever climbed with his father: it was the highest he ever climbed at all.

Later, when Max was in his seventies and a back problem meant that he could no longer climb or ski, a small hotel at Feder in the Italian Dolomites became the favourite summer destination, long walks being substituted for the more strenuous activities he had formerly enjoyed. These holidays included as many members of the family as he could recruit: always his cousin Alice and her partner Mac, and once his brother Franz who moved back to Vienna in the 1980s. By now, Robin and his wife Sue had two children. Introducing his grandchildren Timothy and Marion to the mountains was a particular joy for Max:

> This has been the most successful holiday ever. Feder turned out to be an idyllic village ... There is a variety of walks through forests up to lush meadows, through valleys, along streams from our base at 1200 m up to about 2600 m, all spectacularly beautiful ... To me, introducing Timothy to the Alps was a dream fulfilled. He delighted me by his enthusiasm and stamina and alert observation of everything around. On one steep and hot climb I kept him going by telling him stories when he suddenly said to me 'It is difficult to listen to you when there is so much to look at.' ... Marion also thrived, but was obviously too small to walk up for any length of time ... There was no mishap, no one got ill, no one quarrelled, not even Franz.

Max was perfectly capable of articulating the central role of mountains in his life, health and happiness, and did so in a memoir he dictated in the final weeks of his life. I can do no better than to quote his recollections of his youthful expeditions in full:

> You emerged from the magic snow-covered woods and there in front of you was a glorious snowscape. In the early morning the light of the sun on the snow and the long shadows were overwhelmingly beautiful; every time you looked round the view thrilled you. As you climbed higher, more and more mountains became visible until you finally reached the top and there was a cloudless sky and all around you was a glorious view of snow-capped mountains ... It was a wonderful experience. It filled you with overwhelming joy. Then you started on the downhill run; often there were no signposts, no tracks, you had to find your own way. Skiing in the soft, deep snow made you feel as if you were a bird and had wings

... As you skied down little whiffs of powder snow cooled your face and there was complete silence, the stillness broken only by a swish of the skis through the soft snow. Mountains instilled a great sense of beauty in me. Faust in his pact with Mephisto said:

> If ever I said to the passing moment,
> 'Tarry a while, thou art so fair',
> then you bind me in chains.

But what I said to myself when I recalled that moment: 'I shall be happy forever.'

Later in the summer, rock-climbing became my passion. I would experience the same beauty. In addition, there was something which determined my attitude to life. As you approached the mountain, a formidable rock-wall faced you and you said to yourself: 'Good Lord, I shall never get up there!' So finally you arrived at the wall and you found a system of cracks and fissures and chimneys, tiny bands of rock on which to balance. There were no markings but you had a book which described roughly the way up, so gradually and with absolute concentration you slowly made your way to the top until again you were faced by the beauty of the Alpine landscape all around you and you were totally happy. This time the happiness with which the beauty filled you, was compounded with the happiness that came from the skill, the physical effort, the sense of companionship with your friends, the views of the abyss, which had inspired no fear but only thrills, the sense that your courage had paid off and that you had experienced a very great adventure.

Skill, persistence in the face of difficulty and, above all, the absolute confidence that the summit, when it came, would make the hard work and frustration worthwhile, were all qualities that Max would need in abundance as he moved towards the peak of his scientific achievements: the solution of the structure of the protein molecule haemoglobin.

6

How haemoglobin was not solved

Why grasse is green, or why our blood is red
Are mysteries which none have reached unto
In this low forme, poor soule, what wilt thou doe?
John Donne, *Of the Progresse of the Soule*

The 1950s were a golden age for scientific research in Britain. The war years had seen science applied to real-world problems with astonishing speed, and (in many cases) greatly to the advantage of the war effort. Wartime research produced penicillin, radar and the beginnings of electronic computing, to name but three of the discoveries that would have valuable applications in peacetime. The powers-that-be realised that putting money into science was a good investment for society – something that Bernal and other left-wing scientists had been arguing for almost two decades. Government funding agencies such as the Medical Research Council and the Department for Scientific and Industrial Research suddenly found their budgets greatly increased. Meanwhile the wartime 'boffins', who had done so much to avert the Nazi threat, returned to civil research full of enthusiasm to attack new problems. The combination was to pay off spectacularly by the end of the 1950s, with Cambridge playing a leading role.

Having spent almost a full year on the abortive Habbakuk project, Max reappeared in his lab in the Cavendish at the beginning of 1944. Thanks to extra funds from the Rockefeller Foundation, which was still paying his salary, he had begun to make the move from solo researcher under Bragg's wing, to research group leader with assistants of his own. In the summer of 1942, he had hired two young women, Joy Boyes-Watson and Edna Davidson as his assistants. They had been left to carry on as best they could while he toiled in the Smithfield cold store or mingled with the servicemen and bureaucrats in Combined Ops. Their task was painstakingly to index and measure the intensity of more than 7,000 reflections on 135 X-ray diffraction pictures of crystals of horse methaemoglobin. Max had taken the pictures – each needed a two-hour exposure – during the long nights he had spent fire-watching in the Cavendish after he came back from internment. He hoped that the mass of data generated from the images would reveal a pattern that would provide the key to the haemoglobin structure.

The goal of X-ray crystallography is to construct a three-dimensional model of a molecule, showing exactly how each atom relates to its neighbours in space, based on the position and brightness of the spots in an X-ray diffraction photograph. As mentioned previously, however, you also need to know the phase of each reflection: the part of the cycle from peak to trough of the waves of X-radiation making it. Once you have the phases, then you can put them with all the other information on position and intensity into a mathematical calculation known as a Fourier synthesis after Jean Baptiste Fourier (1768–1830), the French mathematician who invented a form of analysis known as the Fourier series. A Fourier synthesis of the reflections from a layer in a crystal generates a contour map that shows the disposition of the atoms in that layer. Because most of the volume of an atom consists of a cloud of electrons orbiting its nucleus, this contour map is known as an 'electron density map': the electron density increases as you get closer to an atomic position. The calculation involved in completing a Fourier synthesis consists of long runs of addition sums: with large molecules such as proteins, even with a mechanical calculator it is tedious beyond belief. In practice, it would take the invention of electronic computers to make such calculations a realistic possibility.

It was not just the lack of computing power that held Max up, however. Despite his 7,000 reflections, he could not undertake a Fourier synthesis because at this stage he still had no information about the phases. The pattern he sought to construct instead was known as a Patterson map. Lindo Patterson, a former student of the elder Bragg's who then worked in Philadelphia, developed this brilliant piece of lateral thinking in the 1930s. It was a favourite method of Dorothy Hodgkin's, who had recently used it on the way to her solution of the structure of penicillin (a molecule of only a couple of dozen atoms), for which she later won the Nobel prize. It required the same heavy burden of computation as a conventional Fourier synthesis, but had the advantage that it could make use of the position and intensity data in an X-ray photograph without needing to know the phases of the reflections. Like an electron density map, it consisted of a crazy landscape of contours, hills and valleys drawn out on sheets of tracing paper by Max's assistants. The landscape, however, was entirely abstract. While the peaks on an electron density map represent the actual positions of atoms in a molecule, those on a Patterson map represented the distances and directions – the vectors – between regularly repeating features in a crystal, such as a particularly heavy atom.

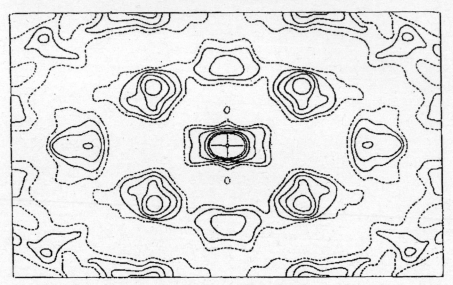

Patterson map of haemoglobin

With a small molecule such as penicillin, this is likely to generate useful information about the possible positions of such features that can be used to find the phases to be fed into the next stage of the analysis. With a molecule as large as haemoglobin, however, Patterson maps are less useful and can be positively misleading. But at the time it was the only line of attack Max knew of and, having spent so much time on it, he had to see it through. In April 1946 he wrote to Franz and Senta, using the slightly grandiose tone he habitually adopted towards his brother:

> I have just completed one of the worst and most tedious jobs that have [sic] ever been done in my subject, a piece of work which just wore me down. Now that it is over I feel much better alround [sic] and life more worth living again. At the moment I spend my days writing comfortably at the first of a grand series of papers on protein structure.

By this time he had lost both his first two assistants. He was sorry to lose Edna Davidson, who already had a PhD, who left in the autumn of 1944 to take up a junior teaching post at the Royal Free Hospital in London. In her place he took on the far less experienced Olga Weisz, a Hungarian-born physics student, who at that time had completed only the two-year Part I of her degree in Natural Sciences. In one of his periodic reports to the Rockefeller Foundation, he expressed frustration at the difficulty in recruiting research assistants in wartime: 'I should have preferred a graduate, and, even better, a man, because I find that the lack of initiative and originality of most women workers (there are some notable exceptions!) is rather a drawback.'

Much of the day-to-day work of crystallography did not require 'initiative and originality': it consisted of the mind-numbing tedium of measuring and indexing reflections, a task Max and his assistants undertook together. Joy Boyes-Watson had come to Max with a First in physics, but very quickly realised that her work would not provide sufficient material for a PhD thesis. She lost interest and motivation (provoking Max to threaten to dismiss her on more than one occasion); by the middle of 1945 she had left to get married.

Olga Weisz (now Olga Kennard FRS) had a similarly disheartening experience. Though she had just missed a First in her Part I examination, she came with recommendations from Henry Lipson and Arthur Wilson, both distinguished crystallographers in the Cavendish, and had acquired the theoretical background to X-ray crystallography as part of her degree. Max gave her a problem of her own, the structure of a small molecule, in order to train her in the basics of practical crystallography. Having spent the whole summer calculating Fouriers by hand, she successfully solved the structure. This was the first complete structure solution to emerge from Max's tiny group, though Kennard herself now dismisses it as 'a ridiculous little structure', and it led to a joint publication.

She went on to work with him on an X-ray study of human haemoglobin. For the first time they were able to grow useful crystals of an oxygen-free form – potentially of great interest, as Max had been unable to get such crystals of horse haemoglobin. One of the main goals of his research was to look for differences between forms of haemoglobin with or without oxygen that could account for the changes in crystal structure observed by Felix Haurowitz in 1938.

However, Weisz did not have the opportunity to pursue this potentially interesting line. Gradually she came to realise that her hopes of doing a PhD in Max's lab would not be fulfilled. She completed a preliminary study on human carboxyhaemoglobin, which she and Max wrote up for publication in *Nature*: yet before it was published she had moved on. As soon as the opportunity arose to recruit better qualified assistants, Max terminated her contract, telling others he did not think she was capable of work as demanding as protein crystallography. Weisz felt crushed by his lack of confidence in her, though she was able to find temporary work with another crystallographer in the Cavendish while she finished her degree. The first thing she did was to develop and publish a reliable method of setting crystals – an idea Max had dismissed – which continued to be cited for more than a decade.

In spite of this setback, Kennard made her career in crystallography, first in London and then, from 1962, in Cambridge where she headed a small chemical crystallography group at the Chemistry Department. Soon afterwards, she became founding director of the Cambridge Crystallographic Data Centre, a national resource. She was elected a Fellow of the Royal Society in 1987. Max visited her lab occasionally and, she says, was never less than encouraging. (One of the most charming letters he wrote to his grandchildren tells how he fell into an ornamental pond – it was on the first floor, where he least expected it – at the opening ceremony for the CCDC's new building in 1992.)

Despite these instances early in his career, it is not true that Max had a low opinion of women scientists. He was a devoted admirer of Dorothy Hodgkin, whose crystallographic insight he always recognised was greater than his own and whose opinion on his papers he often sought. Then, at the end of the war, he met Kathleen Lonsdale for the first time, a crystallographer and conscientious objector who had been imprisoned for refusing to register for war work or pay a fine. He wrote to Lotte: 'She has just been elected FRS, first woman in 300 years ... She is a most resolute little woman of about 45 ... with an amazing output of interesting research.'

Later in his career, he had a number of close female collaborators, and there are several women scientists today who testify to the support and encouragement he gave them when they were young. It could of course have been just a coincidence that those who had to cope with his first attempts at managing a research group were all women. However, when he recruited two young men after the war, they had quite a different experience.

The first was Herbert 'Freddie' Gutfreund, another ex-Viennese who had just finished his doctoral work under Gilbert Adair, and who came ostensibly to replace Joy Boyes-Watson in 1945. Gutfreund had attended Max's Saturday morning lectures on crystallography in the Molteno Institute, and thought he would like to learn more about the subject; Max took him on without hesitation, his salary paid from the Rockefeller grant. However, unlike those of Boyes-Watson, Davidson and Weisz, Gutfreund's name does not appear as a co-author on any of Max's papers. He solved a small structure to learn the technique, attended physics lectures 'to make up for [his] neglected education', and did some routine calculations. Other than that, he spent a considerable proportion of his time sorting out some problems with the work he had done for his PhD thesis. His own assessment of his time with Max was that he was 'more useful to him and Gisela as a babysitter than as a research assistant'. There is no evidence in the files that Max ever complained about this. On receiving his PhD in 1947, Gutfreund moved to a University post in the neighbouring Department of Colloid Science. He retained a close connection with the work of the crystallography group and, as it continued to expand, took part in many of its lively discussions.

The next addition to the lab arrived with the explicit status of PhD student rather than research assistant. His name was John Kendrew, and he was clearly never going to occupy the role of amanuensis to Max. Olga Kennard remembers that, as he walked into the lab for the first time, his new black notebook at the ready, his whole air said: 'I know what I'm doing and I'm getting what I want.' Kendrew had begun doctoral research in physical chemistry in Cambridge in 1939, having previously graduated in natural sciences. The following year, like so many of his scientific colleagues, he was sent to the Telecommunications Research Establishment to join the radar research team. By 1944 he had moved into operational research and was scientific adviser to the commander-in-chief of the air forces in South East Asia, with the honorary title of Wing Commander. In Ceylon he had met Bernal, who was there as adviser to Mountbatten. Among other things, they were testing the effectiveness of bombs in the jungle, which involved a trip into the forest by elephant to observe the impact of depth charges (being used on land because the Air Force had temporarily run out of fragmentation bombs) on 100 caged rats placed at various distances from the explosion. The rats were unharmed, as Kendrew and Bernal had predicted, but they themselves had a narrow escape when Bernal (having drawn out his slide rule with a flourish) miscalculated their own safe distance.

In between discussions of military matters, Bernal persuaded Kendrew of the huge promise of structural work on proteins. On his way home, Kendrew visited Linus Pauling's protein research laboratory in California, and became even more firmly convinced that this kind of work would satisfy his ambition to use insights from physics and chemistry to answer questions in biology. The scientific civil service was reluctant to lose him, offering him at least one post at twice the salary he could expect as a junior research scientist, and the military authorities dragged their heels over releasing him. Finally, however, in January 1946, having arranged everything in advance with Bragg, Kendrew switched the topic of his PhD thesis to protein structure and came to join Max.

He arrived equipped with his own funding: he still had the remains of his pre-war scholarship at Trinity College, and in the meantime Bragg had obtained for him a grant from the Department of Scientific and Industrial Research. He was upright, confident, well organised and efficient, and younger than Max by only three years. All this lent him a certain authority that complicated the student/supervisor relationship (although formally it was not Max but William H. Taylor, head of the Cavendish's Crystallography Division, who was to supervise Kendrew's thesis, presumably because Max himself did not have a permanent University post). In August Max sent his parents-in-law a glowing assessment:

> He knew no X-ray crystallography when he came last January, learnt the elements of the subject in two weeks, set to work on his problem which required great experimental skill, and solved it in a few months' concentrated work. Besides, he is a charming fellow, and it is a pleasure to talk to him; so I consider myself very lucky to have him with me.

By the middle of 1947 Max was describing him to the Rockefeller Foundation's officers as 'probably one of the brightest chaps we have had in the laboratory'. This opinion seemed to be widespread. Around the same time, Kendrew was invited to apply for a research fellowship at Peterhouse – the college Max had belonged to as a graduate student – to which he was duly elected on the strength of his first year's work with Max. From this point onwards (before he had even completed his PhD), he received a steady stream of attractive job offers from both sides of the Atlantic.

Kendrew began by comparing fetal and adult haemoglobin of sheep. At that time, the Cambridge physiologist Sir Joseph Barcroft was making fundamental discoveries about development in the fetus, especially in relation to blood flow and respiration. He made his studies on ewes and lambs, and had observed that the haemoglobin of fetal lambs had a higher affinity for oxygen than that of their mothers. He suggested to Max that he might look for the reason in the structure. It seemed like a suitably self-contained PhD topic for Kendrew, but after six months he realised that it would not be enough to establish his reputation. He made the crystals and took the photographs, but could do no more than make a preliminary description of their structure. The only conclusion he and Max could draw was the erroneous one that, while fetal haemoglobin was composed of four 'identical or almost identical subunits', there was no evidence for the existence of such subunits in the adult molecule.

Kendrew completed his thesis on fetal and adult haemoglobin, earning his PhD in 1949; he kept up his interest in the subject until as late as 1953, but in the meantime he had set off in a new direction. There were other projects that Max had in his bottom drawer that might prove more amenable to analysis, and would give his colleague the chance to work independently of him. In August 1944, Max had written in his report to the Rockefeller Foundation:

> I am still very anxious to extend the work to proteins which are smaller and therefore probably simpler than haemoglobin, notably myoglobin . . . So far it has not been possible . . . because it would take a trained biochemist several months to isolate and purify them.

By July 1946, Kendrew was telling a Rockefeller officer, who visited while Max was still away on his extended Swiss holiday, that he 'hoped soon to start work on myoglobin'.

Myoglobin is a protein that does the same job in muscle as haemoglobin does in blood: it stores and releases oxygen. Like haemoglobin it contains iron, which is why muscle tissue (think raw steak) is red in colour. However, its molecules are only a quarter of the size of haemoglobin, and contain just a single haem group with its iron atom at the centre. The relative sizes of the two proteins later turned out to be extremely significant, but at the time myoglobin's overwhelming attraction was that it was small (for a protein), with 1,200 atoms (not including hydrogen atoms, which are so light that they hardly affect the diffraction patterns) to haemoglobin's 4,800. (Bear in mind, though, that the largest molecules that had then been solved by X-ray crystallography contained a few dozen, not hundreds, of atoms.) Kendrew began, naturally enough, with horse myoglobin: it was reasonable to assume that there might be some relationship between haemoglobin and myoglobin, and so it made sense for Max and Kendrew to work on proteins from the same species. Rather confounding Max's predictions about the need for a 'trained biochemist', Kendrew succeeded within a few months in obtaining crystals from extracts of horse heart. The crystals were disappointingly small, but nevertheless he set to work to take photographs and make the measurements. He was encouraged by his initial results, reporting to the Master of Peterhouse that 'The structure of the protein does appear to be very simple and easily related to that of haemoglobin.'

Max was pleased that his new colleague had made such a good start, but had begun to feel anxious about his own position. He was still on what today is known as 'soft money' – short-term research fellowships from outside organisations with no guarantee of renewal. Since before the war ended, Bragg had been trying to obtain a University appointment for him. This was clearly the most desirable of the options Max sketched out in a letter to his professor in July 1944:

> If I am going to continue my work here for several years to come I should like to be as closely as possible associated with this University and to have a recognised position within it ... it is only the people who are regular members of the University staff and fellows of colleges who have a say in the affairs of the University, are consulted on matters which concern them and are able to get good research students, while holders of fellowships, grantees etc. are just part of the general crowd of MA's [sic] whose position is vague and uncertain.

At the same time he hoped for a College fellowship, though acknowledging the difficulties:

> Owing to the time I lost with my various trips to North America – voluntary and otherwise – I shall not be in a position to submit a fellowship thesis before next spring ... Even then I doubt Peterhouse would have me, because they are almost exclusively specialised on historical subjects. It is hard for me even to approach my college on this question, because although the authorities took a kindly interest in my doings as long as I was a student, they have not done so since I graduated. I don't believe the present Master ever heard my name ... I believe that Burkhill, the tutor, thinks that I should have joined up. I would much prefer becoming a fellow of a more scientifically-minded college ... I really don't know how to tackle this matter.

Even with the progress Max made with his research in the post-war years, he never received any encouragement to apply for a fellowship and did not submit a thesis. (That Kendrew had acquired a Peterhouse fellowship before he had even finished his PhD must have been particularly galling.) Later, he thought his lack of a Cambridge undergraduate degree might have counted against him. Neither had Bragg's efforts to secure a University post for him met with success by 1947: Max's expertise in a field that straddled physics, chemistry and biology did not fit neatly into the structure of the University's courses, and the authorities were not prepared to create a lectureship in such an ambiguous field. He seemed destined to depend on a series of short-term fellowships, and temporary funding would mean permanent insecurity for himself and his little family. Haunted by this prospect through the dark and terrible winter of 1946–47, a despairing Max confided in David Keilin that he would have to find a job in industry.

Instead, Keilin came up with a radical proposal: that Bragg apply to Sir Edward Mellanby, Secretary of the Medical Research Council, for funds to set up a research unit within the Cavendish under Max. Keilin had been on the MRC's council, and had heard the discussions leading to the funding of a new Biophysics Unit at King's College in London under John Randall. Biophysics was flavour of the month, and protein crystallography fell comfortably under that heading. He also knew that the MRC's budget was now between two and three times higher than it had been pre-war, and was set to continue rising. Bragg was still thinking about trying to get funding for a University post such as 'Assistant Director of Research', but Keilin talked him round. Bragg wrote to Mellanby in May, asking for support for a unit consisting of Max, Kendrew, two assistants and two or three research students (yet to be recruited). He warmly commended Max's progress so far and likely success in future. The unit would be well placed in Cambridge, he said, because it would supplement work on similar problems going on in other departments. Keilin also wrote in support:

> The unit exists already and originated about 10 years ago from the collaboration between the Cavendish Laboratory and the Molteno Institute with Dr Perutz acting as a link ... [T]his collaboration ... has opened up several lines of research of fundamental biological importance.

Mellanby met Bragg at the beginning of June, recording in a note of the meeting that, although he had the highest respect for Keilin's view of scientific subjects, he always felt he needed a second opinion 'when it came to a matter of subsidising foreigners of his own race'. Bragg confirmed that he would support Max without hesitation.

Mellanby was sufficiently impressed to ask Bragg to draw up a detailed plan with costings, to be presented to the Council 'for a preliminary run' on 17 October 1947, the first time it would meet after the summer break. The outcome was everything Max could have hoped. Mellanby wrote that 'Rather to [his] surprise', the Council accepted the proposal in full and funded the unit for five years in the first instance. If it was a risk, it was one that was to pay off in the future to a degree that none of the Council members could possibly have imagined.

Bragg was delighted, and responded with a paean of praise to his two protégés, although couched in interestingly different terms:

Perutz is very clever and a beautiful worker. As a refugee he sometimes finds it a little difficult to adjust himself to the way things are done in this country, but I have no real trouble, and he is so brilliant in his own special line and so helpful to other people in the laboratory that I count myself very fortunate to have him here ... Kendrew ... I think is a coming man ... I believe him to be of the type that will make a real leader in the future ...

United in their preference for linguistic precision, Max and Kendrew chose the cumbersome handle 'Medical Research Council Research Unit on Molecular Structure of Biological Systems' for their new outfit. (They later upset the MRC by omitting it from their publications in favour of 'Cavendish Laboratory', because they thought it was too long!) Horace Judson has written that: 'In the English way, it became clear gradually, as the Unit grew, that Perutz was its chief.' However, all the correspondence about the establishment of the Unit makes it clear from the start that it was to be 'under Perutz', and Max himself was never in any doubt on the point. He wrote confidently to Felix Haurowitz in January 1948: 'I have been made head of a research unit on "Molecular Structure of Biological Systems" ... and have been given ample funds to build up a large research team and get better equipment.'

The 'large research team' was perhaps a little strong: in its first year the Unit consisted of Max, Kendrew and a new research assistant, K.K. Moller from Denmark. (Clearly Max had disregarded Mellanby's advice that he should 'get Englishmen and not foreigners' as assistants wherever possible.) The next recruit was Hugh Huxley, who came to work for a PhD under Kendrew in 1948. Huxley was another refugee from physics whose undergraduate degree had been interrupted by war research on radar. He returned to Cambridge to complete his Part II, but seeing photographs of the effects of the atom bombs dropped on Hiroshima and Nagasaki had extinguished his earlier ambition to do research in nuclear physics. Instead, he looked towards biology or medicine. Max told him they could get him a research studentship from the Medical Research Council as long as he got a First in his final exams. In July 1948 Huxley set off for a bicycling holiday in France, not knowing his results but convinced, as he had confided to his sister, that he had done badly. A week or two into his holiday, she telegraphed tersely: 'FIRST CONGRATULATIONS IDIOT', and by the time he got back to England, Max and Kendrew had sorted out the studentship.

At first, Huxley revisited the work Max had done on swelling and shrinkage in haemoglobin crystals, developing apparatus that allowed him to hold constant the temperature and humidity in the tiny tubes that held the crystals, to measure the humidity levels at which the crystals clicked from one stage to the next, and to identify more stages between wet and dry than the four Max had described in his 1942 paper. Max took a new set of photographs of the ten stages, hoping that comparisons between them would help to solve the ever-present phase problem. Huxley then turned his attention to muscle, wondering how the structure of muscle fibres gave them the ability to contract.

A year later the Unit added to its small staff an electrical engineer, Tony Broad. With the well-staffed and equipped Cavendish workshop close by, it might seem odd for the Unit to feel the need for its own engineer, but this was a far-sighted and extremely astute appointment that was to have huge benefits in the years to come. Broad was hired to design and build a new kind of X-ray tube called a 'rotating anode' tube. X-ray tubes generate a lot of heat, along with the X-rays: if not turned off in time, their anodes can burn out. Now, in the late 1940s, tubes were produced in which the anode rotated, so that no single part of it became hot enough to incur damage. These tubes were capable of producing X-ray beams twenty times as strong as conventional tubes, making it possible to work with smaller crystals and to obtain sharper diffraction images. Broad's designs were extremely successful, eventually being produced commercially (they are still in use today): more importantly, they meant that from the early 1950s Max's unit was using the best X-ray equipment in the world.

With his Unit established, Max continued writing his 'grand series of papers'. The first had appeared in the *Proceedings of the Royal Society* in 1947. It summarised all of the work he, Boyes-Watson and Davidson had done since he began the study on swelling and shrinkage of crystals in 1941. As in his wartime *Nature* paper, he described how layers of molecules were interleaved with layers of water, and how drying the crystals removed the water layers without altering the structure of the molecules themselves. This was a critically important finding. One of the main reasons that biochemists doubted that crystallography could tell them anything interesting was that proteins in the body are always in solution, bathed in the body fluids. If each haemoglobin molecule changed its shape on wetting and drying, then the likelihood that molecules in crystals had the same structure as molecules in red blood cells would be called into question.

In the same paper, Max dared to speculate publicly about the size and shape of the molecule. He tentatively proposed that it was a squat cylinder, 34 angstrom units (ten-millionths of a millimetre) high and 57 Å in diameter, made up of four layers: he called it his 'hatbox model'. The implication was that there would be a haem group in each layer, although he gave no evidence for this.

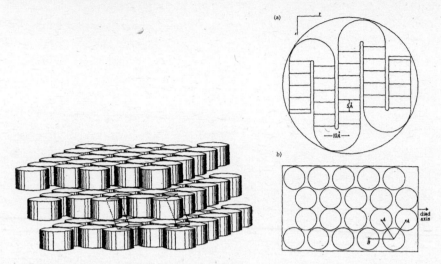

(Left) Max envisaged 'hatbox'-shaped molecules of haemoglobin arranged in layers in the crystal; (right) inside each 'hatbox' were four layers of parallel protein chains.

Having done the best he could with his tiny crystals, John Kendrew performed the same type of analysis on his myoglobin photographs and concluded that the myoglobin molecule was a flat disc, like a penny: perhaps haemoglobin was like a stack of four of these pennies. To find out more about the nature of the layers would take an even more heroic calculating job: a three-dimensional Patterson synthesis, which involved finding not just the superimposed vectors seen along one direction in the unit cell, but in all directions. It was far too much to do with a mechanical adding machine, as they had done so far. Instead, Max sent the data to the Scientific Computing Service in London, who did the calculations using Hollerith punched card machines, electromechanical devices capable of adding and sorting at what then seemed to be incredible speed, and at the time the latest technology in office computing. The bill came to hundreds of pounds, more than twice what Max had estimated, but he felt that the ends justified the means. As the results began to come in, he and Kendrew convinced themselves that they could see features suggestive of parallel 'rods' lying side by side in the layers, as in a packet of cigarettes. At last, they thought, something new to report! Max was uncharacteristically curt in refusing to mount an exhibit on his past work for the Chemical Society centenary in July 1947 when invited to do so by Bernal:

I have shown my old work a hundred times and I am sick and tired of producing the same old results of myself and Dorothy's over and over again. There is no doubt that a new and decisive step forward will come from my three-dimensional Patterson, but the results will hardly be ready in a presentable form in July ... I have the data for the complete Patterson now and they shall be plotted by the end of next week. There is no doubt whatever about the 'rods' ...

There were numerous delays in completing the three-dimensional synthesis and its write-up. Power cuts during the viciously cold winter of 1947 shut down the Scientific Computing Service for a while (as well as freezing the gas pipes at the Green Door, so that Max and Gisela could not cook or warm themselves). Later that year, Sir Joseph Barcroft died, and both Max and Kendrew were heavily involved in organising a conference on haemoglobin in his memory to take place in June 1948, to which they each contributed a long paper. Throughout this time, Max was also organising his glacier flow study, which took him to Switzerland in the autumn of 1947 as well as during the experiment itself in the summer of 1948. However, he felt more and more confident as the final results emerged from the Hollerith machines.

Max continued to correspond with Felix Haurowitz, who had started him on his haemoglobin journey. Since his move to the University of Istanbul in 1939, the extraordinarily resourceful Haurowitz had changed fields, learned to lecture in Turkish, and written a widely-acclaimed biochemistry textbook in German. In 1948 he accepted the chair in biochemistry at Indiana University in Bloomington, and Max wrote to invite him to participate in the Barcroft Symposium on his way to the US – Haurowitz had been high on the wish list of international speakers drawn up by the organising committee. Max also brought him up to date with the latest on his work:

Our work here has more than justified the labour spent on it. The fact is that my work on haemoglobin combined with Kendrew's recent results on myoglobin have solved the riddle of globular protein structure, at least as far as the general layout of the polypeptide chains is concerned. From the three-dimensional Patterson projection of haemoglobin the chain direction, the repeat pattern along the chains ... and the general packing of the chains in the haemoglobin molecule can clearly be recognised ... The myoglobin results confirm the general picture of globular protein structure derived from haemoglobin and show them in a delightfully simple model ...

Delightfully simple it may have been – but it bore no relation to reality. Driven by what? – The need to succeed? The need to have something to publish? – Max had seen meaning in his Patterson maps where there was only ambiguity. What led him astray was his 'great and childish faith' (his own words) that protein structures would have some kind of neat regularity that could be revealed by Patterson analysis, when in fact they did not. Fatally, none of his mentors knew enough to put him right. Haurowitz told him it was 'the most important contribution to protein chemistry', though as one who spent most of his time thinking about the forces that hold molecules together, there were some things that bothered him: 'It would be very important to know, whether the myoglobin disk, which you drew in your letter, maintains its oblate structure, when dissolved . . . Would not such a disk break down owing to thermic oscillations of the peptide chains?'

There is no evidence from the correspondence – and it is remarkable how carefully Haurowitz preserved Max's letters over fifty years and during moves between three countries – that Max replied to this point. He submitted his manuscript in June 1948, and presented its contents to the Royal Society in person in December of the same year. 'The photographing, indexing, measuring, correcting and correlating of some 7000 reflections,' he wrote, 'was a task whose length and tediousness it will be better not to describe' – a personal aside rather refreshing to find among the dry prose of the *Proceedings*. He went on to augment his earlier report on the shape of the molecule with the new 'evidence' that the chains looped into parallel rods in regular layers stacked on top of one another. He hedged his bets only slightly, admitting that 'the details [in his two diagrams of the layout] are, of course, purely imaginative', but asserted that 'the general lay-out which they indicate follows from the vector structure'.

The beauty of the scientific method is that wrong ideas cannot survive for long. Sooner or later someone will discover the flaw, like a wrongly-placed jigsaw piece, and the whole picture has to be adjusted as a result. The person who was soon to rearrange Max's jigsaw was Francis Crick.

Crick was the son of a Northampton boot and shoe manufacturer. Neither of his parents had a university education, but his mother harboured ambitions for her son and encouraged his restless curiosity. He won a scholarship to Mill Hill School, a fee-paying school in London, where he acquired a thorough grounding in physics, chemistry and mathematics and an unshakeably self-assured manner. It was a natural progression to take a degree in physics at University College London. Then, under the professor of physics Edward Andrade, he began a PhD on 'the dullest subject imaginable' – the viscosity of water – but fortunately the outbreak of war cut short his studies and he moved into the much more interesting field of designing non-contact mines for the Admiralty.

By the time the war ended, he had decided to switch from physics to biology, a decision only strengthened as he began to read up on the subject. (At this stage he barely knew what a protein was.) One of the books that impressed him was *What is Life?*, a book based on a series of lectures that the physicist Erwin Schroedinger gave at Trinity College Dublin in 1943 that foresaw a time when life would be understood in terms of its physics and chemistry. Like Kendrew, Crick was excited by the idea of exploring what went on at the boundaries between living and non-living matter. Advised by the physiologist A.V. Hill to go to Cambridge, where he would 'find his own level', he initially joined the Strangeways Laboratory, headed by Honor Fell, where he worked for two years on the physical properties of cytoplasm (the material that surrounds the nucleus inside cells) as a member of the Medical Research Council's research staff. The subject did not enthral him, but it did give him the opportunity to learn some biology and to begin to formulate his own ideas about what the important questions were. By his own account, these centred on genes and the importance of molecular structure.

While visiting the Secretary of the MRC, Sir Edward Mellanby, to report on his progress, Crick expressed his desire to do research at the molecular level. Mellanby revealed that on his desk at that moment was an application for funding for a new unit on that subject at the Cavendish Laboratory, headed by Max Perutz. Despite his junior status, Crick clearly already had the air of someone whose opinion was worth having, and Mellanby asked him if he thought the MRC should support 'these foreigners'. Rather horrified at the question, Crick replied that of course they should support Max. Soon afterwards, Max received a visit from a young Austrian mathematician, Georg Kreisel, who had become friends with Crick at the Admiralty and was now doing a PhD at Cambridge. According to Max, Kreisel cryptically asked if they had room for a physicist in the lab, not mentioning any names. Max wondered who the shy and retiring individual behind this embassy could be, but Crick has denied that he sent Kreisel to spy out the land, saying his friend had acted on his own initiative. Soon after, Crick turned up in person and talked to both Max and John Kendrew. 'I impressed Perutz sufficiently for him to welcome the idea of my joining him, provided the MRC would support me,' he later wrote. 'He just came and we talked together and John and I liked him,' was Max's recollection.

Crick arrived in the autumn of 1949, having just returned from a honeymoon in Italy with his second wife Odile. Max and Gisela immediately offered the couple the tenancy of the Green Door. After an interval of five years, they themselves were expecting another child. The Green Door was impossible for a family of four, and so they had taken the daring step of buying a small, three bedroomed semi-detached house in Eachard Road on the northern outskirts of Cambridge, to which they had moved that summer. As they had little in the way of savings, their former landlord Mr Kitteridge generously provided them with a loan of £1,200 to help them bridge the gap between the £3,000 price tag and what the bank was prepared to lend as a mortgage. Gisela's parents, who had been appalled when they visited for the first time earlier that year and saw how their pregnant daughter had been living, offered to cover the interest on this second loan. While house buyers today think nothing of taking out loans that are much larger multiples of their income, the whole notion of borrowing went very much against the grain for Max, who wrote rather melodramatically to his parents-in-law: 'Even with your generous offer . . . our debts will still hang like millstones around our necks for the next twenty years. If we buy unwisely, the education of our children will have to suffer . . . You must realise the enormous responsibility which I have to shoulder.'

After they moved into the new house, Gisela's parents continued to hint that Max was not maintaining her in the style to which she was accustomed. He felt that they suspected him of 'keeping her short', an accusation he hotly denied.

Gisela gets all the money I earn. This literally means all. I don't buy the most essential scientific books, I don't buy new clothes unless the old ones are in rags, I don't take any holidays unless they are financed from an outside source . . . As I neither smoke nor drink nor spend my money by betting on horses, my vices, if any, cost me no money. All the money I earn is spent on our household [plus the £50 annuity to his parents] . . . You may conclude from all this, that obviously my income is quite inadequate and that I should never have married Gisela anyway. It was small indeed in 1942. But it is really quite substantial now. During the past financial year I have earned *1100 pounds* . . . Our difficulties are due almost entirely to the very difficult times we live in. With income tax rarely less than one quarter of one's income, with prices soaring far ahead of salaries I think you could look far and wide to find a young couple who have done better than we.

This outburst began as a letter of thanks for £10 that the Peisers had sent for the new arrival, which Max decided to invest in an educational insurance policy for the new baby: Vivien already had a similar one, a gift from Gisela's brother Steffen. The Peisers had done much, ever since Max and Gisela married, to help out financially: there was a constant stream of small gifts, together with generous presents for Christmases, birthdays and new babies, and a substantial donation of their own furniture. Max seems both to have appreciated their generosity and been embarrassed that it was necessary. One way and another, Robin Noel was born on 27 December 1949 into a home that had more comforts than the family had previously enjoyed, with a nurse and an au pair girl on the payroll to make life as easy as possible for Gisela.

Robin was a fat and happy baby who laughed all the time. Gisela commented to Evelyn Machin that 'second children are luckier than first', and it was perhaps not entirely a coincidence that the little boy began to refer to himself as 'Lucky Robin' almost as soon as he could speak. Meanwhile Vivien seemed enthusiastic about the school she began to attend in January 1950, though Max complained that it was 'terribly churchy, with the result that we hear [her] chanting cheerfully "The best book for me is the Bible" and developing a passion for praying.'

While Max and Gisela were hugely relieved to have a proper house at last, Francis and Odile Crick were only too pleased to take on the tiny furnished flat with its enviably central location. The Crick household immediately became one of the social hubs of the lab and its associates. Crick's impact was much more than social, however. His stentorian voice and loud falsetto laughter echoed round the lab. He knew little of X-ray crystallography on arrival, though he had taken the precaution of reading Max's haemoglobin papers. Max taught him the basics of practical X-ray work: growing, selecting and mounting crystals, setting up the X-ray apparatus, and measuring the intensity of the reflections caught on film. He recommended Charles Bunn's textbook on the theoretical background, but Crick found it unsatisfactory: he began to rethink the subject for himself. 'I soon found I could see the answer to many of these mathematical problems by a combination of imagery and logic, without first having to slog through the mathematics,' he reported. Both Max and Bragg encouraged him to re-register for a PhD, despite his advanced age (he was now thirty-three), and he began the work that would eventually form the basis of his 1953 thesis, which bore the catch-all title *X-ray diffraction: polypeptides and proteins*.

Meanwhile Max's publications, which put him clearly in the forefront of the protein structure field, considerably raised his international profile. In the spring of 1950, he made a six-week trip to the US, something he had hoped for since his visit in 1943 when he first established contact with American researchers. One object of the trip was to learn about a machine for carrying out crystallographic calculations called the X-RAC (X-Ray Automatic Computer), built three years earlier by Ray Pepinsky and now at Pennsylvania State College in Philadelphia. Crystallographers from both sides of the Atlantic had visited to run their data through it, and had been pleased with the results. In April 1950 Pepinsky held two conferences designed to show off its paces to a wider group, in the expectation that further copies of the machine (a large, cumbersome and somewhat temperamental device covered in screens, dials and switches) might be built elsewhere. Max was going to stay on after the conferences to analyse some of his own and Kendrew's data.

During the rest of the visit, Max gave several lectures at leading American universities. He began by visiting Harvard, where the audiences showed great interest in his work – he found that 'most of what I had to say was new to them', as few, apart from his host John Edsall, seemed to have read his recent papers. His reception was almost overwhelming: he was showered with invitations by everyone he met, so many that he could not accept them all. Lotte came to stay for a few days, and was swept up in the social whirl (though she later complained to Gisela that she had wanted to have Max to herself). Max was delighted to receive so much attention and felt on top form. A note of insecurity crept in later, however, when he commented to Gisela that no one had yet offered him a professorship, 'either temporary or permanent'. She reassured him that England was the best place for him to be, but she had rather missed the point: he wanted the offers not so that he could leave Cambridge, but to increase his market value and cement his position there.

Chairs or no chairs, there was no doubt that he had begun to make his name in the US, and his lectures in Boston and Philadelphia increased his visibility. He heard on the grapevine that Linus Pauling's lab at the California Institute of Technology had held colloquia on his two most recent haemoglobin papers, and had 'considered them sound'. 'Yet,' he confided to Gisela, 'how I wish I could get convincing proof for my conclusions.' The conferences in Philadelphia brought together representatives of every group from both sides of the Atlantic who were interested in protein structure, but Max was inclined to dismiss the American contribution.

There is a sharp contrast between the English crystallographers who do beautiful work in solving ever more complex structures by improvements in existing methods and the American School whose thought is concentrated on the discovery of new methods which they try to test on the simplest kind of structure, or sometimes on the development of purely abstract theories.

One American about whom Max seemed particularly exercised was David Harker, a former student of Linus Pauling's who was rumoured (correctly) to have been given a million dollars to set up a new protein structure group at Brooklyn Polytechnic. Harker was one of those trying to develop what were known as 'direct methods', which aimed to reach a solution purely through applied mathematics. He had already had some success on small molecules. Max was dismissive of his prospects, consigning him to 'the lunatic fringe of American crystallography'.

Harker had kept quiet during the conference and not mentioned a word of his new method of protein analysis. I was determined to bring this out into the open and to force a discussion, so that we could all judge the validity of his claims. I did this with great success. Harker had to explain his method to us and it became perfectly clear that it was devoid of any physical basis. I had the entire meeting behind me ... I expect Harker will still get his million dollars but his shares have dropped a lot in value.

Harker did get his million dollars, set up his group and went on (in 1967) to solve the structure of the enzyme ribonuclease – although he was beaten to the punch earlier that same year by Hal Wycoff and Fred Richards at Yale. Max later realised that his early assessment of Harker had been too hasty – he himself used a mathematical method developed by Harker as an important step towards the solution of the haemoglobin structure. Two other American crystallographers, Herbert Hauptman and Jerome Karle, went on to develop direct methods further and won the Nobel prize in 1985. It is difficult to avoid the conclusion that Max's suspicion of the direct approach stemmed partly from his own lack of mathematical insight. He was right, however, that direct methods would only be of limited usefulness in solving protein structures, which were too large and too irregular to offer many neat mathematical short cuts.

Pepinsky's X-RAC was another matter. It did not involve any theoretical advance: it simply represented all the terms in a crystallographic calculation as waves and added them up, thus saving all the effort in punching a manual calculator or producing piles of punched cards for a Hollerith machine. You could try different phases by setting dials and flicking switches, the result appearing as a contour map on an oscilloscope screen. Max called it 'miraculous'. He told Gisela it had yielded 'quite promising results' on Kendrew's myoglobin data, and that he thought it had immense possibilities. But the X-RAC was almost immediately eclipsed by the infinitely greater possibilities of the electronic computers that were becoming available for the first time.

Max brought the first news of these to Philadelphia: it caused a 'sensation', though he himself may not have fully appreciated its significance. To the great advantage of the MRC Unit, Cambridge was in the forefront of electronic computing – and, as so often happens, it was one of the more junior members of Max's group who first saw its immense potential for crystallography. One day over dinner at Christ's College, soon after he joined the Unit, Hugh Huxley was moaning to an Australian graduate student, John Bennett, about the tediousness of the calculations he had to perform. Bennett was doing his PhD in the Mathematical Laboratory, which had been set up in the 1930s to develop large-scale computing methods that would be a resource for all the University's science departments. As soon as the war ended, the new director, Maurice Wilkes, began to build an electronic computer, one of the first in the world that could store programs and run a wide range of applications. By 1949 the Electronic Delay Storage Automatic Calculator (EDSAC) was complete, and in 1950 it came into regular use as a service to University members. Bennett was convinced that Huxley's problem could be programmed.

The EDSAC's very limited memory – it could not store all the terms at once and so had to do the calculation in stages – meant that the program Huxley and Bennett produced was slightly cumbersome, but it was still a huge advance on doing it all by hand. By the time the program was working, Huxley had moved on from protein crystallography to muscle fibres, but he told John Kendrew what he and Bennett had been up to, and Kendrew took up the challenge 'like a duck to water'. With Bennett he improved the program, which they eventually published in 1952. In an early progress report, Max told the MRC that a computation that had previously taken six months and cost £450 could now be done in thirty hours at 'negligible cost'. Others around the world soon saw that electronic computers would become an essential adjunct to protein crystallography. Pepinsky's machine gradually became obsolete, and plans to build further models were dropped.

Max never transferred his personal enthusiasm for the X-RAC to the EDSAC. Programming was not the kind of intellectual exercise that appealed to him (it might have been significant that Bragg, too, regarded computers with suspicion). Once it became obvious that crystallography and computing would forever be linked he was happy for his haemoglobin data to be analysed in that way; but he left it to others in the group – Kendrew being only the first of these – to develop new and better programs, and to supervise the work of the young women who were gradually recruited to prepare the data for the computer.

As well as improving his scientific standing, Max's 1950 trip to the US had helped to put him and his family on a much healthier footing financially. While he could hardly be described as rich, Max received a substantial salary increase in 1950 and earned fees for other activities, including his lectures in the US, his talk on glaciology for BBC Radio and the preparation of an exhibit for the Festival of Britain in 1951. He wrote to Lotte complaining (not without some self-satisfaction) that he was constantly in demand:

> I have had work crowding in on me from all sides . . . I cannot work more than 8 or 9 hours a day, and I do want to enjoy having a wife and children. Life is worth living *now*, rather than being endured for some imaginary paradise in the future. My trip did me a lot of good in many ways, not least because it relieved me of my terrible financial worries.

He told Lotte that he had been able to pay off his debts both to Heini Granichstaedten and to his 'English friends' (presumably the Kitteridges), and now owed only the mortgage on his house which was 'quite bearable'. He even suggested increasing the contributions he, Franz and Lotte made to their parents' support. For some time now his father had been unable to work, although he benefited from a small state pension as a result of the jobs he had held during and after the war. Out of kindness, Gisela's uncle had given him a less strenuous job in London than the one he had struggled with in Cambridge, but even that became too much. Dely too had given up the part-time handicraft work she had dabbled in, in order to look after her increasingly ailing husband. She greeted Max's offer, characteristically, with 'curt thanks and brusque rejection' before accepting it at his persuasion, while at the same time Hugo wrote to tell Lotte how grateful they were as money was very short. For Max, the greatest satisfaction was the freedom to concentrate on research:

> As a result of my salary increase I decided to drop practically all my subsidiary activities, such as writing and broadcasting, and much of the private tuition I do, and concentrate all my energies once more on

research. It was nice and reassuring to find that I was successful as a writer and broadcaster, and to have a horde of editors at my tail clamouring for more and asking me to write books on all sorts of things, but having once proved to my own satisfaction that I could do it I might as well leave off for a few years and concentrate all my energies on research – the most satisfying of all my activities.

7

Annus mirabilis

If you go on hammering away at a problem, it seems to get tired, lies down and lets you catch it.

W.L. Bragg

Of the fifty or so protein molecules that had been crystallised, photographed and measured worldwide by the early 1950s, around half were accounted for by haemoglobin and myoglobin from different species studied by Max, Kendrew and their assistants. Max had also become interested in the molecular basis of the haemoglobin defect that causes sickle cell anaemia, an inherited blood disorder (see Chapter 11). In the first five years of its existence, Max's MRC unit had established itself as one of the two leading centres in the world for the study of protein structure (the other being Linus Pauling's lab at Caltech).

However, as far as his studies of the haemoglobin structure were concerned, Max himself had reached an impasse. Without a means to find the phases of his 7,000 reflections, he could not hope to reveal the arrangement of atoms in the molecule that they so tantalisingly represented. All he had to go on were his Patterson maps, which were perfectly correct but equally inscrutable. He put his faith in his hatbox model, and hoped for the best.

One day in 1950, Bragg appeared in the lab with a broom handle into which he had hammered corks to form an extended spiral, or helix. He challenged Max and Kendrew to use such models to work out from first principles what possible configurations a protein chain could adopt. To recap, a protein is made of one or more strings of amino acids: one amino acid links to another to form a protein fragment called a peptide, and the join is called a peptide bond; a chain of amino acids is a polypeptide. In order to compress themselves into the globular shape of a protein molecule the polypeptide chains must fold, for at least part of their length, in some regular fashion, held in place by the bonds that form between successive loops. Knowing how they did this would be an important step forward in understanding protein structure. William Astbury at Leeds, who had pioneered the study of protein fibres such as hair and wool, suggested that the chain folded back and forth like a ribbon. However, this model did not fit the data convincingly, and more recently an American researcher called Maurice Huggins had suggested that the chains might adopt a helical structure, like coiled springs.

Max and Kendrew took all they knew about the sizes, shapes and bonding preferences of amino acids, which would constrain the number of possible configurations, and used them to construct a whole family of helices, varying the number of amino acids per turn and the steepness of the rise at each turn. The approach was exactly right, but it failed because they clung to one unnecessary constraint, and failed to observe an absolutely necessary one, of which more below. It was ultimately a depressing exercise. None of their model structures fitted the data any more convincingly than Astbury's ribbon, but they published the paper anyway, in the *Proceedings of the Royal Society* in 1950. 'It was one of those papers you publish mainly because you've done all that work,' Max told Horace Judson years later. Bragg called it 'the most ill-planned and abortive [paper] in which I have ever been involved', again with a considerable amount of hindsight.

Much of the information they used about the sizes of the amino acids and the lengths and angles of the bonds between atoms had come from the work of the only lab in the US that was anything like their equal: Linus Pauling's group at the California Institute of Technology. Unbeknown to the Cambridge researchers, Pauling had also been thinking about protein chain configurations. Two years before, he had spent six months at Oxford as a visiting professor. In April 1948 he had fallen victim to a British cold and retired to bed with a pile of detective novels. The next day, bored with such unchallenging material, he began to think again about protein chains. Taking a piece of paper, he drew out, from memory, the zig-zag sequence of atoms that form the continuous part of a chain of amino acids, ignoring short chains that protrude to the side. He drew it flat, then rolled up the piece of paper so that, by adjusting the angle at which it was rolled, he could bring the atoms from either side neatly to connect with one another. The result was a helix with 3.7 amino acids per turn.

It looked right, but Pauling was anxious. Astbury's early X-ray photographs of hair gave a prominent reflection indicating a repeating feature every 5.1 Å, which he took to mean the rise between one turn of the chain and the next. The helix that Pauling had created had a 5.55 Å rise (later revised to 5.4 Å). Pauling was even more worried when he went over to Cambridge and heard from Max that haemoglobin also had a reflection that supported

Astbury's 5.1 Å repeat. He said nothing as Max showed him his Patterson maps, pointing out the 'rods' of density that he thought represented parallel chains in the molecule. Pauling privately thought the rods would have the form of his helix, but because of the discrepancy in the height of the rise he was reluctant to commit himself. He was also worried that if he embarked on a discussion with Bragg and his team, he might inspire them to come up with the correct solution themselves: he was shaken to see how much better the X-ray facilities at the Cavendish were than those he had at Caltech. He returned to California to do more work on the problem with his colleague Robert Corey, but published nothing for almost three years.

Pauling had what his biographer called a 'vertiginous moment' when he saw the title of the 1950 Bragg, Kendrew and Perutz paper on the config-uration of polypeptide chains. Had his hesitancy about publishing his new helix let them get ahead? But he had no cause to worry. Their most serious error was a failure to understand that the set of three chemical bonds including the link from one amino acid to the next had to be 'planar', and rigid. Pauling had published a discussion of this point almost twenty years previously: it was something that any well-read chemist should have known. Their second mistake was to assume that there had to be an exact number of amino acids in each turn of a helical chain. So they had tried chains with two, three and four amino acids per turn, none of which sat comfort-ably with no strain on the bonds.

One fine Saturday morning at the beginning of June 1951, Max dropped into the Cavendish library to catch up on his reading. Picking up the two most recent issues of the *Proceedings of the National Academy of Sciences*, the top science journal from the US, he discovered not one but eight papers from Pauling and his colleague Robert Corey on protein chain configurations, the highlight of which was the structure that they chris-tened the alpha helix. Max was 'thunderstruck'. He could see at once that the structure, chemically speaking, must be right. The bonds were all in the right position to hold the structure in shape without strain, unlike the wretched attempts he and Kendrew had spent so long constructing.

The structure looked dead right. How could I have missed it? Why had I not kept the amide groups planar? Why had I stuck blindly to Astbury's 5.1 Å repeat? On the other hand, how could Pauling and Corey's helix be right, however nice it looked, if it had the wrong repeat? My mind was in turmoil. I cycled home to lunch and ate it oblivious of my chil-dren's chatter and unresponsive to my wife's inquiries as to what the matter was with me today.

By the time he had finished his lunch, Max had realised that if Pauling's structure was right, then there should also be a reflection on a protein fibre X-ray photograph that represented the rise of *each amino acid* round the axis of the fibre – like the height of a single step on a spiral staircase, rather than the height between one turn of the stair and the next. This height was 1.5 Å: he was sure there had been no 1.5 Å spot on any of Astbury's photographs of alpha-keratin in hair, but then he realised that such a spot would be so far from the centre of a conventional photograph that it would fall off the edge. To show up, it would also require the fibre to be set at a different angle to the X-ray beam from the one Astbury had used. Max rushed back to his lab and rummaged in a drawer for a horse-hair he remembered keeping there. He set it up in the X-ray apparatus at the correct angle, and used a cylindrical film to catch reflections over a much wider range than Astbury had managed. You can't rush these things: it took two hours to take the photograph, but finally he had what he wanted and feverishly developed the film. There was the 1.5 Å spot! Of all the models of protein chains that had been put forward, only the alpha helix would have produced it.

Max was jubilant. Pauling may have had the imagination to infer the correct structure, but he, Max, had provided the proof. On the Monday morning, he burst into Bragg's office with his photograph and showed him what he had found. When Bragg asked him what had made him think of the experiment, he said it had been 'sparked off' by his fury at not having thought of the alpha helix himself. Bragg's dry rejoinder, 'I wish I'd made you angry earlier', became the title of the volume of essays on science and scientists that Max published towards the end of his life.

Max had missed the alpha helix not because of a failure of technique or inadequate data, but simply because of a failure of imagination, and not having Pauling's bonding rules at his fingertips. He had allowed himself to be corralled by received wisdom, instead of opening his mind to all possi-bilities. Once he had the result of his horsehair experiment, however, he was inspired to go further. He went on to find the same 1.5 Å reflection in other structures made of orderly arrays of protein molecules, including porcupine quill, muscle fibres and (somewhat looking with the eye of faith, as the resolution of his images was very poor) haemoglobin. Immediately, he wrote the results up and despatched them as a letter to *Nature*, arguing that the presence of the 1.5 Å reflection left 'little doubt about [Pauling and Corey's] structure being right', and that it provided the first 'simple and decisive criterion' for testing the validity of a proposed model of protein structure. He also cited it as evidence in support of bundles of parallel rods he believed he could see in his Patterson maps. Max was in no doubt that this was the most important discovery he had ever made, as he boasted to his parents-in-law:

> I have made a discovery of decisive importance which changes the whole outlook of our work . . . This discovery vindicates my interpretation of the haemoglobin data . . . and dispells [sic] all doubts about the hours of labour spent on what seemed to practically all my friends and colleagues a problem of hopeless complexity. The field of protein structures is thrown wide open and the next few years will see the solution of many of these structures. By my new method one can find the chain direction in any protein and given that the structure can be solved . . . The greatest share of the glory will of course go to Pauling whose solution of the chain problem is a true stroke of genius. It is a terrible shame that we missed this model through a combination of unfortunate circumstances and blunders for which we shall never forgive ourselves . . . But all this disappointment is forgotten over the intense thrill at having removed all doubts and having my ideas about the structure of globular proteins vindicated.

When Max wrote to Pauling to congratulate him on the alpha helix and to mention his own contribution as 'the most thrilling discovery of my life', Pauling graciously told him that it 'seems to me to be very important'. Max had managed to share the limelight and so survived Pauling's bombshell with his self-esteem intact. Bragg was not so fortunate. With Pauling's alpha helix paper clutched in his hand and agitation in his breast, he had crossed the courtyard from the Cavendish to the University Chemical Laboratory, presided over by Alexander Todd. This, according to Todd's account, was his first visit, even though members of their respective departments had collaborated on studies of simple peptides. Todd rather patronisingly explained that Pauling's structure must be correct, because of the planarity of the peptide bond, adding moreover that any organic chemist could have told him that. Had he not sought advice from a chemist? Bragg muttered that yes, they had spoken to Charles Coulson – a theoretical chemist from Oxford who had, it turned out, given them misleading information. To be beaten by Pauling (not for the first time in his career), and then shamed by Todd, was too much. Bragg resolved that henceforth no structure from his group could be published unless it had been personally vetted for chemical plausibility by Todd.

Crick, meanwhile, had tried to find a protein of his own to work on, but of the three or four he chose, none had suitable crystals. So he turned his attention to the tentative conclusions that Max and Bragg had drawn from their Patterson maps, now authoritatively published in the two Royal Society papers. He did what Max afterwards kicked himself for not doing: he calculated from first principles what the Patterson peaks should have looked like if the haemoglobin molecule had indeed had the 'hatbox' structure Max had suggested, with most of the protein consisting of layers of parallel rods. He immediately found that they would have been several times stronger than the peaks Max had actually recorded. Max's results were not consistent with his own model – haemoglobin could contain only much shorter stretches of folded polypeptide chain.

Crick announced his findings in public at a research seminar in the Cavendish in July 1951. This was one of the earliest gatherings of British protein crystallographers: Bernal was there, as were Dorothy Hodgkin and most of her team from Oxford, and Astbury from Leeds. Bragg presided, proud that his little group was leading the field in protein structure. In his turn, Crick rose to speak; it was only the second research paper he had presented, the first being on the work he had done on cytoplasm at the Strangeways Laboratory. He had gone over what he was going to say with John Kendrew, and asked him what title he should give his presentation. The literary Kendrew suggested 'What Mad Pursuit!' from John Keats's *Ode on a Grecian Urn*.

Crick not only pointed out the glaring flaw in Max's model: he told the assembled company that 'Broadly speaking . . . they were all wasting their time and that, according to my analysis, almost all the methods they were pursuing had no chance of success.' David Davies, then a graduate student in the Oxford crystallography lab, remembers the scene vividly:

> Max proposed a structure for haemoglobin which was a number of helices in a bundle, parallel to one another. Then Francis got up . . . He demolished Max. He had a little wooden ball, and his wife Odile had painted on it the contours of the Patterson function at 5.4 Å resolution, and if all these helices were parallel to one another, you would expect to see an enormous peak. And there was no such thing, so Francis showed pretty conclusively that Max's model couldn't be correct.

> Bragg, who had already found Crick's brash personality and inescapable voice hard to take, was furious. He accused him of 'rocking the boat', and soon after let him know that he wished him to leave Cambridge as soon as he had finished his PhD. Max, perhaps surprisingly, took Crick's criticism with much better grace, and never held this public demolition of his ideas against him: indeed, he was instrumental in quietly interceding with the MRC in Crick's defence so that his future in the Unit was assured.

Crick's evidence was undeniable. He published the ideas presented at the seminar only with the throwaway line: 'Rough calculations show that the vector rods to be expected from such a model would be more dense than those observed'; the sentence appeared in a paper of 1952 in which he gave accurate figures for the Patterson peaks that could be expected from various simple models. Some of the same material featured in his PhD thesis. Max had to live with the unpalatable fact that the first two papers in his projected 'grand series' were seriously in error. Trying to recover some dignity with hindsight, years later he wrote:

> Perhaps I had suspected the weakness of the vector rods even before Crick . . . proved it to me, but I had pushed this thought aside, because I could not face the stark truth that the years of tedious labour, the many nights of interrupted sleep and the appalling strain of measuring the intensities of thousands of little black spots by eye had brought me no nearer the solution of the structure of haemoglobin, and that I had wasted the best years of my life trying to solve a seemingly insoluble problem.

Crick's presence in the unit acted as a magnet to others. Whereas Max liked to work quietly at his bench, Crick liked to talk, to examine ideas critically, to float balloons. Bragg may have groaned and covered his ears (he said Crick made them buzz), but others were stimulated by his intellectual energy. Freddie Gutfreund dropped in daily, and crystallographers in the Cavendish who had an interest in the theory of diffraction by different structures, notably Bill Cochran and Eric Howells, also joined in the discussions on protein structure. Crick's demolition of Max's hatbox model made it clear that the problem was too complex to be solved by more and better photographs. There was room, too, for back-of-the-envelope calculations, for well-informed conjecture and, above all, for discussion: all of which Crick was prepared to contribute in abundance, whether or not it was asked for. Hugh Huxley was also one of those who valued his presence.

> He was extraordinarily clever at calculating things that you might think were too difficult to calculate. He had a tremendous grasp of the theory and logic behind things. In any discussion he would analyse it all so clearly and deeply and lucidly – he was marvellous to talk with about any topic in science, he certainly clarified my thinking and others too. He was also very jolly and cheerful and funny.

The discovery of the critical 1.5 Å reflection in haemoglobin did confirm that a proportion of the protein took the form of alpha helices, though

Crick's calculation showed this proportion to be much less than Max had originally claimed. Crick now attacked a problem that had long been waiting for a solution. If a biological structure was helical, what X-ray pattern would it give? Working with the theoretician Bill Cochran in the Cavendish, and following up some earlier work by Vladimir Vand at the University of Glasgow, he published a general theory of diffraction by helices. He and Cochran then went on to demonstrate why Astbury's 5.1 Å spot had misled people for so long: it turned out that in protein fibres such as keratin in hair, three alpha helices wound round each other like the strands of a rope – Crick called it a 'coiled coil' – shortening the pitch of each turn. These papers finally persuaded Bragg that Crick might actually be useful after all. He himself decided to take a grip on the haemoglobin project by becoming much more actively involved.

Crick's next trick removed any remaining doubt about his incisive intelligence. The story of the DNA double helix has been told so often that I will retell it only briefly here. James Watson was a tall, gangling and very bright young geneticist who had entered the University of Chicago at the age of fifteen and had completed a PhD at the University of Indiana at Bloomington by the time he was twenty-two. There he became the newest and brightest recruit to the circle, led by Salvador Luria and Max Delbrück, working on the genetics of phage, a form of virus that infects bacteria. The chief lesson he absorbed was the overriding importance of understanding the nature of the gene. In the summer of 1951, Luria met John Kendrew at a conference in the US, and decided that Watson should go to Cambridge.

So it was that one day in September 1951, Watson stuck his crew-cut head round the door of Max's office and asked, without preamble, if he could come and work there. Max, according to Watson's own account, could not have been more affable. He assured him he would have no difficulty learning protein crystallography, introduced him to Bragg and obtained his formal permission for Watson to join the lab, took him to find lodgings in the town and showed him King's College Chapel. Watson was entranced by everything he saw, except the dingy state of typical Cambridge lodging houses. In the years that followed, he adopted the Cambridge lifestyle enthusiastically: he let his hair grow, affected tweed jackets and V-necked pullovers, and engaged in conversation on every possible subject over lunches in the local pub and bibulous dinners in the homes of his friends, who mostly represented a heady mix of science and gracious society.

Soon after term began he wrote to tell his sister that he found Max 'very pleasant in a reserved way', and that the group was 'quite small though it appears quite intelligent'. He also mentioned the presence of a 'very likeable research student' who had filled him in on the social niceties of lab procedure: this is almost certainly his first reference to Francis Crick. Formally, Watson was to work under Kendrew, but as soon as he discovered that Crick shared his obsession with the nature of the gene, the two of them began to spend most of their waking moments arguing, debating, drawing diagrams and making models in an effort to divine the structure of the genetic material: deoxyribonucleic acid, or DNA.

Two vital discoveries had laid the groundwork. First, in 1944 Oswald Avery and his colleagues had published a paper proving, against all expectations including his own, that the 'transforming principle' that could confer heritable characteristics on a colony of bacteria was made of DNA. Chemically speaking, DNA is very simple: it consists of only four components, the nucleotides adenine, thymine, guanine and cytosine, strung on a backbone of sugars and phosphates. Because it seems so simple compared with proteins, most biochemists were very sceptical that it could do a job as apparently complex as making genes. By 1951 it was still not universally accepted that genes were made of DNA, though Watson was a believer and so was Crick. Second, Maurice Wilkins at King's College London had begun to take X-ray photographs of DNA fibres. Watson had seen one of these photographs at a talk Wilkins gave in Naples in the summer before he came to Cambridge. Although he knew nothing about X-ray crystallography, he immediately took Wilkins's point that the distinctive pattern he had obtained must mean that the fibre had a regular structure and so could in principle be solved. There was also a third critical discovery, whose significance all the key players were extraordinarily slow to recognise: the Columbia University chemist Erwin Chargaff's rule that in any preparation of purified DNA, there were always the same number of adenines as thymines, and guanines as cytosines.

After their first, embarrassingly unsuccessful, attempt at modelling the structure in the winter of 1951, Bragg told the two men to leave DNA alone. There was an unwritten code among British crystallographers not to trespass on one another's territory – and DNA 'belonged' to Wilkins and his new colleague at King's, Rosalind Franklin. Indeed, all the X-ray evidence Watson and Crick had for their DNA model came from the King's lab – they themselves never took any X-ray photographs of DNA. For a few months, Crick returned to his work on helical diffraction, and Watson made a successful X-ray study of tobacco mosaic virus, showing that it had a helical structure. In late January 1953, however, a manuscript arrived in the lab that changed everything. The manuscript was from Linus Pauling, whose son Peter was in Cambridge (somewhat lackadaisically working on a PhD with John Kendrew). Peter had already hinted to Watson and Crick that his father had a structure for DNA. Now at last they could see it – and see that it was just as wrong as the one they had produced more than a year earlier. Watson took Pauling's manuscript to London to show to Wilkins, and during the same visit Wilkins showed him a new photograph that Franklin had taken of a form of DNA she called the 'B' form.

By this time, Watson had absorbed much of the theory of helical diffraction that Crick had developed, and he saw instantly that the photograph, with a distinctive X-shaped feature at its centre, suggested that DNA was a helix. On his return to Cambridge he explained the potential significance of this photo to Max and to Bragg. As Bragg put it in his Foreword to Watson's book *The Double Helix*, 'When competition comes from more than one quarter, there is no need to hold back.' Convinced that if they did not work on the structure, Pauling would get there first, he put his scruples aside and encouraged Watson and Crick to pursue the DNA problem. There was more unpublished DNA data, though not photographs, in a report on the work of the King's lab that Max had received as a member of the Medical Research Council's Biophysics Committee. When, a couple of weeks later, Crick asked to see it, Max saw no ethical difficulty in showing it to him, as it was not marked 'Confidential' – a decision he would find himself having to defend more than fifteen years later (see Chapter 12). The data allowed Crick to make the crucial inference that two chains of DNA ran in opposite directions.

Too impatient to wait for the metal components from the workshop, Watson made cardboard cutouts of the DNA bases, and in only a few more weeks arrived at the now iconic double helix structure. Whereas previously they had put the nucleotides on the outside, now he and Crick put the sugar-phosphate backbone on the outside and the nucleotide bases on the inside forming complementary pairs: adenine with thymine and guanine with cytosine. He and Crick excitedly built the tall wire model that features in the classic photographs taken soon afterwards. Needless to say, Max was thrilled beyond measure that members of his unit had made such an important discovery. He immediately wrote to Harold Himsworth, Secretary of the Medical Research Council, to tell him of the result.

> They used . . . a certain amount of unpublished X-ray data which they had seen or heard about at King's. All these X-ray data were either poor, or referred to a different form of structure, and while they indicated certain general features of the structure of DNA they did not give a guide to its detailed character.

While Watson and Crick were building their structure here, Miss Franklin and Gosling at King's obtained a new and very detailed picture of DNA. Watson and Crick only heard of this photograph when they sent the first draft of their paper to King's, but it now appears that this new photograph confirms the important features of their structure.

The account that Max gives in this letter of the role of the King's data in Watson and Crick's success is inaccurate in many respects, particularly in regard to the timing of their knowledge of Franklin's picture. He knew that Watson had seen it weeks before: the only explanation for his highly uncharacteristic departure from strict truthfulness must have been the need to conceal Wilkins's role in revealing Franklin's photograph from the director of the biophysics lab at King's, John Randall. Watson and Crick themselves struggled with the wording of their acknowledgement to King's. On 25 April 1953, their one-page paper 'A Structure for Deoxyribose Nucleic Acid' appeared in *Nature*. They settled on the formulation that they had been 'stimulated by a knowledge of the general nature of the unpublished experimental results and ideas of Dr M.H.F. Wilkins, Dr R.E. Franklin and their co-workers', a form of words suggested by Wilkins himself. They also stated that when they built their structure they were 'not aware of the details of the results' presented in papers by the King's groups back-to-back in the same issue of *Nature*, one of which included the B form photograph, and that the double helix rested 'mainly though not entirely' on published experimental data and stereochemical arguments. According to Franklin's biographer, although she was on friendly terms with Crick until the end of her life, Franklin herself never knew just how 'stimulating' her B form photograph had been.

No one in Max's unit revealed publicly that Watson had seen the photograph until he himself published his explosive book *The Double Helix* in 1968. Possibly they feared a dispute with King's about credit, though this possibility seems to have been effectively neutralised by the agreement to publish simultaneous papers, in which all comment on the fit of the model to the experimental data was left to the King's group. Wilkins's longstanding friendship with Crick appeared undamaged by the Cavendish coup, to which he responded with surprising sang-froid that although they were 'a couple of old rogues' there was 'no good grousing' and 'who the hell got it isn't what matters'. Franklin, meanwhile, had left King's and was starting a new project on the structure of viruses in Bernal's department at Birkbeck College. In the event there was no row until many years later, after the publication of *The Double Helix* provoked a debate about the ethics of all those involved that is still not finally resolved.

Watson and Crick's *Nature* paper famously concluded with the deceptively, casual remark: 'It has not escaped our notice that the specific pairing we have postulated immediately suggests a possible copying mechanism for the genetic material.' In another paper a month later they added:

> . . . in a long molecule many different permutations are possible, and it therefore seems likely that the precise sequence of the bases is the code which carries the genetical information.

Watson and Crick had discovered the secret of life. There would be other secrets to discover – there still are – but none that would have such a profound impact on our understanding of the living world, from the theory of evolution to the causes of cancer. It would be some time before the full significance of their discovery was generally understood and accepted, but no one could be in any doubt that the little group over which Max presided had fully justifed the MRC's investment. The original five years' funding had been extended by only a further year from October 1952, while the MRC waited to find out what Bragg was going to do about his retirement. The discovery of the double helix in 1953 was only one of several outstanding advances in that remarkable year that ensured the funds would continue to flow.

While the DNA saga was unfolding, Bragg had thrown himself into the protein structure problem with renewed vigour, with Max enthusiastically at his side. Together they published four papers in 1952 alone: a publication rate, as Bragg's biographer Graeme Hunter noted, higher than the 62-year-old Bragg had achieved since the 1920s. The papers steered clear of the dangerous area of chemistry in which Bragg had found himself to be so woefully lacking, and concentrated on strictly crystallographic considerations in which he felt quite at home. He returned to the data Max had collected on the swelling and shrinkage of the crystals at different levels of moisture, and more recent experiments Max had done varying the amount of 'salt' (actually ammonium sulphate) in the water layers between the molecules of haemoglobin. Most significantly, he drew on new measurements Max had made of the intensities of the spots on his photographs. Conventionally, crystallographers measured these intensities by eye, by comparing them with an arbitrary reference set of spots ranging from strong to weak. For the new study, Max measured what are known as the 'absolute' intensities. This was a method of accounting for the proportion of the

X-rays in the incident beam that were actually diffracted by the crystal, rather than passing straight through it.

Max calculated the absolute intensities by comparing the reflections on his haemoglobin pictures with spots given by a crystal of anthracene, a hydrocarbon with a very regular crystal structure. He looked at how the absolute intensities of the 'low order' reflections – those due to the positions of the molecules in the crystal, rather than the atoms in the molecules – changed with different levels of salt concentration. These differences allowed him, in some cases with more confidence than others, to obtain information about the phase angles of these few reflections. Because a molecule of haemoglobin is made up of two identical halves, the job of finding some of the phase angles is much easier: instead of each angle being anywhere between 0 and 360 degrees, it only has to be given a plus or minus sign.

Max was able to find the signs for low order reflections in a photograph of haemoglobin taken at right angles to the axis that split the two symmetrical halves. We are talking about fewer than a dozen signs out of thousands of reflections in total, but it was a real step forward. Bragg used the information to calculate, without worrying too much about the detail, what shape of molecule would give those figures, and came up with the answer that it would be an elongated sphere, or ellipsoid, measuring about 55x55x65 Å. These turned out to be very close to the true dimensions. In a further paper he went on to show that they fitted well what was known about the crystal structure of several of the different forms of haemoglobin Max had studied over the years. The work shed light only on the external form of the molecule, not its internal structure, but it was a modest breakthrough none the less. For the first time since Crick had demolished his 'hatbox' model, Max had something positive to go on. It was a rewarding and stimulating exercise, but at the same time a humbling one. He wrote later:

> The papers on these findings, authored by Bragg and myself, were largely written by Bragg. They mirror his originality, his lucid arguments and his profound understanding of diffraction, but they made me feel that I ought to have thought all this out myself.

A third 1952 paper looked at the internal structure of haemoglobin, trying to find an alternative to Max's discredited 'hatbox'. A graduate student in the Cavendish, Eric Howells, had worked with Bragg on some of the protein data. Bragg, Howells and Max examined the effect of packing the polypeptide chain, now known to include some proportion in the form of an alpha helix, into an ellipsoid of the size Bragg had calculated. They began with the assumption that 'the protein molecule consists for the greater part of parallel polypeptide chains in a hexagonal packing'. But Howells's analysis, included in the paper, showed that the absolute intensities of the reflections supposed to be due to the end-on view of such chains were 'only one third of what would be expected'. The paper concluded with some speculation about why this might be, focusing on the known fact that haemoglobin consisted not of a single chain but two identical halves, and also airing the possibility that 'the chains are very far from straight'. The paper is interesting in that it reports what is essentially a negative result: the model proposed, based on knowledge of the size and shape of the molecule, the length of the polypeptide chain, the arrangement of atoms in an alpha helix and the apparent distance between neighbouring chains suggested by Max's Patterson diagrams, could not be reconciled with the data on absolute intensities. It was, in effect, a not too embarrassing way of saying their previous assumptions were quite wrong.

Max's measurement of the absolute intensities, however, had an unintentional consequence of much further-reaching significance. In the course of his 'What Mad Pursuit!' seminar, Crick had conceded that the one method that might offer some hope of solving a protein structure was the method known as isomorphous replacement. If you can introduce into a molecule an atom much heavier than all the others, without changing the overall arrangement of the atoms, then X-ray photographs of the two forms (with and without the heavy atom) will show measurable intensity differences that allow you to solve the phase problem. This was not a new idea. In 1936, the year Max came to Cambridge to learn crystallography with Bernal, John Monteath Robertson at the Royal Institution had solved the structure of the pigment phthalocyanine using isomorphous replacement. He had grown crystals of the pigment with and without an atom of nickel nestled among the carbons, nitrogens and hydrogens that make up the rest of the molecule: his solution of an organic molecule with 40 non-hydrogen atoms was a breakthrough at the time.

Phthalocyanine was uniquely appropriate for this technique: not only was it accommodating in accepting the nickel atom at a fixed location, but its structure was beautifully simple and symmetrical, greatly reducing the number of parameters that needed to be found for its solution. Proteins were very different, hundreds of times the size and much more complex. Optimistic as ever, Desmond Bernal mentioned isomorphous replacement as a possibility for determining protein structures as early as 1939, and even earlier than that he was encouraging Dorothy Hodgkin to try and grow insulin crystals, which naturally contain zinc, with other metal atoms. But in general protein crystallographers had dismissed the idea. They believed that in such a large molecule no single heavy atom would have a measurable effect among the many thousands of lighter atoms. In their paper to the Barcroft Symposium in 1948, Max and Kendrew had written that 'Even the heaviest of heavy atoms would make a negligible contribution to the reflexions from so large a unit cell.' Bragg, with his felicitous way with words put it more succinctly in a *Nature* article the following year: 'No heavy atom could stand out in such a crowd.'

Crick claimed that when he suggested isomorphous replacement in 1951, he had already calculated on purely statistical grounds that a heavy atom would indeed have an effect.

> Essentially the fraction from the atoms in the protein as a whole add up in a random manner and . . . the effect of an added heavy atom, assuming it is a proper isomorphous replacement, is proportional to what you add. In other words the small number of heavy atoms all work together, while in the rest of the protein [the atoms] interact at random, so if you have 100 more atoms you only get ten times the effect. The general level due to the protein is lower than you'd expect compared to the level you'd get from a heavy atom and that's a straightforward bit of crystallography.

While Crick's statistical argument must have lodged in Max's mind, it was the absolute intensities that convinced him that isomorphous replacement might work after all. When he did these measurements for the papers with Bragg, he was surprised to discover that the fraction of the X-ray beam that was diffracted by the protein was much smaller than he expected. Far more of the diffracted beams cancelled each other out by interference than reinforced each other and produced a spot on an X-ray photograph. Under those circumstances, as Crick had predicted, the coordinated diffraction from a small number of very heavy atoms would have a clear impact on the intensities. This provided a theoretical route to the Holy Grail, the solution of the phase problem, but the practicalities of how to achieve it were not obvious.

It was an absolute requirement of the method that the introduced heavy atom should not jostle the atoms in the protein molecule out of place. 'Isomorphous' means having the same shape. The heavy atom-containing molecule had to have exactly the same atomic arrangement as the natural protein, otherwise nothing could be learned. Max had no idea how this might be achieved. It was a job for a good biochemist: and it happened that by great good fortune Max had recently acquired one. The story is a wonderful illustration of how much great scientific discoveries owe to the informal, international network of scientists that hummed with scientific and personal information long before the invention of the World Wide Web.

Vernon Ingram was born in Breslau and came to England as a schoolboy refugee in 1939 (his family name was Immerwar: Ingram was picked at random from the telephone directory). Towards the end of the war, he studied organic chemistry at Birkbeck College, where he met Desmond Bernal. When Bernal discovered that Ingram was interested in proteins, he advised him to go to the Rockefeller Institute in New York State to study enzymes with the eighty-year-old Moses Kunitz. After a year there, Ingram moved on to Yale for the second year of his fellowship. What he might do once his two years were up was becoming a worry. He had sent off more than a hundred letters of application for academic jobs in the UK, but without success. Then, just as he was about to despair, Freddie Gutfreund, who had close associations with Max's lab, arrived at Yale as a Rockefeller Fellow. As fellow ex-pats with similar histories, they naturally found common ground. One day, Gutfreund received a letter from John Kendrew, asking him if he would be interested in helping with heavy atoms when he came back to Cambridge. As Gutfreund recalled later, he suggested Ingram instead: 'I said I'm not a green-finger biochemist, but there's someone here looking for a job in England and he's just the kind of guy who'll cook the thing for you.'

Ingram wrote offering his services, was immediately accepted and arrived to work in the autumn of 1952. When Gerard Pomerat of the Rockefeller Foundation arrived for one of his regular visits a few months later, he noted in his diary that: 'Perutz says that I[ngram]'s training . . . with Kunitz makes him an enormously valuable member of the group and he cannot understand how they ever got along without a really competent biochemist.'

Ingram was working on an idea that had just arrived in the post from the US. Austin Riggs, a biochemist at Harvard, had read the paper Max had written a couple of years previously about haemoglobin in sickle-cell anaemia (discussed later, in Chapter 11). Riggs had investigated the effect of blocking certain groups of atoms in the haemoglobin molecule on its capacity to absorb and release oxygen. These are the so-called sulphydryl groups, each of which contains a single sulphur atom linked to a hydrogen atom, that form part of the amino acid cysteine. Sulphur forms exceptionally strong bonds with mercury, so Riggs used a mercury compound (called paramercuribenzoate) to block the sulphydryl groups of human haemoglobin. He found that in both normal and sickle-cell haemoglobin it changed the characteristic S-shape of the relationship between oxygen pressure and the affinity of the molecule for oxygen. He thought Max might be interested in this result, and sent him a copy of his paper.

To say that Max was interested was an understatement. Although flattened, the curve still had its S-shape, which meant that the molecule was doing its job as an oxygen-carrier. Max later told Horace Judson: 'I jumped when I saw that, because it was clear to me that if it left the biological properties intact, then it would also leave the structure intact.' And if that was the case, crystals of haemoglobin could grow with mercury atoms incorporated. This was the challenge he now set Ingram. Ingram's first job was to find out how many sulphydryl groups per molecule of horse haemoglobin would take up a mercury atom. Although there are more than two such groups in the molecule, he found he could reliably get exactly two atoms of mercury attached, one in each half molecule, which was very convenient from the point of view of the analysis. The biochemistry itself was simply a matter of mixing the haemoglobin and the mercury compound together in the right proportions. Crystals of the mercury haemoglobin grew in a little over a week: Ingram says it was easy.

Although Ingram created these compounds he is adamant that Max, not he, had had the idea. And it was Max who conducted the final, critical experiment: to photograph the crystals with and without mercury. He knew exactly what he was looking for. If the crystals were isomorphous – like a pair of twins who differ only by the presence or absence of a mole on the cheek – the pattern of spots should be the same in each photograph. The presence of the mercury atoms, however, should alter the intensity of some of those spots. With mounting excitement, Max developed the photographs in the basement of the Cavendish. There are not many eureka moments in protein crystallography – or at least there weren't in those days – but when he saw the image emerging in the developing bath, Max knew that his eureka moment had arrived.

I found the exact intensity changes which knowledge of the absolute intensities had led me to expect. I raced up three floors to Bragg's office and asked him to come down to the darkroom. As we looked at the two pictures, we realised that the phase problem which had baffled us for the past 16 years was at last solved, and Bragg went around generously telling everybody that I had discovered a goldmine.

In later accounts of his life and work, Max always presented this moment as a highlight, if not *the* highlight: Horace Judson commented on the 'buttery glow of satisfaction' that Max's voice took on some seventeen years later as he described 'the most exciting moment in all my research career'. He told Bernal's biographer, Andrew Brown, that isomorphous replacement in proteins was 'My idea and my discovery. If you like, it's what I am famous for.' Why such a relatively simple piece of work should have caused him such excitement might seem puzzling to an outsider, but it was the excitement of anticipation: as soon as he saw those pictures he believed that he would have the solution to the structure within a year. Not for the first time, his optimism would prove to be misplaced. Nevertheless, his demonstration that isomorphous replacement could work in proteins would ultimately prove to be the breakthrough that opened up the whole field of protein crystallography.

Max was due to make another extended trip to the US between August and October 1953, the highlight of which would be a conference on protein structure organised by Linus Pauling in Pasadena, California at the end of September. Working through the long vacation with his student David Green, he fixed the positions of the mercury atoms in the molecule, and assigned signs to the phases of many more reflections. To his delight, they confirmed the accuracy of the few signs he had already managed to determine after years of work on the data from swelling and shrinking. 'It was a triumph to find that there were no inconsistencies,' he wrote in the paper he published with Ingram and Green in 1954. The total of 150 signs he now felt confident about was enough to calculate a Fourier synthesis of haemoglobin – a set of figures that when plotted as a contour map would reveal a real picture of the arrangement of the atoms in the haemoglobin molecule. He had not got far enough at this stage to be able to produce a three-dimensional model – his picture would be a projection, looking at the molecule from above and showing all the atoms superimposed – but he hoped that would at least give him clues about the whole structural arrangement. To be able to present this picture at Pauling's conference in September – the first electron density map of a protein ever seen – was a prospect he looked forward to with eager anticipation.

Meanwhile Kendrew had also made a breakthrough. He had discovered that while myoglobin produced from horse muscle was unsuitable for his experiments, the muscles of marine mammals were an ideal source of the protein. Because they spend so long under water, dolphins, whales and seals need to store a great deal of oxygen in their muscles: there is so much myoglobin in whale muscle that it looks almost black. With post-war Britain still under the privations of rationing, frozen whale meat was being imported in large quantities as a cheap form of protein, and Kendrew was collecting any other whale or seal species he could get his hands on. Max wrote enthusiastically to the Rockefeller Foundation to tell them of this development:

One of our most exciting events lately has been Kendrew's discovery of a magnificent source of myoglobin crystals ... We have a supply in Cambridge of six different kinds of whales ... We recently managed to collect a porpoise which had got stranded in Scotland ... Kendrew is now getting a dugong from Ceylon.

Eventually, Kendrew settled on sperm whale myoglobin as most suitable for his purpose. Naturally, he also wanted to use the isomorphous replacement technique to solve the phase problem, but the myoglobin molecule lacked the sulphydryl groups that would allow Ingram to attach mercury atoms: all such attempts failed. It seemed that, despite its larger size, haemoglobin was much nearer to a solution.

Even with its mercury atoms attached, however, haemoglobin would not give up its structure easily. The tidy accounts of published scientific papers inevitably conceal weeks, months or years of false steps, doubts and corrections of errors. Max's first isomorphous replacement experiment was no exception. An almost daily record of his hopes and fears emerges from the letters he wrote to Gisela at the time. She had taken the children to her mother's in Switzerland in mid-August and did not return until after his departure for the US: altogether they were apart for seven weeks, the longest Max had gone without seeing his wife and children since his US trip in 1943. While every letter declares his love and longing to see her again, they also reassure her of his continued health and happiness: it is hard to avoid the conclusion that two months in which he could focus almost entirely on haemoglobin without distraction outweighed the disadvantages of waking up every morning alone. Indeed, he briskly turned down her plea that he join her in Switzerland for a few days before leaving for the US, on the grounds that it would not leave him enough time to spend with his brother and sister.

Max had always discussed his work with Gisela, and his letters relate the advances and setbacks in some detail:

22 August 1953

David Green has worked out the exact mercury positions now, so that preparation for the main haemoglobin Fourier can start on Monday ... Let us hope it will show something interesting when we get it, the Fourier that is.

I had a good night's rest. This and the certainty of success have much improved my state of health. While on Thursday I still felt as though I had a gastric ulcer combined with angina pectoris and a disintegrating colon, today I felt fit to climb Mount Everest . . . However, the intensity of the emotions felt when Nature reveals to you one of its great secrets is quite overwhelming for a poor mortal and perhaps I shall be forgiven when I go quite off my head at times.

28 August 1953

I do hope the present exciting days will not be unique in my life. The present discovery should start a new era in Crystallography of Proteins; ie in future I hope one will actually try to determine structures, instead of trying to make reasonable (and often erroneous) guesses, and being harrowed by doubts from within and criticisms from without. If this hope is fulfilled there are many exciting discoveries yet in store for us.

After all the Fourier, which I may get tomorrow, will have only a low resolution and its interpretation may be difficult. It is by now almost certain that a Fourier at full resolution can be obtained, and this will be a much greater event than the present one. There is about a term's work required to get that out. The really thrilling discovery would be if we get structural information which explains the biological activity of proteins, and that I fear may be a long way off yet . . .

The thing to do when one is worried about too big a pile of work in front of one is to do it, and one is surprised how soon it gets less.

30 August 1953

The Fourier appeared yesterday morning. On checking we found it to be riddled with errors and spent all day correcting these . . . This job was finished just in time for me to take the sheet with the numbers to [the] Braggs for dinner, where we plotted the usual contour map afterwards. The picture of the molecule is a crazy-looking object, quite unlike anything we imagined it to be, and certainly not something that can be interpreted in terms of polypeptide chains and haem groups at this stage . . .

I rang Dorothy [Hodgkin] tonight and asked her if she would

like to come round and see the Fourier ... [she] asked if she could spend the night here ... I hope the neighbours won't be scandalised.

1 September 1953

After days of brooding over my results and trying to do his worst, Bragg came in this morning, just as I was explaining results to Dorothy [to say] that he found my arguments inescapable. The mercury positions I had found are the only possible ones, he found, and we have to accept the Fourier even if we don't like it.

Dorothy gazed at it for an hour last night, without being able to get any sense out of it ... I felt rotten yesterday when the whole correctness of the Fourier was in doubt again and thought I was getting a fever, but now that the doubts are stilled I feel quite well once more.

8 September 1953

Lunn's secured me a passage for tomorrow night ... I shall travel with my bags full of nice results, confident that they are correct except for one or two details.

The conference in Pasadena was a remarkable gathering, though just how remarkable did not become apparent until later. Of fewer than fifty participants, seven would eventually join Bragg as Nobel prizewinners – it should have been eight, but Dorothy Hodgkin was refused a visa to attend the conference because the American authorities ruled her 'statutorily inadmissible' for her Communist affiliations. (She was not a party member, though her husband and many of her friends either were or had been.) Apart from Hodgkin and Bernal (they went to Moscow instead), all the famous names of protein science were there: Astbury, Bragg with the crack squad from the Cavendish, Patterson, Edsall, Pepinsky, Harker, and of course Pauling himself and his Caltech colleagues.

Pauling had organised the conference as a celebration of the alpha helix, but in the event he was trumped by his Cambridge competitors. That year, 1953, had truly been an *annus mirabilis*: for most Britons the ascent of Everest and the coronation of Elizabeth II were the high points, but for Max the wonders were all scientific. As well as his isomorphous replacement and Watson and Crick's double helix, there was Hugh Huxley going much of the way to explaining, at the level of individual fibres, how muscles contract. Huxley attributes his success to the philosophy of the Unit, which was that 'you weren't going to get anywhere with theories until at least you knew what the structure was.' Using a very fine focus camera, he was able to obtain diffraction diagrams from live frog muscles that suggested that muscle tissue consisted of two different types of filament, tiny fibres laid along the length of the muscle. In 1952 he had gone to MIT to learn how to use electron microscopy to study the same problem. There he joined up with Jean Hanson, another British scientist visiting from London, who was looking at muscle with the light microscope. Between them, they confirmed that muscle consisted of two sets of partially overlapping filaments, with a protein called myosin forming the dark-coloured A-bands seen under the microscope.

Meanwhile, Francis Crick was beginning a rather miserable and unsuccessful year in David Harker's lab at Brooklyn Polytechnic, working on ribonuclease, and Jim Watson had taken up a fellowship at Caltech: as they were all reunited in Pasadena to present their work to the cream of the protein science community they were in high spirits.

Stopping in Berkeley on his way to Pasadena, Max had been anxious about the reception they would receive. He wrote to Gisela:

I gather that we shall have a hard job convincing biochemists of our X-ray results, because they stoutly maintain that they cannot understand how these results are derived. 'Do you really believe that globular proteins consist of alpha-helices?' Watson and Crick is regarded as an interesting hypothesis, but the power of the X-ray and stereochemical proof eludes people. Even Edsall cannot really take it in yet.

But once he got to Pasadena he was reassured. The small size of the conference allowed plenty of time for discussion. Max told Gisela that it was 'most interesting and certainly not a thing which I could have afforded to have missed'. He went on:

My paper was well received and most people realised the achievement as well as the potentialities of the new approach . . . Cambridge made a brilliant show I think and I was very proud of us . . . John K gave a beautiful paper even if he had no conclusions at the present stage. Bragg introduced the discussion of globular proteins by comparing haemoglobin to a mountain climb on which he and I had been firmly roped together and taken the lead in turn, with myself having just led the way up the last pitch. It was charmingly put and a very fair description of our respective parts. My only quarrel would be with describing my present struggle as the final pitch . . . The Paulings went out of their way to be nice to me.

Max kept a note of caution, but he was unquestionably upbeat about his prospects for success: elsewhere he talked of interpreting his electron density map, the crucial step before the structure became clear, in a matter of months. At Pasadena, in a glow of approbation as warm as the Californian sunshine, he could not possibly have foreseen the combination of problems, personal, professional and scientific, that would still stand in his way.

8

In search of solutions

Honours thrive
When rather from our acts we them derive
Than our foregoers.
William Shakespeare, *All's Well That Ends Well*

While everyone at the unit celebrated their success and the triumphant reception they had received in Pasadena, a cloud on the horizon began to cast an increasingly threatening shadow. In May 1953, Sir Lawrence Bragg, who had given Max a lifeline in 1939, championed his cause with the MRC to get the unit established, and worked side by side with him on the haemoglobin structure, had accepted the post of Director of the Royal Institution in London from January the following year.

Founded at the dawn of the nineteenth century, the Royal Institution was a private members' club in Mayfair, which had the twin aims of conducting excellent research and promoting enthusiasm for science. The chemist Humphry Davy, appointed in 1802, established a tradition of popular public lectures. His assistant Michael Faraday, discoverer of electromagnetic induction, went on to become Director and introduced annual lectures for children at Christmas. He also inaugurated the Friday Evening Discourses, at which the cream of London society mingled while attending a scientific lecture. Both traditions have continued to this day. Meanwhile, the work of its Davy Faraday laboratory, principally in chemistry, put the RI at the forefront of British research institutions.

Accepting the directorship was a good move for Bragg: while he would have to retire as Cavendish Professor in a few more years, the RI managers had extended the Director's retirement age to seventy-five. His father had been a legendary Director until his death in 1942, since when the institution had been in political turmoil: the previous Director, Edward Andrade, had been removed after a vote of no confidence and a legal battle. The finances were in a parlous state, but if the constitutional and financial problems could be sorted out, there was a golden opportunity to restore the RI as a beacon for public understanding of science, a cause close to Bragg's heart. The RI had also been a centre for research under W.H. Bragg, and his son was equally keen to see the Davy Faraday Laboratory restored to pre-eminence – and to continue his own involvement in the protein structure work.

It was less clear what the move would mean either to Max personally or to the Unit as a whole. Since 1939, Max had relied heavily on Bragg's support and, although by 1953 he had clearly established himself as an independent researcher, his gradually acquired self-confidence owed much to Bragg's unwavering encouragement. His professional position was by this time secure, but his medium-term prospects were uncertain. In 1949 Edward Mellanby had been succeeded as Secretary of the MRC by Harold Himsworth. Himsworth was a brilliant clinical researcher, who had conducted the studies that ultimately led to the distinction between Type I (insulin-sensitive) and Type II (insulin-insensitive) diabetes. He had reached his position despite having left school in Huddersfield at sixteen and putting himself through night school while working in a woollen mill.

Himsworth later became one of Max's most loyal supporters and friends, but when Bragg had written early in January 1950 to ask for a permanent MRC staff position for him, he replied that Max's work might not be sufficiently close to medicine to justify MRC support in the long term. He argued that it might be easier for the MRC to help Max if it were bearing only part of the cost, and suggested that the University should take him on to its staff. Bragg assiduously lobbied the General Board until it created a Lectureship in Biophysics jointly between two faculties, which Max would take up in October 1953. Although seven years previously it had been his dearest wish to have a University appointment, by now he had very mixed feelings about this, as he wrote to Lotte in June 1952:

> I would prefer on the whole to remain on the MRC staff to becoming a University lecturer . . . The difficulty is that due to my rapid advancement by the MRC, my present salary exceeds that of a U lecturer . . . I may be faced with a substantial drop of income . . . Everything remains in the air, and as I do not know whether my new appointment is a good or a bad thing I have not told people about it, especially not the parents . . . I am not at all worried, and think that if I continue to produce interesting work everything will turn out all right in the end.

His fears were confirmed. After a visit in April 1953, Gerard Pomerat of the Rockefeller Foundation recorded that 'poor Perutz has rather had to take a financial beating to get onto the academic ladder'. It was nothing new for Max to be worried about his income, but there were new pressures: he had already decided that three-year-old Robin was to be privately educated, at the academically excellent boys' school The Leys in Cambridge. He and Gisela were planning a third child, but Max believed they first needed a bigger house. He had hoped in vain for a legacy that would make this possible when Gisela's father died at the end of December 1951. The Perutzes did not move to a larger house until 1960, and they never had a third child. On the positive side, Hugo and Dely had at last received some compensation for their losses from the Czech government, which meant that Max and his siblings no longer had to pay them a pension. The elderly couple, who had moved to London at the end of the war when Hugo went to work for Gisela's uncle, moved back to Cambridge and settled in a rented flat in Harvey Road, with a landlady who became one of Dely's greatest friends.

One uncertainty was cleared up in December 1953 when Cambridge appointed Nevill Mott, head of the department of physics at Bristol University, to succeed Bragg as Cavendish Professor. Max told Himsworth that 'Everyone here is very pleased with the choice', though he added, 'I hope that he will treat the Unit kindly.' Mott was a solid-state physicist, interested in the properties of metals and semiconductors. He had no interest in protein crystallography. Despite the successes of the MRC Unit, and of the Cambridge radio-astronomers whom he had also encouraged, physicists had not generally regarded Bragg as a good head of department: he had no taste for administrative work and little understanding of the Byzantine politics of Cambridge committees. Mott wanted to set about restoring something of the prestige Cambridge physics had enjoyed under Rutherford. He wrote to Max in February 1954, to say he thought it 'doubtful' that the Cavendish should continue to house indefinitely an MRC unit of the present size, and that he could not commit himself to keeping it beyond July 1955.

Up to that point, Bragg had led Max to believe that he had 'settled it all' with Mott, and there was no need to worry. Bragg, however, now had another agenda. He wanted to establish protein crystallography at the RI, and the easiest way to do it would be to move some or all of the MRC Unit to join him. Knowing that Max was opposed to moving, he told Himsworth that he thought Max ought to stay in Cambridge, though he would like 'continued contact'. On the other hand, he would be able to get started more quickly if, for example, 'Kendrew and some of his people' came to join him in London. Throughout 1953, he pressed the case, arguing that it was 'fair to say that the protein research was pretty well stuck in a sand bank till I floated it off again'. He proposed that the work of both labs should be under his direction, and began to negotiate with the Rockefeller Foundation to fund the new development. Warren Weaver advised Himsworth that 'even if the team had to be split (which he hoped would not be the case) the main reliance should be placed on the part containing Bragg'.

Max was opposed to either moving the Unit, or splitting it. Neither was he happy about being under Bragg's direction. He told him in no uncertain terms that he regarded himself as being in an independent position at Cambridge: were a joint scheme to be developed, he would want to be joint director. 'This attempt to set himself up as Director of Research over us all is too much for me,' Max confided in Gisela. He had some reason to claim his independence. As well as having presided over the staggering array of brilliant results produced by the Unit the previous year, he had been elected a Fellow of the Royal Society in March 1954. Election meant admission to the elite of British science, a mark of recognition by one's peers valued at least as highly by those chosen as many apparently more valuable honours. The initiative for this undoubtedly came from Bernal, who had started trying to enlist Bragg's support two years previously. Bragg had been reluctant to give it until Max had achieved something concrete, a condition that had now been fulfilled by his electron density map of haemoglobin.

In an otherwise difficult year, Max's election was a high point: literally, in that the bulk of the congratulations that poured in from around the world reached him while he was on a ski-mountaineering expedition in Val d'Isère. As a total of sixty-five letters arrived he wrote to Gisela: 'When I see the FRS on the envelopes I still find it a bit incongruous after my name, something that hardly belongs to me. But now that it's known among the crowd I expect you had better put it on [your letters] as it is considered the proper thing to do.'

His father's response was disappointingly equivocal. Told of Max's election, he said: 'Aber dafür zahlen sie dir doch nichts.' [They don't pay you a thing for that.]

On his return from his holiday he had immediately to plunge back into the problems of ensuring the future of the Unit. Bragg now launched a full-scale charm offensive against John Kendrew, inviting him to bring the whole myoglobin programme to London and to take a leading role in setting up a new protein crystallography lab at the RI. 'You can play such a vital part in this that I am prepared to go to the limit in meeting any condition you like to make,' he wrote.

Kendrew had no desire to move. He was by this time well established in Peterhouse, and played a full part in college life. His home was in Cambridge. Like Max, he placed a high value on the resources of the Cavendish. Yet Bragg continued to plead with him, alternating letters with lavish lunches at the Dorchester, eventually wringing from Kendrew an agreement to accept a part-time readership carrying an honorarium of £100 per year, with no specific time commitments. Sensitive to Max's feelings about Bragg's attempts to split the Unit, Kendrew made it a condition of his acceptance that Max should have an identical offer. Kendrew wrote to tell Jim Watson about all this, adding: 'The only good reason I can think of for going would be to escape from Max, which would greatly please me, but is not sufficient to compensate for the loss of Cambridge.'

This letter is a rare piece of direct evidence that Kendrew did not find his working relationship with Max wholly satisfactory. The reasons are not obvious from the archives. Max was never less than supportive of Kendrew and, although they all shared somewhat cramped conditions, Kendrew had everything he needed for his research. However, although they accorded one another civility and respect, they never became friends, and having to share an office was a trial to them both. Colleagues from the time were struck by the differences in their personalities. Vernon Ingram thought the Unit benefited from their contrasting approaches to science.

John was exceedingly well organised and very intelligent, very bright. The organisation of the research was what really interested him, not just the scientific knowledge. Max was much less well organised, but driven by the scientific content and how that affected adjacent areas of science, like physiology, evolution, and diseases. Max was very much a European all of his life, John was very much an upper class Englishman with all that implies. But they worked very well together.

Max, meanwhile, in an equally rare admission, later told the writer Horace Judson that Kendrew's jealousy of his own success strained their relationship, and also told his daughter that Kendrew had been a 'thorn in his side'. Whatever the reasons for their differences, circumstances during 1954 were not conducive to a relaxed and fruitful working relationship.

The long-term future of the lab remained uncertain. Mott had suggested that the Unit might move to the former linear accelerator building in Madingley Road, about two miles from the centre. Max had a look at it: it had possibilities as it was already equipped as a lab, but he found it 'somewhat derelict and surrounded by the depressing remnants of war'. He told Himsworth that 'All the members of this Unit very much dislike the idea of being isolated so far from the centre of scientific activity and are there-fore anxious to find some alternative solution.'

By the middle of 1954, the University authorities had at least committed themselves to keeping the Unit in Cambridge and providing accommoda-tion, but the question of the location had still not been solved as University departments bickered jealously over the space available.

On top of all this, Max's research had stalled again. The electron density map he had presented in such triumph at Pasadena was a genuine break-through, but it was impossible to interpret. A two-dimensional map gives an image of the molecule projected on a flat surface – in other words, the crystal structure looks as it would if you ran over it with a steamroller. To infer the lengths and angles of all the bonds between the atoms in such a picture is impossible: it was not even clear where one molecule ended and the next began in Max's projection. When Dorothy Hodgkin had come over to see it, she gave him a tip that might provide a way out: she suggested he read a paper published a few years previously by the Dutch crystallog-rapher J.M. Bijvoet on *double* isomorphous replacement. By using two different heavy atom derivatives plus the native crystal, Bijvoet suggested, you could calculate an electron density map in three dimensions. While this was a great idea, Max could not see how to put it into practice. Neither he nor Ingram could immediately think of another way of getting heavy atoms into haemoglobin.

To make matters worse, Max could not even repeat the success he had had with the first batch of Ingram's paramercuribenzoate compounds. The crystals he made had a fault in them that meant they were no longer precisely the same shape as the native haemoglobin crystals, and so any comparison would be meaningless. Max was stuck.

He was worried about his income, worried about the future of the Unit, plagued with problems in the lab and threatened by Bragg's desire to lead the protein work, and his indefatigable courting of Kendrew. Then, in September 1954, Gisela's mother fell ill and Gisela had to rush to Switzerland to be with her. Thereupon Max's own health broke down, with a serious recurrence of the 'tummy trouble' that had plagued him for years, and feel-ings of weakness and depression. The local hospital in Cambridge could not find out what was wrong, and in December he was admitted to St Bartholomew's hospital in London for tests under a leading gastroenterol-ogist, E. Cullinan. He diagnosed irritable colon and low glucose threshold (though Max was not diabetic), and prescribed a mixture of belladonna, kaolin and bismuth. If anything it made him worse. As far as diet was concerned, Cullinan said vaguely that he should 'find out what suits him'.

Frustrated with the lack of improvement, Max discharged himself after ten days and went home to Gisela's care. The family went to Kitzbühl over Christmas; as usual, the mountain air improved his spirits, though he felt too ill to ski. On his return he relapsed completely, doing nothing but sitting in a chair and staring into space. Although a letter to the Rockefeller Foundation in March states that he is 'very much better' and has started working in the lab again, other correspondence suggests that he did not return to work in earnest until the autumn of 1955, after a restful and sunny family holiday in Devon. Within days he was off sick again: there was flu going round, but Jim Watson wrote to his girlfriend Christa Mayr: 'Max Perutz's psychosomatic illness is back and he has already stopped lecturing. Noone knows how to cure him as he thinks that his major symptom – weakness – is organic.'

Watson was not alone in thinking Max's increasingly chronic ill-health was psychological. His preoccupation with his health breached an unwritten code of British society, in which it is considered rather bad form to talk about illness, and which requires one to grit one's teeth and struggle on in the face of all but the worst afflictions. Rather than attracting sympathy from his colleagues, his sickness tended to increase his isolation.

While Max recognised that he was under a great deal of stress, later telling Horace Judson that 'Bragg's departure was the beginning of an exceedingly difficult time', he discounted this as a contributory factor in his illness and continued to make strenuous efforts to find out what was wrong with him. He visited the eminent nutritionist and professor of experimental medicine at Cambridge, R.A. McCance, 'thinking naively, as we were both Fellows of the Royal Society, that McCance would try and help me'. This august figure shouted at Max and threw him out for wasting his time. In desperation, in May 1955 Max consulted a medical researcher in Cambridge called Werner Jacobson, who told him it sounded as though he had coeliac disease.

Coeliac disease is an intolerance to the protein gluten in wheat flour. Its cause was first recognised in children in the Netherlands in the postwar years, who got better when there was no wheat available to make bread: by the time of Max's conversation with Jacobson the same condition had just been recognised in adults. The symptoms include inflammation of the small intestine, diarrhoea, and malnutrition due to poor absorption of vitamins and other nutrients from the gut. The only cure is to stop eating anything that contains wheat. At the time the 'gluten-free' diet was almost unheard of, but Max took Jacobson's advice, and slowly began to recover. Over the following decades, and by trial and error, he eliminated more and more foods from his diet, including all cereals, peas, beans and most fruits, in a desperate effort to avoid 'tummy trouble'.

Max's absences from the lab over a period of almost a year from September 1954 had little effect on its activity and continued vigorous growth. In October 1954 a new research assistant for Max, Ann Cullis (now Ann Kennedy), made a considerable impact by arriving in Cambridge on an old barge she had rescued from a half-sunk state in the Thames. At her interview some months previously, she had immediately gained Max's admiration by saying that she had arranged the interview for 9 a.m. because she had tickets for an opera at Glyndebourne afterwards. (He even asked if she could get him tickets too.) Cullis was an Oxford graduate who had been taught by Dorothy Hodgkin, and had previously worked in another MRC lab. She was an ideal addition to Max's team: bright, competent, vivacious and good company, but, by her own admission, content to be Max's assistant rather than wanting to establish an independent scientific career. For the first month or two she was somewhat at a loss, as Max fell ill just after she arrived, but in time he came in and taught her the basics of protein crystallography. She soon graduated from interminable intensity measurements to growing and mounting crystals, photographing them and supervising the work of two 'human computers' – young women who came in part-time to do the tedious calculations. As the lab at this time lacked any secretarial assistance – Max wrote his own letters on an ancient type-writer – she also kept the accounts.

The publicity surrounding the advances made by the lab during 1953 attracted people from further afield. Altogether, Max's report to the Rockefeller Foundation for the period 1953–56 lists nine members of staff and twelve visiting and associated workers, many of whom – including Jerry Donohue, James Watson (who came back in 1955), Howard Dintzis, Alex Rich and Linus Pauling's son Peter – came from the US. Most of the newcomers were highly productive. Finding a large enough nest for what must have seemed to Nevill Mott like a voracious cuckoo chick was proving ever more of a problem. In 1956 a partial solution emerged when the University decided that instead of demolishing the Metallurgy Hut, a 'temporary' building next to the Cavendish Laboratory, they would make it over to the Unit. The Hut was basic but quite large, and such was the reputation of the Unit by this time that its rather rustic accommodation did not deter good researchers. Francis Crick had for some time been agitating to get Sydney Brenner, an extremely bright South African geneticist, to come to Cambridge: Brenner declared he would be happy to work in a cupboard if necessary.

They did slightly better than that. Brenner came in January 1957, as a full member of the MRC staff. Crick had proposed that he might be given a visiting fellowship, but when Max went to visit the Deputy Secretary of the MRC, Sir Arthur Landsborough Thompson genially asked, 'Why don't we take Brenner on our staff?' As Max later commented, 'No panel, no referees, no interview, no lengthy report, just a few men with good judgement at the top.'

The Unit moved into the Hut, plus some extra rooms in the old Anatomy Department for Crick and Brenner, in October 1957, the tenth anniversary of its creation. At the same time, Max proposed to Himsworth that it should change its name:

> The old name 'Molecular Structure of Biological Systems' no longer covers all our activities and the name 'Molecular Biology' has grown up in recent years for a science which looks at biological events from the point of view of the large molecules involved in them. I suggested to my colleagues that we should be renamed Molecular Biology Research Unit and they have indicated their unanimous approval.

Max's Unit became the first in the world to put its allegiance to molecular biology on its nameplate. The same year, John Kendrew accepted the first editorship of the *Journal of Molecular Biology*, which continues to flourish as the leading journal in the field. From its modest beginnings, the Unit had in ten years established an international reputation and a clear identity at the head of a new and rapidly-moving field – even if they were still working in a glorified shed. A senior official from MRC head office in London remarked that the accommodation and facilities compared 'very unfavourably' with those of the Council's other units. He contrasted this with the very high quality of the staff:

> I thought that Perutz probably exercised a great deal of quiet personal control and had his finger closely on all aspects of the Unit's activities . . . I got the impression of a closely integrated team with Perutz, Kendrew and Crick as its nucleus. They were all bursting with ideas, seemed to know exactly where they wanted to go, and were very confident of getting there. There is an exciting atmosphere of discovery about the place and I found it altogether a most stimulating visit.

Max delighted in telling the story of the delegation from the Soviet Union that came to Cambridge for a congress in 1959 and asked to see his 'Institute': 'When I took them to the hut they looked perplexed and went into a huddle. Finally they asked me "And where do you work in winter?"'

*

Max felt strongly that his personal status and income had failed to keep pace with the burgeoning reputation of his Unit. In 1956 Cambridge University turned down his application for a Readership. Max told Himsworth that he wanted to resign and re-apply for a post on the MRC staff: 'I am unhappy about the idea of being indefinit[e]ly penalised by my transfer from the Council to the University staff which has put me into the position where I can only improve my salary by leaving the Unit which I created and seeking my fortune elsewhere.'

The Council of the MRC agreed to take him back, though the salary they offered, linked for rather spurious political reasons to University pay scales rather than their own, higher, scales, still left him suspecting that his colleagues earned more than he did: 'I believe that Kendrew, with his various part-time activities, has already reached that stage. Crick and Ingram have reached it by spending periods of work in the United States.'

His position as head of the Unit carried with it less power and influence than might at first appear. Research there was conducted very much according to the Cambridge tradition of recruiting excellent people and letting them do more or less what they wanted. Deference had to be earned: it could not be demanded. At this stage in his career, it was not by any means obvious that Max stood out from his colleagues, though his understated leadership style was much more effective than they probably realised. He lacked the brilliance and charisma of a Bernal, or the gravitas of a Bragg. His own haemoglobin research group – mostly Ann Cullis and one or two graduate students, with assistance on specific projects from Vernon Ingram and Howard Dintzis – remained small, with himself very much a hands-on participant. For five long years, between 1953 and 1958, the work plodded on with little to show in the way of progress – there was still no obvious way to solve the problem of how to interpret the first haemoglobin maps.

Other members of the Unit, notably Crick and Kendrew who had their own groups of students, assistants and post-doctoral workers, were enjoying more success. Crick was by this time a key figure in a transatlantic effort to solve the structure of RNA, the molecule that transcribes and translates the DNA message into proteins, and to crack the genetic code. Kendrew had – thanks to Howard Dintzis – at last got several heavy atom derivatives of myoglobin and was close to a structure.

The most important result from this period was not part of the protein structure project at all. Vernon Ingram was using a version of the protein sequencing technique developed by Fred Sanger in the Biochemistry Department to study the difference between haemoglobin from normal human blood and from the blood of people with the genetic blood disorder sickle cell anaemia. His discovery that the sickle cell mutation caused a single amino acid difference in the haemoglobin sequence demonstrated for the first time the impact of a faulty gene on a protein. In the twenty-first century, an international research effort involving thousands of scientists and billions of dollars of research money is dedicated to tracing the molecular basis of disease, and all of this can be traced back to Ingram's little experiment (see Chapter 11).

Ingram's discovery is a fine illustration of the way the lab worked. Fewer than twenty people, crammed into a small space, all passionate about science and working on different but related problems: it was inevitable that new possibilities would emerge from the enthusiastic exchange of ideas. There was no need for a director to write a five-year plan and allocate specific tasks to his workers. Max's role was to make sure that everyone had what he or she needed. Long after the MRC had taken the Unit on, he retained a warm relationship with the Rockefeller Foundation that ensured he could buy all the latest equipment, and he set great store by having a well-qualified engineer such as Tony Broad on the staff, as well as adequate numbers of more junior technical and computing staff. Ingram, who left for a career at MIT soon after making his discovery, still regards Max as a mentor in his approach to scientific management: 'One of the things that I learned from Max was a proper disregard for authority, for the bureaucratic aspect of being director of a very lively, very creative research unit.'

Socially, too, the lab worked well, despite its rather high ratio of prima donnas to ordinary mortals. While Crick had irritated Bragg beyond endurance, Max liked and admired him, recognising, wisely, that he was one of the Unit's greatest assets. Crick credits Max with creating the generally harmonious atmosphere in the lab.

> Max's great talent was that he got everybody happily working together . . . I don't think [the differences] were very great, but in a situation like that you can have various groups in rivalry with one another . . . I think Max made a major contribution by the tactful way he administered.

If Max did not take the lead intellectually in the lab, neither did he socially. He would arrive on his bicycle at 9.00, having taken one or other of the children to school, and leave at 5.00 to go home for dinner. He did not join the pub lunches where Crick, Watson, if he was around, and others would converse raucously on every topic from science to girls. Entertaining his colleagues was largely limited to tea after a day in the lab on Saturday afternoons. Sundays were reserved for his parents, and Max's daughter reports that Gisela found dinner parties so stressful that Max was 'reluctant to impose them on her'.

Ann Cullis (now Kennedy) recalls that there were 'lots of parties' at the time, but as far as she could remember Max was never invited. Regular musical evenings took place in John Kendrew's room in Peterhouse, where he had the latest hi-fi equipment and a superb record collection. Hugh Huxley was a frequent guest, along with the Cricks and several others, but he doesn't remember ever seeing Max at these occasions, and doubts he even knew about them. Francis and Odile Crick gave many dinner parties at their house ('The Golden Helix') in Portugal Place, where they moved after returning from the US in 1954. These parties were not the formal kind intended to ingratiate themselves with the great and good of Cambridge, but purely social events for their friends – but not Max.

Cullis had become the latest tenant of the Green Door, and it too became a social hub. One Christmas she threw a party there and didn't invite Max. On that occasion, she remembers, he heard about the party afterwards and was upset, much to her mortification. It was not that nobody liked him: Max's colleagues generally held him in some affection, though he couldn't be said to be personally close to any of them. Cullis herself was very fond of him and regarded him as a father figure, to the extent of inviting him a few years later to make the 'father of the bride' speech at her wedding. Today she struggles to explain why Max was so often excluded from the social life of the lab.

> Max was very friendly to me at all times . . . He did live incredibly simply. He probably didn't drink [he didn't] . . . Max invited me to tea at Eachard Road soon after I arrived, and that was very nice, but Francis invited me to dinner . . . Francis's house was ideal for parties and he and Odile were such sociable people. Perhaps it was partly that Max seemed rather delicate at that time, and that he would bicycle off home at five.

Max's close family life did set him apart from most of his colleagues, though not all: the Cricks also had young children. It was probably more significant that his delicate health left him little energy for socialising. Others have suggested that he found it difficult to keep up with the quick-fire punning and repartee that characterised any conversation in which Crick was involved, his own sense of humour having a less subtle and more child-like quality to English ears. He liked to tell what he called 'funny stories' and was good at it: but attracting the attention of an audience when Crick, Brenner and some of the 'young Turks' were around might have been difficult. He loathed gossip, and would certainly not have been interested in the high-testosterone atmosphere captured by James Watson in his autobiographical books.

Sociable and conducive to good work as it was, the Hut was only ever meant to be a temporary expedient. The wheels began to turn again thanks to another serendipitous meeting early in 1957. Possibly travelling up to London on a train – he didn't remember the details – Francis Crick ran into Fred Sanger. Small, self-effacing and modest in the extreme, Sanger had determination, perseverance and the scientific equivalent of green fingers for developing new lab techniques and getting them to work: he was the kind of scientist who thinks with his hands. He worked independently in the Biochemistry Department, funded by the MRC but without a University teaching post. A few years earlier, he had been the first to unravel the complete amino acid sequence of a protein, the hormone insulin. This was an alternative, and complementary, approach to understanding proteins to that of the crystallographers: rather than trying to discern their three-dimensional shape, Sanger successfully read off, one by one, the sequence of amino acids along each of the two polypeptide chains that comprise the insulin molecule. It took him years and involved inventing a variety of techniques to cut up the chain, separate the pieces, place them in order and analyse their content. It was a landmark discovery. The secret of life, at its simplest, is that genes direct the production of proteins. Sanger's technique revealed that each protein has a unique amino acid sequence, with no discernible pattern of repetition. With the structure of DNA also solved, the challenge was now to understand how the sequence of nucleotides in the double helix encoded the amino acid sequence of a protein.

Though his Nobel prize did not come until 1958, Sanger's discovery had gained him international recognition. He had had some contact with Max's Unit since at least 1950, but attempts to forge closer links had so far not succeeded. Until Vernon Ingram found the amino acid difference in sickle cell haemoglobin, protein chemists such as Sanger had had little to do with geneticists, a group to which Crick was drawn through his interest in the coding problem and his collaboration with Sydney Brenner. As Sanger and Crick got talking on their shared train journey, Sanger said he'd like to learn more about genetics. Crick invited him to attend a series of weekly meetings at The Golden Helix, where people working on genes and proteins could learn about each other's work. The newly-arrived Brenner also began to spend time in Sanger's lab, learning his techniques.

Sanger had no room for expansion in the Biochemistry Department. Learning that Max's unit had been promised the Hut for only five years, he suggested that they might jointly apply to the MRC for the funds to establish a free-standing institute. After hearing from Sanger, Himsworth assured Max that he regarded his Unit as 'holding the key position in this matter'. It therefore fell to Max, working with Crick, to set out the case for what they were now calling the Laboratory of Molecular Biology. His statement, later published in the magazine *Endeavour*, was breathtaking in its scope. Careful to assume no previous knowledge on the part of his readers, Max described how in a very few years the structure of DNA had been discovered, the first steps had been taken to discovering the structure of proteins and small plant viruses, and the link had been made between mutations, abnormal protein sequences and disease. While his review was international, the reader could not fail to notice that the bulk of the work Max described had been carried out in Cambridge, including Kendrew's recent solution of the structure of myoglobin (of which more below).

On 18 April 1958 Max presented his case in person to the MRC Council. He was nervous about facing the massed ranks of such a powerful body, but gave a charmingly personal spin to the powerful arguments in his paper:

> If such a proposal had been put before the Council ten years ago it would certainly have been turned down, and rightly so. In making it I would have acted like a nineteenth century engineer forming a company to build aeroplanes. Five years ago I would not have felt quite such a fraud any longer, because I thought I had discovered a method to make my plane fly. This year Kendrew has actually built a plane and made it fly a little way. This makes us confident that our plane can be developed into a means of transport . . . [M]edicine and biology stand to gain from this new approach in the next few decades as much as they have gained from pure biochemistry in the past.

Max need not have worried. Even before his turn had come to speak, several members of the Council had told him that his paper was 'the most exciting document they had ever read'. Himsworth wrote to Bragg after the meeting and told him that he had never seen the Council so enthusiastic about anything.

Persuading the MRC of the scientific case for a new laboratory was only the first step, however. The problem of where to put it was much harder to solve. All the discussion about the ultimate home of the Unit had come to nothing, and the larger group now envisaged would present even more of a challenge. Max was careful to keep his options open, even to the extent of telling Himsworth that they would 'certainly be willing to move elsewhere' if a site could not be found in Cambridge. This willingness to consider all options was perhaps more apparent than real, part of a negotiating position designed to bounce the University of Cambridge into making a decision: Max had also told Himsworth that Kendrew was 'dead against the entire scheme' unless it could be part of the University. But if Max had thought the University would be eager to embrace the new Laboratory, he was mistaken. The head of Biochemistry, F.G. Young, actively opposed it on the grounds that it would have an unfair advantage in recruitment: MRC staff would not have to devote time to teaching students.

A way forward suddenly appeared in March 1959. The MRC had received a generous legacy from a wealthy Indian entrepreneur, Cusrow Wadia, and said it was willing to use the money to build a new laboratory if the University could provide the land. At the same time the University hospital, Addenbrookes, was due to move out of the city to a site two miles to the south on Hills Road. Joseph Mitchell, Regius Professor of Physic, was to have a new Department of Radiotherapeutics adjacent to the hospital; he suggested that the molecular biology laboratory should also be built there. Although they all felt that the distance from colleges, the computing lab, the Cavendish workshops and, most of all, colleagues in other departments was a serious disadvantage, Max and his fellow researchers knew they had exhausted all other avenues. Their plans had continued to expand: Hugh Huxley, who had moved to University College London in 1955, was to return to carry on his electron microscopy work, and Max had also invited the virus group from Birkbeck College, led by Rosalind Franklin, to join the new lab. Franklin's life was cruelly cut short when she died of ovarian cancer in the spring of 1958, aged only thirty-seven, but her colleague Aaron Klug would be bringing the rest of the team.

The latest plans for the laboratory envisaged a building of 24,000 square feet. After six years of wrangling, it was clear to the molecular biologists that, however successful they were (Sanger's Nobel prize was announced in October 1958), the heads of the science labs in central Cambridge were never going to concede such a large amount of space from their fiefdoms. Less than ten days after Max had told Himsworth that Crick would 'prefer to accept a post abroad' than move to the outskirts of Cambridge, he and his colleagues agreed to move to Hills Road. On 24 June 1959, Himsworth's assistant Miss Potts wrote to Max on his behalf to say that the Minister of Health (whose department owned the land) had approved the proposal and so he could feel 'safely at home'.

*

Miss Potts's words must have had a particular resonance for Max, who had felt for so long that he was tossing on stormy waters without a port in sight. Now, not only was the future of the Unit secure, but the landfall that for so long had seemed to recede as fast as he gained on it – the haemoglobin structure – was clearly visible on the horizon. The break-through of the isomorphous replacement technique in 1953 had been followed by years of frustration as Max and his colleagues failed to create suitable crystals.

> I got quite desperate about it. There was Kendrew going ahead, things were going splendidly, he was probably solving his structure, and I, after having produced the method, was getting absolutely nowhere! . . . Perhaps my unhappiest years were from 55 to 57 when I knew I had this marvel-lous method but could make no headway.

In 1957 a Norwegian visiting scientist, E. Alvor, discovered that increasing the acidity of the solution in which the mercury haemoglobin was kept cured the fault in the crystals that had bedevilled Max's attempts to repeat his 1953 experiment. The next problem was to find a second heavy atom derivative that would enable him to work out the structure in three dimensions.

For two years, Howard Dintzis had soaked haemoglobin crystals with a huge range of metal compounds in the hope of getting some atoms into different positions, but he returned to the US in 1956 without apparently having succeeded. However, he had had much greater success with myoglobin, and by the time he left, he had provided several derivatives. With the help of the Cambridge computer EDSAC I, Kendrew was on course to produce a structure for the molecule.

A first, two-dimensional analysis had had the same problem that Max had encountered with haemoglobin: with the atoms all superimposed, it was impossible to disentangle their positions. Moving to three dimensions was a challenge: it required several different heavy atom derivatives. Wisely, Kendrew chose not to try for a resolution that would show the positions of individual atoms, but to start with a simpler analysis, at a resolution of 6 Å, that would show how the protein chain folded in the molecule and its relationship to the single haem group. The lower resolution had the great advantage that it required him to measure only 400 reflections per deriva-tive, not the tens of thousands he would need to resolve individual atoms.

In the late summer of 1957, EDSAC ran the seventy-minute program that for the first time revealed the three-dimensional structure of a protein molecule. The result was a complete surprise. Since the earliest discussions on protein structure in the 1930s, most people had believed that the molecules would have a pleasingly regular shape. Along with the assumption that once you had found one protein structure it would be easy to get all the rest, this proved to be a consoling fantasy.

The computer chuntered out all the figures representing the density of the electrons throughout the molecule, his assistants plotted maps of the density at different depths, and Kendrew made a three-dimensional representation of it out of modelling clay. As a visiting MRC official remarked, it looked like nothing so much as 'an anatomical model of abdominal viscera', bending back and forth apparently randomly. (An unnamed cartoonist at the University of California at Berkeley lampooned it as 'Frankendrew's Monster'.) Nothing could be further from the neat bundle of parallel chains he and Max had previously envisaged. Kendrew expressed his own astonishment in the paper published in *Nature* the following year:

> Perhaps the most remarkable features of the molecule are its complexity and its lack of symmetry. The arrangement seems to be almost totally lacking in the kind of regularities which one instinctively anticipates, and it is more complicated than has been predicated by any theory of protein structure.

Max wrote to Gerard Pomerat at the Rockefeller Foundation that 'we are all very thrilled with [the structure]'. It was naturally a feather in his cap, as director of the Unit, and the timing could not have been better as it gave him a world-leading result to put into his application to the MRC for the new laboratory. His pleasure, however, must have been mingled with frustration. He had kicked himself that Pauling and not he had discovered the alpha helix: now Kendrew had beaten him to the first protein structure, even though he had been working on the problem for longer and had made the first technical breakthrough that showed it would be possible one day. However, all that others saw was his excitement at the discovery: Vivien, who was twelve at the time, remembers him taking her to see the model. Even his closest colleagues, such as Ann Cullis, detected no sign of jealousy: 'He may have been depressed at home, but he was the most equable person I ever met, and it was only in old age that he mentioned that he had been desperately keen to get there first.'

Max might have been more cast down were it not for the fact that, by the time the myoglobin structure came out, things were also looking up for haemoglobin. Having solved the problem of growing crystals with mercury that consistently kept the form of the native molecule, Max had also found the second derivative that would be vital to a three-dimensional analysis.

When he went back to Harvard, Howard Dintzis had left behind dozens of test tubes. In desperation, Ann Cullis was looking through these again when she struck gold. Mercury can bind to particular locations on the haemoglobin molecule where there are sulphur atoms. One of Dintzis's experiments had involved blocking these sites with another compound, iodoacetamide, and then adding another mercury compound, mercuric acetate. When Cullis came to look at the resulting crystals, she found that this had allowed the mercury to bind in places different from those in Ingram's paramercuribenzoate crystals. At last they had what they needed.

A newcomer to the group would play a crucial role. Michael Rossmann was born in Frankfurt and came to England as a schoolboy refugee in 1939. His early scientific education was unconventional, but he had managed to complete a PhD in crystallography with Professor J.M. Robertson at the University of Glasgow while teaching physics at the nearby technical college. Rossmann found that the mathematical rigour of crystallography suited him perfectly. On a two-year fellowship in the US, he heard Dorothy Hodgkin speak about John Kendrew's work on myoglobin and decided that protein crystallography offered the challenge he wanted. He wrote one letter to Hodgkin, and another to Max Perutz, asking if he could come and work with them: Hodgkin never replied, but Max invited him to come to the Molecular Biology Unit.

By the time Rossmann arrived in the summer of 1958, Max, Ann Cullis, and various assistants had collected data on several different haemoglobin derivatives to a resolution of 5.5 Å: as in the case of myoglobin, this would be good enough to see the overall architecture of the protein, but not enough to see the positions of individual atoms. In order to calculate the phases of all the reflections, they needed a good method of finding the positions of the heavy atoms, not just as projected from above, but in all three dimensions. Within three months of his arrival, Rossmann had devised a new and much more powerful method, which he programmed on Cambridge University's new EDSAC II computer. In October 1958, he was able to show Max contour maps of the haemoglobin derivatives that clearly showed the heavy atom positions. He remembers that Max was 'absolutely delighted', and invited both Bernal and Bragg to come up and see the maps. Being young and impulsive, Rossmann thought they now had all they needed to move on to calculate a three-dimensional Fourier synthesis – the final calculation that would enable them to recreate for the first time for human eyes the route taken by the coiled chains in haemoglobin.

Max, however, was more cautious: he had made mistakes before, and knew the humiliation of having to acknowledge an error: 'I had no idea

what the molecule ought to look like and I wanted to be sure that whatever emerged could be proved to be right beyond any possible doubt.' He
would not let Michael run the calculation until he and Cullis had collected
data on more derivatives, so that they could be sure that the phasing was
as accurate as possible. Eventually, they had six heavy atom derivatives,
plus the data from the native haemoglobin, to feed into the program. By
August of 1959, they had recorded 40,000 reflections in total, their intensities measured by an 'army of young assistants'. Rossmann had located the
positions of the heavy atoms. Under Rossmann's supervision, Max's new
research student Hilary Muirhead had written a program to determine the
phases of the reflections using a method developed by the American crystallographer David Harker. They also applied a modification proposed by
Francis Crick and a previous research student of Max's, David Blow, to
compensate for possible errors in the data. Now, at last, they had all they
needed to run the program that would fulfil the promise of the X-ray
method: to turn a pattern of spots on a film into a three-dimensional image,
albeit on the fuzzy side, of the haemoglobin molecule.

Normally Max steered well clear of the computing side of things, leaving
it all to Rossmann and Muirhead – but this time, sensing a historic moment,
Rossmann encouraged him to come over to the Maths Lab and see it
through. All the data had been entered on punched paper tape, which
Rossmann fed into the computer. He invited Max to press the lever that
would start the program running: but Max, fearing that something might
go wrong with the temperamental machine, refused. The program took
two hours to run. Then, at last, it churned out the lists of numbers, and
the young 'computors' transcribed them as contour maps. Everyone could
see that the maps were good: even to a casual eye the haem groups stood
out strongly, and areas of high density were concentrated in 'rods' that
could only be lengths of alpha helix.

Rossmann was all for embarking on a full interpretation of the maps,
tracing the route of these helices through the different map sections that
together represented the molecule, but Max seemed curiously uncertain.
After a week closeted in his room, he gave no sign that he was any further
forward. Rossmann, impatient as ever, found his hesitation incomprehensible and in his frustration did something he later deeply regretted: 'I felt,
probably wrongly, that Max didn't have much trust in me being able to
interpret the maps. I thought he thought I was some kind of computer
expert who didn't understand this kind of thing.' Eager to prove his worth,
he took the maps home with him and did the interpretation himself. Using
coloured pencils, he traced the alpha helices, which showed up even better

than they had on the myoglobin molecule, from one map to another. The next day, he brought them back into the office he shared with a number of others from the haemoglobin and myoglobin groups: Kendrew was in the room, as was David Davies, the former graduate student from Oxford who was now based at the National Institutes of Health in Washington DC and had come to learn some protein crystallography. Picking up the low-resolution clay model of myoglobin, Rossmann turned it in his hands, glancing repeatedly from the haemoglobin map to the myoglobin model. All at once he let out a cry of recognition that brought the others over. Rossmann had seen that haemoglobin looked like four molecules of myoglobin stuck together.

Kendrew urged him to go and tell Max, who was working at the far end of the hut in the biochemistry lab, and Rossmann rushed to show him the maps. He found him trying to crystallise another mercury derivative. Apparently thinking that the maps were uninterpretable, Max was about to embark on another round of experiments to get more data.

To Rossmann's mortification, Max was not delighted that he had successfully interpreted the maps, and indeed that they revealed the first evidence that proteins with related functions might also have similar structures. For a moment, his face darkened with fury. He left Rossmann in no doubt that he was upset to have had the final stage of his epic journey pre-empted by a junior colleague who had worked on the problem for only a year. It was a difficult moment for both of them. Later that evening, Rossmann went round to Max's house to apologise and plead to be forgiven. To this day he bears a burden of guilt about what happened.

Max himself never spoke about this, and we have no way of knowing how he resolved matters in his mind. Clearly, the realisation that the journey really was at an end soon restored his naturally optimistic outlook.

I need not have worried so much about proving that the answer was right, because it proved itself at first sight. There emerged two pairs of almost continuous chains of electron density, each linked to a dominant peak that clearly marked the haem: the conformation of the two pairs of chains in horse haemoglobin was similar and closely resembled that found by Kendrew and his collaborators in sperm whale myoglobin. No conceivable combination of errors could have produced that striking similarity between two independently determined structures.

It takes a rare feat of imagination to 'see' the three-dimensional structure of a protein molecule in the few dozen maps that represent slices through it. Max set about building a three-dimensional model. Rather than using clay as Kendrew had done, he traced the areas of high density on each map on to sheets of thermosetting plastic about a centimetre thick, cutting out the pieces and assembling them layer by layer. Knowing that there were two each of two types of chain (he called them alpha and beta), he made two black chains and two white. Each chain harboured a haem, which he represented as a flat disc, in a pocket on its outer surface; it was easy to see how the four chains fitted neatly together. Although each chain, like the myoglobin molecule, looked quite irregular, the four of them packed together into the flattened sphere Bragg had described in 1952, 'like Spanish chestnuts in their case' in his own apt phrase.

For so long haemoglobin had been a will o' the wisp, just beyond reach, its form unknown: now Max could hold it in his hands, turn it over, send pictures of it to his friends as he did those of his children. There would be later and better models, but none would match this one for the simplicity and directness with which it revealed one of life's most intriguing secrets.

The 'striking similarity' between haemoglobin and myoglobin had momentous implications. At the time Max's group solved the haemoglobin structure, they still did not know the sequence of amino acids in the molecule – but if the four haemoglobin chains folded the same way as myoglobin, then the likelihood was that the sequences would also be similar. And that meant that, despite coming from species as different as horses and whales, they had a common evolutionary origin. Max wrote excitedly to Himsworth to explain the significance of what they had found:

> If the myoglobin of a whale is like the haemoglobin of a horse, then the structure of these two proteins is probably much the same throughout the animal kingdom. There must be certain standard sequences of amino acids which all these proteins have in common and which determine the characteristic loops and turns of the chain. These must have developed from a common primeval gene which provided the physiological basis for the development of the higher animals, by making possible the storage and transportation of oxygen.

A century after the publication of On the Origin of Species, Max and Kendrew had provided the first evidence that Darwin's ideas held true at the level of individual molecules. In so doing they launched the new and fascinating field of molecular evolution, which traces the unity of life through the family relationships of protein molecules (see Chapter 11).

By the time Max and his team had completed their haemoglobin model, Kendrew and the myoglobin group had taken the next step. Benefiting again from the much more powerful EDSAC II, and the programs Michael Rossmann had developed for it, they had taken the resolution of their myoglobin structure from 6 Å to 2 Å. It was a tremendous feat of data collection and analysis, requiring 9,600 reflections from crystals of five different isomorphous heavy atom derivatives. By this time, however, the computer could do much of the work: the days of poring over tiny black spots and measuring them by eye had long gone. A structure at 2 Å is almost good enough to 'see' individual atoms. What they could certainly see was that the rods of density they had previously identified were hollow cylinders, helices that exactly matched the dimensions of Pauling's alpha helix. David Davies remembers the moment one hot day when most of the others had left for summer holidays and he and Kendrew sat down to interpret just one small section of the 2 Å map: 'There was an alpha helix: it was the first time it had been directly visualised. Francis had a room across the hall – he came rushing in and within minutes he had a theory about how proteins folded.'

By the end of 1959, Max and Kendrew had each quickly written a short paper for *Nature*, announcing the low resolution structure of haemoglobin and preliminary conclusions about the high resolution structure of myoglobin. Max was so delighted with the photographs of his model that he had several extra printed so that he could stick them in his Christmas cards. 'This is our Happy Christmas', Gisela wrote in their card to her brother and sister-in-law, Steffen and Primrose Peiser, and doubtless many others. The paper would not come out until February, but Max made sure that key people had advance copies. On Boxing Day he set off for his usual winter skiing trip, meeting the Weissels and other friends in Bad Gastein, before going on to give a lecture in Munich that, he wrote to Gisela, caused 'quite a stir'. On his return home, he found that the news of the structures had focused attention on the lab: 'When I came back,' he told an interviewer years later, 'I suddenly realised that I had become famous.'

'Famous' is a word that Max used frequently, and mostly without irony – though not always. On setting off for a congress in Munich in July 1962, he wrote his first letter to his children 'as I am now v. famous' in the form of three pages of spoof notes by a journalist 'Anthony Scooper ... who follows my every step'.

MFP settled in empty carriage, immediately unpacked various coloured files and proceeded to work on manuscript of world shaking discovery relating to <u>Secret of Life</u> ... Ventured a few words by offering to help with his luggage. Declined charmingly. Delightful personality ... On arrival at Harwich MFP had correct passport and tickets actually ready, not absent-minded professor after all ... Look for next report 'On the Spot with MFP' or The Secret of Life in tomorrow's column ...

While the column is a good illustration of Max's child-like sense of humour, it is even more revealing of his child-like desire for recognition. He wanted scientific success, but he also wanted to be admired for his success, and not only by his scientific colleagues. Though he was never pompous or grand, he was not without a touch of vanity and eagerly welcomed opportunities to place his work – and himself – in the public eye. (The same is true of most scientists, but by no means all: Fred Sanger, one of only four scientists ever to have won two Nobel prizes, could not be more self-effacing.)

The name of Max Perutz has never registered in the public consciousness as did that of Albert Einstein, or even those of Watson and Crick; but from 1960 onwards he certainly began to make a wider impact. That year BBC television's science series *Eye on Science*, presented by one of the pioneers of popular TV science, the former fighter pilot Raymond Baxter, made a whole programme on Max and Kendrew's work. Entitled 'Shapes of Life', the programme consisted mostly of a studio interview between Baxter and the two scientists. To the viewer today, watching it on video, the contrast between them is striking. Max, dressed casually in a V-necked pullover, is clearly enjoying himself and explains with animation what haemoglobin is and does, with the aid of the models on a table. Somehow the BBC allowed the suited Kendrew to spend most of the broadcast with his back to the camera: much of his contribution was in the form of a scripted voice-over for a filmed insert showing the whole process, from crystal mounting to computer printout.

Unbeknown to Max, Sir Lawrence Bragg had begun to canvass support for Nobel prizes for him and Kendrew almost as soon as the models were built. He proposed that Max, Kendrew and Dorothy Hodgkin should share the 1960 prize for physics. Nothing happened that year, or the following year, but the Nobel committees in Stockholm decided that 1962 was to be the year of molecular biology. In October, news reached the newly established Laboratory of Molecular Biology that James Watson and Francis Crick had been awarded the prize for physiology or medicine for the discovery of the DNA double helix, together with Maurice Wilkins of King's College, London. Two weeks later, while Max was in the middle of preparing models of his latest work for an important conference in New

York, he received a call from a London journalist saying he and Kendrew had won the chemistry prize. He went home that night with his feelings in turmoil, but said nothing to Gisela and the children: he knew only too well that both Dorothy Hodgkin and Linus Pauling had received similar calls that turned out to be false alarms, and didn't want his family to be disappointed. On 31 October he received another call from a journalist, but this one was from Stockholm. Once again he bicycled home; this time he could contain himself no longer and revealed to the children that their father might be going to get the prize.

The next day the lab was buzzing, but the uncertainty continued. Suddenly two telegrams arrived, addressed to himself and Kendrew. His secretary rushed into his office waving them excitedly – 'This must be it!' thought Max. Tearing his telegram open with trembling hands, he found it was from the Pontifical Academy of Sciences, asking how many reprints he wanted of the paper he had presented at a conference the previous year. But eventually confirmation arrived, first in the form of a call from Associated Press in Stockholm, then, finally, the official telegrams from the Royal Swedish Academy of Sciences. Max Perutz and John Kendrew were to be awarded the 1962 Nobel Prize in chemistry 'for their studies of the structures of globular proteins'. Max immediately commandeered the lab van – he did not own a car – rushed home to collect Gisela, and went straight to his mother's to break the news. After thirty years, his decision to defy his parents and study chemistry had been proved justified beyond doubt, and the fear of being a disappointment to them had been finally put to rest. Dely greeted the news with tears: she was overwhelmed with grief that her husband had not survived to witness their son's triumph. In poor health since his arrival in the UK, Hugo Perutz had died in 1958, quietly and without fuss as he had lived.

The avalanche of congratulations began at once. Robin, then aged twelve, enthusiastically undertook the task of logging them: Max's files contain a frequently-updated list in coloured crayon and a childish hand that eventually records a tally of a hundred telegrams from sixteen countries. There were even more letters. The senders represent the full spectrum of Max's lifetime acquaintance: Viennese schoolfellows, teachers and fellow students, plus one of the textile workers from his father's former factory; team members from his glaciological expeditions and ski courses; fellow internees, including the Schubert scholar Otto Deutsch; there was a personal note from Lord Mountbatten recalling the work on Habbakuk; scientists from all over Europe and the US wrote offering warm and genuine praise and clearly basking in the reflected glow; and the great and good of British science saluted the achievement with one voice.

Lawrence Bragg had continued to lobby the Nobel committees on behalf of all the Molecular Biology Laboratory scientists since 1959. At the time of the announcements he was seriously ill in hospital, recovering from an operation for prostate cancer. His wife Alice wrote to tell Max that after the first bulletin about Watson, Crick and Wilkins he asked every day 'What about Perutz and Kendrew?', and that once he knew their prizes were also in the bag he insisted on telling 'the glory of the protein story' to every doctor and ward sister in the hospital. She was still very anxious about his state of health, but hoped 'this will be a tonic'. He went on to make a full recovery.

Desmond Bernal, who was generally agreed to have launched the field with his studies of pepsin, was naturally delighted to see it reach its goal, though his feelings must have been mixed as he saw his successors romp away with the glory. His letter contained a slight rewriting of history: he remarked that it seemed 'only a few months ago that I gave you the first haemoglobin crystal . . .' As Max could not possibly have forgotten, Bernal had little if anything to do with his introduction to haemoglobin, though Max would have been the first to acknowledge his critical role as mentor and inspiration. (Bernal was to make a similar adjustment to history two years later when Dorothy Hodgkin won her own Nobel, managing to imply in a *New Scientist* article that she had conducted her ground-breaking work on the penicillin molecule under his direction.) He went on: 'You know better than I do that what you and Kendrew have done is only the first swallow but what a summer it is going to be.'

Felix Haurowitz, who really did introduce Max to haemoglobin, wrote a letter full of humility, acknowledging how wrong he had been to doubt that the structure of such large molecules would ever succumb to crystallographic methods:

> I want to say that I have a great admiration for your work, for your patience and your persistence, and for the continual improvement and refinement of your methods which finally led to the high resolution and the results published during the recent year . . . You and Kendrew accomplished much more than I ever had considered as possible.

At the Laboratory of Molecular Biology, the champagne flowed once more, at a 'great party' where Max revelled in the attention. The LMB has continued this tradition of celebrating good news with spontaneous and informal parties at which everyone has a chance to feel that they share in the success. At the party for the protein Nobels, there were formal presentations: Max received a toy horse and a bottle of blood, and Kendrew a 'stuffed whale' and a lump of frozen whale meat.

The Nobel ceremony took place on 10 December. The Royal Swedish Academy of Sciences goes to enormous trouble to ensure that the whole experience is as enjoyable as possible for the prizewinners, beginning by encouraging them to bring as many family members as they wish. Max took Gisela, Robin and Vivien, his mother and his mother-in-law; his cousin Alice Frank came from Paris through her connection with the Swedish Technical Academy. (Although the Royal Swedish Academy covered all their travel and accommodation costs, Vivien remembers that Max was so short of money that he had to take out a loan to buy the new clothes they all needed.) They were royally entertained in Stockholm for more than a week, with a constant succession of parties, dinners, balls and outings. From the moment he arrived, Max at last knew what it really meant to be 'famous':

> We were, as always, the last to leave the plane. As I stepped, unsuspecting, out of the door, I found myself blinded by a host of floodlights and heard camera shutters clicking all round. When I arrived at the bottom of the steps I was overwhelmed and deeply moved by friendly faces giving me a tremendous welcome, as though I were a traveller returned from space, or the saviour of their country come back from a victorious campaign. I soon realised that here in Sweden my friends and I would be national heroes.

The children had also very quickly got used to the idea that while in Sweden they would be treated like 'fairy princes'. On visiting the ballet with his grandmothers and sister but without Max, Robin was disappointed to discover there were no press photographers to greet them. 'Without Daddy we are just nobody,' he remarked sadly, though Max noted that he and Vivien had their fair share of press coverage.

Three days after their arrival came the prize ceremony itself. Dressed in tails and a starched white shirt (the same suit he had brought from Vienna in 1936), Max processed with the other prizewinners onto the stage of the Concert Hall in Stockholm.

> I felt gay and happy as a lark, and was surprised to find Gisela in the front row of the stalls with tears streaming down her face as though she was attending my funeral. Robin luckily was unaffected by all this solemnity and exchanged some cheerful winks with me.

Each presentation was preceded by a speech in Swedish which, Max noted, 'judging by the faces of the audience were unrelieved by humour'.

Because the prizewinner for physics, the Russian Lev Landau, was too ill to come to the ceremony, Max was the first to receive his medal and certificate from the King of Sweden.

> He did it in a charming, informal and fatherly manner, talking not to the public but only to me; he congratulated me, told me how pleased he was to be able to give me the Prize and wished me every success for my future work . . . I felt cheerful and extremely happy throughout . . . What made me happy was the warmth and friendliness of all around me . . . And finally the thought that the Prize was only the final expression of the great enthusiasm aroused all over the world by John's and my research.

Max's unalloyed enjoyment was obvious for all to see. Years later John Kendrew described his state to Horace Judson: 'Max was in an absolute mood of euphoria, leaving the floor, you know – he was so pleased with the whole thing. I suppose we all were – but Max wears his heart on his sleeve.'

That evening Gisela led the procession into the Great Golden Hall for the Nobel banquet on the arm of the king, 'looking beautiful and regal in a magnificent black velvet gown'. Max himself sat with Odile Crick on one side and Princess Christina, a girl of about Vivien's age 'with much the same interest and little more sophistication' on the other. Rising in his turn at the end of the meal to make the required two-minute speech, which he had carefully prepared and learned so that he could deliver it without notes, Max spoke simply and from the heart:

> I was deeply moved by the Royal Swedish Academy's decision to elevate me, a man of modest gifts, to the Olympian heights of a Nobel laureate . . . Unravelling the anatomy of the haemoglobin molecule may have needed much perserverance and hard work, but comparatively little of the imaginative power which made the giants of the scientific revolution create the world we live in today. I stand in awe of the company which I am now supposed to join.

In conclusion he reminded his listeners that the task of explaining the physiology of breathing in terms of the architecture of the haemoglobin molecule had only just begun, and quoted the prayer by Sir Francis Drake that he took as his personal motto:

> When Thou givest to Thy servants to endeavour any great matter, grant us also to know that it is not the beginning, but the continuing

of the same until it is thoroughly finished, which yieldeth the true glory.

For Max the Nobel prize was indeed a new beginning. Many of his colleagues have commented on the new confidence it gave him: while he had long faced doubts that what he was doing was worthwhile, now his persistence had been utterly vindicated. National recognition quickly followed the accolades of his scientific peers. In 1963 he was awarded a CBE, which he happily accepted; in 1969 he politely turned down an offered knighthood: 'If I were titled this would put me on a pedestal and set me apart from my young colleagues; I fear that this might spoil the happy atmosphere we now enjoy.' He told friends that he thought a title would make him look a fool among his colleagues in continental Europe, and that it would make a barrier between himself and others in the lab. However, it was the title to which he objected, not the honour. He accepted the higher distinction of Companion of Honour in 1975, and in 1988 he was admitted to the most select body in British public life, the Order of Merit, which is in the personal gift of the Queen and limited to twenty-four members at any one time.

Another distinction that meant at least as much was his election, on winning the prize, to an honorary fellowship at Peterhouse, his old college. Saturday lunches in Peterhouse became a regular fixture in his week. Soon afterwards, another Cambridge college, Trinity Hall, invited him to apply for the Mastership. Later, Max denied he had ever thought seriously about accepting, but Vivien remembers that he was tempted. It was Gisela who immediately saw that such an administrative burden, with the obligation to do a lot of wining and dining, would not suit him: she talked him out of letting his name go forward.

For none of Max's fellow prizewinners in 1962 could the prize have meant so much. Until 1962, letters to Gisela occasionally included the comment that new acquaintances in other countries 'were very nice to me': the remark carried faint connotations of surprise or relief, as though his experience of life had led him to expect a less friendly reception. His account of the Nobel ceremony is full of references to the warmth and friendliness of the Swedish scientific hosts, noting that they 'had been just as nice to me on my first visit in 1951 when nobody could have foreseen that I would win the Nobel prize'. Now the outsider's anxiety was finally dispelled.

Belonging to the 'club' of Nobel prizewinners brought with it the opportunity for a higher public profile, which Max would use in pursuit of causes close to his heart. His first action, though, was to nominate his dear friend Dorothy Hodgkin for the prize, feeling embarrassed that she had been passed over in his and Kendrew's favour. She was duly announced as the sole winner of the 1964 prize for chemistry, the first (and so far only) British woman to win a science Nobel.

9

A structure for science – the LMB

I am just a chap who messes around in the lab.
Fred Sanger (double Nobel prizewinner)

Max and his colleagues moved into the Medical Research Council Laboratory of Molecular Biology (LMB) in March 1962. In fifteen years, the original Perutz-Kendrew partnership, occupying a couple of rooms in the Cavendish, had metamorphosed into a 90-strong institute with its own five-storey brick building. Among the twenty-five independent research scientists, one was a Nobel prizewinner (a figure that rose to four within months of the official opening), and four (soon to rise to six) were Fellows of the Royal Society. How such a concentration of prima donnas should be managed was a question that exercised the Secretary of the MRC, Sir Harold Himsworth, from the moment the lab was first mooted.

In 1957 he had a meeting with Kendrew at which it emerged that the latter's reservations about the plan had less to do with the distance from the centre of Cambridge than with difficulties that he foresaw over 'organisation and personalities'. Himsworth's diary note of the meeting records that he thought the best idea might be to have a series of sections coordinated by 'a kind of Soviet of directors'. Two years later, he put these ideas to Max, emphasising that each of the senior people should have responsibility for the scientific policy of his own section or division, and that Max might rely on 'propinquity and joint round-table meetings' to secure coordination.

Max was enthusiastic about this proposal, as long as the number of divisions was kept within reasonable bounds. He drafted a constitution for the lab that established three autonomous divisions: Protein Crystallography (later Structural Studies) headed by John Kendrew, Protein Chemistry headed by Fred Sanger, and Molecular Genetics headed by Francis Crick. Max himself was to be Chairman – deliberately not Director – with Kendrew as Deputy Chairman. These four were to constitute the Governing Board of the lab. Divisions would not have their own budgets, but would put forward proposals for spending that Max and the board would consider and unite into a single budgetary request to Head Office. Each division was to have its own floor in the building, with a common workshop and stores in the basement. The workshop, stores and any particularly expensive pieces of equipment were shared among all the divisions to minimise duplication. The individual heads of division would have the right to appeal directly to MRC head office if there was any dispute over sharing of resources. 'We could approach the MRC without going through Max,' says Crick, 'although as a matter of courtesy we almost always did.'

This constitution was a departure from MRC practice. The usual philosophy behind its units was to give long-term funding to a single researcher of proven ability to build a group that he or she would direct. However successful the group, the retirement of the director would usually terminate the life of the unit. The MRC's Council was clearly slightly dubious about the LMB's constitution, but accepted it on condition that it should in no way be regarded as a precedent.

Notwithstanding this bureaucratic caution, the MRC was justifiably proud of its new creation. To Max's transparent delight, it invited the Queen to perform the official opening ceremony in May 1962. Not all members of the lab had the same reaction: both Francis Crick and Sydney Brenner deliberately arranged to be away on the day of the ceremony. Jim Watson, in contrast, flew over from the US specially to be presented to her. On the day, Max was in his element, using his models to explain to Her Majesty what the research of the lab involved.

> The Queen was so interested in everything we showed her and had such a straight, natural and warm approach to everybody that even the anti-royalists were charmed by her; one of her accompanying ladies, when shown our models, exclaimed: 'I had no idea that we had all these little coloured balls inside us.'

For his part, Max there and then became a staunch admirer of Queen Elizabeth, whom he was to meet several more times. The reply he received from her office, after he wrote to thank her for her visit, thanked him for the 'charming and lucid way' in which he had explained the 'thrilling and complicated' subject of his research, and asked to be kept in touch with future developments.

A constitution was one thing, but how would Max's unorthodox management structure work out in practice? All the evidence is that it worked superbly well. The divisional heads had what they wanted, which was complete freedom to conduct their research. Hugh Huxley was initially disappointed not to be included on the Governing Board: he was bringing a new and very successful area of research to the Laboratory, and was already an FRS, the youngest in the country at that time. However, within a few years he was made joint head (with Aaron Klug) of a sub-department within Structural Studies and took his seat. Sydney Brenner, who shared the headship of Molecular Genetics with Crick, also joined, and subsequently Max decided that the sole criterion for board membership should be election to a Fellowship of the Royal Society. In practice, the board rarely held meetings. As Himsworth had predicted, 'propinquity' – bumping into each other in the corridor or the cafeteria – was usually enough to sort out any problems. Max would circulate the heads of division with anything he felt they needed to know, but paperwork was negligible – there are no files of governing board minutes. This suited him down to the ground: he hated administration and had every intention of spending as much time as before on his own research. At the same time he was never less than conscientious, and made it a point of honour that all letters should be answered within forty-eight hours of receipt.

He also recognised that he needed someone to keep track of the pennies and see that the building was maintained, and that therefore he had to have some administrative help in addition to his personal secretary. The MRC found him Audrey Martin, who came from its Dunn Nutrition Laboratory nearby and, with two assistants, took care of book-keeping and personnel matters.

For anything to do with the building and equipment, Max put forward someone he had known for a long time. He had recruited Michael Fuller as a fifteen-year-old school leaver to an apprentice technician's post in 1952. Fuller's interview with Max, John Kendrew and Tony Broad had been perfunctory. What, Max asked him, was sodium chloride? 'Salt,' replied the boy, quick as a flash, and he was in. His first task for Max was to go to the slaughterhouse and get a few pints of horse blood. By 1961, Fuller's job was still officially to maintain the somewhat temperamental rotating anode X-ray tubes on which all the protein crystallography depended, but Max had clearly seen that there was more to him than this, and he was soon proved right. On his own initiative, Fuller had produced a minutely detailed master plan for establishing the infrastructure of the new lab, which involved both ordering new equipment and moving purpose-built machines such as the X-ray tubes from the Hut to the new quarters. Once the move had taken place, he became the presiding genius of the stores, making many of the spending decisions on his own initiative. It was a considerable increase in responsibility, but Max just left him to it. 'Once

he realised that you could do something, he would let you get on with it,'
says Fuller.

He was given the title Laboratory Steward, which he kept until his
retirement, though MRC rules kept him on the technicians' salary scale.
(His 'retirement' was nominal – he is still to be found in the LMB most
days, with the title of Special Projects Coordinator, and works with the
lab archivist on preserving its history.) Fuller took pride in sorting out practical problems for Max. If Max was travelling to speak at conferences, Fuller
would make sure he had everything he needed for his talks and demonstrations in a specially-fitted case. When Max injured his back, Fuller took
on the job of driving him to and from the lab each day – on at least one
occasion stretched out flat in the back of the van. He arranged for the
construction of a special high lectern, attached to the wall of his office,
so that Max could work standing up.

The common stores had another function, which was to act as a meeting
point where members of the different divisions would mix as they dropped
by to pick up a stock of chemicals or a new Biro. Max placed a high priority
on this mixing. In the Cavendish and the Hut the group was so small, and
the accommodation so cramped, that mixing was unavoidable. With the
new divisional structure, a larger staff and a much larger building, he had
to plan for it. The key tool for this in Max's mind was a cafeteria where
staff could continue the Cambridge University tradition of coffee, tea and
lunch breaks.

MRC Head Office did not see the point at first. The new hospital that
was going up adjacent to the lab would have a cafeteria that lab members
could use: wasn't that good enough? In Max's eyes this was highly unsatisfactory, especially as the hospital cafeteria would serve food and drink only
at set times. It was Gisela who voluntarily took on the job of setting up a
corner where she could run a modest cafeteria (circumventing the MRC's
financial objections as she was unpaid), which rapidly developed into something much larger built on to the roof of the building. She specified every
detail, from the kitchen equipment to the choice of furniture, hired the
staff and kept the accounts. MRC officers visiting soon after the opening
commented that they had had a good lunch in the cafeteria, which they
found 'furnished with some imagination' and with 'well designed modern
stainless steel cutlery (made in Hong Kong!)'. They noted that two of Sir
Lawrence Bragg's watercolours (he was an enthusiastic amateur artist) hung
on the walls. Gisela had also installed some ceramic reliefs created by her
mother, inset into the pillars.

Coffee, lunch and tea became an essential part of the LMB culture. At the Cavendish there had been separate coffee rooms for clerical staff, technicians and academic staff; at the LMB in contrast, everyone had coffee together. The most junior technician could, if he or she dared, join in discussions with the 'gods' of molecular biology. For his part, Crick, whose table was guaranteed to be the liveliest, made a point of including everyone within range in his remarks. Max himself used the canteen as an opportunity to keep up with what everyone in the lab was doing and offer encouragement, taking his tray to a different table each time he visited. Occasionally, after one of these conversations, he would have to check the name of the new arrival he'd been talking to: to save such future embarrassments, Michael Fuller arranged to have everyone photographed so that a 'rogues' gallery' could be put up on the wall.

Gisela remained as an unofficial and voluntary manager, coming in three times a week even after the cafeteria was equipped and staffed to her satisfaction. She continued to do the accounts weekly until Max's retirement, keeping the prices as low as was possible consistent with breaking even. Although the paid catering staff were mildly exasperated by her tendency to rearrange the yoghurts, they regarded her with affectionate tolerance (and called Gisela and Max 'Grandma and Grandpa' behind their backs). Gisela too would take the opportunity to keep her ear to the ground, generally on personal rather than scientific matters, and she discreetly made sure that Max got to hear about anything that might be troubling one of his staff. When one of the canteen staff fell ill, Max personally intervened to ensure she saw the top specialist at the hospital: Gisela's intercession almost certainly achieved this. Having devoted her married life to making Max's personal and professional life run smoothly – helping him get ready for trips and assisting with correspondence as well as running the home and caring for the children – it was natural for her to offer her services to help with something that Max thought important but was unable to organise himself: 'It became necessary, so it wasn't a question of Max asking me . . . For him, good facilities for working were the prime need for him and for others . . . I saw it as my job to help where I could.'

There was no rule that said you had to go to the canteen, but almost everybody did. Visiting American scholars, used to a more austere work ethic, were astonished at the frequency with which top-class scientists would down tools and head off for a coffee or a plate of pie and chips. They were quickly converted. The opportunity to talk to people who were working in different but related areas could prove extremely fruitful, and the challenge of explaining your work to those with different backgrounds could only help to clarify your thinking. Some tried to establish the same culture on their return to the US, with varying degrees of success. According to one anecdote, the Nobel prizewinning biochemist Arthur Kornberg was so impressed with the LMB canteen that he set one up on his return to Stanford – but even when he laid on free coffee and cookies, few of the senior people turned up. Instead, they would send technicians to collect the free snacks and bring them back to the lab.

Francis Crick introduced a further innovation that formalised the sharing of ideas. At the beginning of each academic year the lab held a one-day seminar – later stretched to a full week as numbers grew – at which research workers stood up and gave short presentations to all the others. Attendance was compulsory. Max enthusiastically participated in these 'Crick weeks': apart from his pleasure in the achievements of his colleagues, he was keen to keep on top of the subject and include the latest advances in the under-graduate biochemistry lectures that he continued to give each year. He made it his speciality to ask questions of the often inexperienced speakers that forced them to clarify their accounts, to their own benefit as well as everyone else's. Academics often ask questions designed to show off their own brilliance: Max's style was to ask questions that made him look naïve or even stupid. Newly-arrived young researchers could be misled into regarding him as something of a figure of fun, especially as the more senior people were not above teasing him: 'Explain it so that even Max can understand it,' Crick would bellow. Later they came to realise that Max's questions were an educational tool, designed to avoid jargon and ensure that everyone – the lecturer included – knew what he or she was talking about. Clarity of presentation was something Max insisted on: he also vigorously edited the written papers of his junior colleagues, and many of them express their gratitude for his guidance on standards of written English.

Everyone who has worked in the lab remarks on its informality. Max insisted on first names, regardless of status. John Kilmartin, who joined the lab as a PhD student in 1965, remembers him reprimanding one of the technicians, Jan Fogg: 'Jan, if you continue to call me Dr Perutz I shall have to call you Mrs Fogg!' Only Michael Fuller never quite managed to shake off the formal ways of the old Cavendish and called Max Dr Perutz until the very end. Max kept his door open and, although time-wasters were given short shrift, he was always accessible if you needed him. Informality and frequent coffee breaks, however, did not imply a casual approach to work: one hapless recruit recalls finding a sharp note on his desk when he failed to turn up by 9.30 in the morning. In general, there was never any need to crack the whip. The lab hummed with the energy of ambitious scientists driven by the desire to understand. Max's main recipe for success was to enlist good people and give them everything they needed. The distinguished careers of many of those recruited under his chairmanship suggest that most of the time he and his divisional heads got it right.

The other notable feature of the laboratory community was the very high number of foreign researchers, mostly from the US, who came for two or three years and usually brought their own funds. The combination of a small number of established researchers with a larger population of keen young people early in their careers made for an energetic and productive environment. Many of the visitors returned to head labs in their home countries.

In a very short time, the LMB had established a powerful group identity, based on a combination of informality, personal responsibility and commitment to the highest standards of research quality. MRC officials on their regular tours of inspection pronounced themselves highly satisfied with their new creation:

> The slightly anarchic way of running the Laboratory seems to work well in practice.
>
> The easy, informal atmosphere of the laboratory was as striking as it was pleasant, and I formed a most favourable impression of the work being undertaken, and of the atmosphere in which it was carried out.

In the decade following the lab's opening, it continued to grow rapidly. Individual research teams remained small, but under Max's leadership new arrivals and existing lab members generated an astonishing burst of scientific creativity.

Since Sydney Brenner had arrived at the Unit in 1957, he and Crick had been key members of an international network of chemists and geneticists who between them cracked the genetic code and discovered how it was translated into proteins. By the mid-1960s, they had discovered that the four letters of the DNA alphabet – A, T, G and C – spell out the sequence of amino acids in a protein using a three-letter code: CAT means histidine, GTT means valine and so on. DNA is in the nucleus of the cell, and proteins are made in the cytoplasm, the fluid surrounding the nucleus and bounded by the cell membrane. How does the code reach the protein-manufacturing machinery in the cytoplasm? The answer is that the DNA message is transcribed into another nucleic acid, messenger RNA, which moves out of the nucleus into the cytoplasm. At specialised structures in the cytoplasm called ribosomes, nucleotide triplets of transfer RNA collect their corresponding amino acids and line them up along the messenger RNA template, forming the protein chain.

With this problem solved and the 'secret of life' now known to all, Brenner was looking for a new challenge. In 1965 he decided the time was right to try to understand a whole animal in terms of its genes – not just its physical form, but even its behaviour. The animal had to be small: the one he chose, the nematode worm *Caenorhabditis elegans*, was less than a millimetre long and made up of fewer than 1,000 cells. Brenner soon built up a small team studying the genetics of the animal, looking at how mutations affected its behaviour, and studying its development from egg to adult so that he could understand how genes controlled growth. In the early days, some saw the project as overambitious. Max himself was doubtful, writing to his son Robin in 1978 that it was 'a pity' that Brenner's work had not been more productive, and describing his results as 'meagre'.

However, the philosophy of backing long-term projects paid off. In the course of painstakingly watching cells in the developing worm larva divide under the microscope, John Sulston and Bob Horvitz in Brenner's group discovered that in the normal course of development some cells are deliberately killed off. This 'programmed cell death' turned out to have much wider implications, for example in understanding why cells that are supposed to die fail to do so in some cancers. Brenner, Sulston and Horvitz were awarded the Nobel prize in 2002 for their studies (mostly completed two decades or more previously) of how genes regulate development in the nematode worm.

During the 1960s, Fred Sanger shifted his attention from protein sequencing, which was by this time a tool that anyone could use, to finding a method of reading the sequence of RNA and DNA. It took him many years, but by 1977 he had hit upon an ingenious solution. A single strand of DNA, provided with a plentiful supply of all four 'letters', or nucleotides, will make a copy of itself, one chain prescribing the sequence of the other as predicted by Watson and Crick. Sanger would take a strand of a few hundred bases and mix it with nucleotides that included a few that were in an altered form. An altered nucleotide would join the chain, but stop it at that point, like a jigsaw piece that had a hole on one side but no corresponding blob on the other for the next piece to fit onto.

In test tubes Sanger made four mixtures, each containing a few copies of the altered form of only one of the four nucleotides A, T, C and G, plus plenty of the normal forms of all of them. The result was that in each tube chains would grow of different lengths, but each chain ended on the same nucleotide when by chance it picked up the altered form. The final step was to run the DNA samples, labelled with a mildly radioactive chemical so that they made an image on photographic film, through a gel: the smaller pieces move faster than the larger ones, so that you end up with a ladder of stripes like a bar code. By lining up the four ladders side by side, Sanger could read off the order of the four nucleotides all the way down the gel. Using an earlier version of this manual method, in which he simply restricted the amount of one of the nucleotides in each tube so that the chain stopped when it ran out, he and his colleagues read the sequence of just over 5,000 nucleotides in a tiny virus in 1977. Sanger won his second Nobel prize in 1980 for this achievement.

A little over a decade later, commercial firms began to build automated sequencing machines based on the same principle, and John Sulston, in Sydney Brenner's lab, bought a couple to read the complete 100-million-letter DNA recipe to make a nematode worm. His evident skill at managing a large sequencing project caught the eye of the Wellcome Trust, Britain's wealthiest medical research charity, who put him in charge of their new automated sequencing centre at Hinxton, just south of Cambridge. The Sanger Centre, as it was named, contributed one-third of the human DNA sequence which was assembled by the international Human Genome Project and finally completed in March 2003 – fifty years almost to the day after Watson and Crick discovered the DNA structure.

Cesar Milstein, an Argentinian biochemist, joined Sanger's Protein Chemistry division in 1963. He studied antibodies – protein molecules that form a key element of the body's self-defence armoury – trying to understand how there seemed to be a specific antibody for every conceivable infectious agent or other 'foreign' tissue. With Georges Köhler, a post doc from Germany, he worked out how to make unlimited quantities of 'monoclonal' antibodies, all recognising the same target. Acknowledging the enormous potential of monoclonal antibodies as tools in both medical and industrial contexts, the Nobel committee awarded them the prize for physiology in 1984. The technology's potential began to be fulfilled soon afterwards when the National Enterprise Board formed a company, Celltech, to exploit the commercial possibilities of monoclonal antibodies from mice. When another of Milstein's colleagues at the LMB, Greg Winter, found a way to 'humanise' the antibodies to make them safe to use in human patients, LMB had to fight to retain control of the intellectual property. Today, the LMB enjoys a steady income (currently £40 million per year) from licence fees, even though some of the best-known monoclonal antibody products, including pregnancy testing kits and the cancer drug Herceptin, do not bring them any royalties.

Aaron Klug, who had worked with Rosalind Franklin at Birkbeck College on the structure of viruses using X-ray diffraction, moved to the LMB on its foundation in 1962. There he found a way to apply the Fourier methods of X-ray crystallography to photographs taken using an electron microscope, to generate three-dimensional images of assemblies of proteins and nucleic acids that could not have been studied with either technique alone. For this achievement he won the Nobel prize for chemistry in 1982. In this golden era of the LMB, Klug found himself in an extremely favourable environment for the development of new ideas and techniques.

In 1965 I introduced the use of optical diffraction to look at electron micrographs. I used some home-made things, and borrowed some from the Cavendish, but then I built an optical diffractometer in a lift shaft, which cost about £1500. It was a lot of money in those days, but there was no trouble in getting it . . . I simply wrote a short note to Max, he understood the importance of it and I got it.

Hugh Huxley had also used the powerful combination of X-ray diffraction and electron microscopy with considerable success in his studies of muscle structure. Max later thought he had made a tactical error in trying to achieve separate Nobel prizes in physiology and chemistry for the LMB in the same year, as had happened in 1962, rather than a joint award in physiology for Huxley and Klug: when only Klug was successful, he was somewhat embarrassed.

Most of the 'stars' of the LMB were there from the beginning, but Max also made at least one significant appointment and proved a shrewd judge. The early molecular biologists tended to be chemists and physicists by training, with some exceptions: Fred Sanger was a biochemist, and Sydney Brenner had trained as a doctor. The MRC was keen to see more biological research. One day, Max read that a young scientist called John Gurdon, in the Zoology Department at Oxford, had successfully cloned adult frogs for the first time. Max decided that Gurdon sounded exactly the kind of person they needed. He made some discreet enquiries before approaching him informally and asking him if he'd like to move to the LMB. Gurdon said no at first, but later wrote to Max to ask if he could come after all. Max replied at once that that would be 'marvellous', with not a hint of irritation at Gurdon's apparent dithering. So, in 1971, Gurdon became the head of a sub-division under Crick and Brenner. Although Gurdon's work on embryonic development was far removed from anything he himself had done, Max took a lot of interest in it and they became good friends, even ice skating together on the frozen fens when the opportunity arose. Today,

Gurdon has his own institute for developmental biology and cancer research at Cambridge University. He tries to emulate Max's style in its management, with informality and interaction between groups the watchwords.

This is far from the only example of the LMB philosophy being exported. After working with Max on the low-resolution structure of haemoglobin, Michael Rossmann had formed a partnership with David Blow, a former research student who had returned to the lab in 1959. Together they developed a set of mathematical tools to help unpick the structures of big molecules composed of several subunits, which could exploit the near-symmetry of the different parts. Rossmann called the technique 'molecular replacement'. They had gone on to work together on the structure of another protein molecule, the digestive enzyme chymotrypsin. In 1963, however, Max told Rossmann that he had not been promoted to a permanent post and would have to find another job. This in itself was not too much of a surprise: what came as more of a shock was that Max was keeping on the other half of the partnership, David Blow. Rossmann believes Max thought he was spending too much time on crystallographic techniques, and that a medical research lab was not the right place for that kind of work.

With hindsight, Max and his colleagues on the board may have made an error of judgement in not keeping Rossmann on, but, if anything, the move acted as a stimulus to his career. He accepted an invitation to set up a new crystallography lab at Purdue University in Indiana, now one of the leading centres in the world for the crystallographic study of proteins and viruses. Within six years, he had solved the structure of a large enzyme; in 1985 he became the first to solve the structure of a rhinovirus, one of the large family of viruses that cause the common cold. Characteristically, Max could not have been more delighted by Rossmann's success: Rossmann is certain that Max was behind his election as a Foreign Member of the Royal Society.

To what extent could Max take credit for this extraordinary run of success, both at Cambridge and further afield? Most agree that his apparently hands-off approach was precisely what was wanted. Max put his energies not into directing research, but into creating conditions in which it could flourish. He trusted his research leaders, he made sure they had what they needed, he minimised paperwork, he encouraged a sense of collaboration between the divisions and of self-worth at all levels. When they needed more space, he got it – one extension was agreed within a year of the lab's opening, and another towards the end of the 1970s. Of course, all the available space was soon filled to capacity, but no one seems to have complained about the cramped conditions: it was a sign of the lab's extraordinary appeal and productivity. John Gurdon comments especially on the care Max took to make sure he knew who everyone was in the lab, and what they were doing: 'It's easy to overlook how influential he was,' he says, 'because it was done in such a mild way.' People often asked Max what was the secret of its success, as though there was a formula that could be applied again and again:

> I feel tempted to draw their attention to 15th century Florence with a population of less than 50,000, from which emerged Leonardo, Michelangelo, Raphael, Ghiberti, Brunelleschi, Alberti, and other great artists. Had my questioners investigated whether the rulers of Florence had created an interdisciplinary organisation of painters, sculptors, architects and poets to bring to life this flowering of great art? . . . Creativity in science, as in the arts, cannot be organised. It arises spontaneously from individual talent. Well-run laboratories can foster it, but hierarchical organisation, inflexible, bureaucratic rules, and mountains of futile paperwork can kill it.

Although as able as any in this community of creative thinkers, John Kendrew was an exception in that after he won his Nobel prize he took virtually no further interest in the practice of science. There are some hints that he felt he would have been a better head for the lab than Max, but Aaron Klug is emphatic that this was not a general view: 'Max was without any doubt the spiritual leader: Kendrew was the perfect staff officer.'

Kendrew's early experience in the scientific civil service had led to a number of advisory posts – and in the early 1960s, he became deputy to the Chief Scientific Adviser to the government, Sir Solly Zuckerman – but much of his effort in the 1960s was devoted to developing the subject of molecular biology in Europe and establishing it as a discipline in its own right. Max could have made a fuss about the fact that one of his divisional heads was essentially an absentee landlord: but he seems to have decided that, since Kendrew was engaged on other worthwhile activities, it was better not to rock the boat.

Kendrew's concern was that, although the discoveries of the previous two decades – many of them by European researchers – emphasised the unifying power of a molecular approach to biology, European universities continued to teach students according to the traditional rigid divisions into botany, zoology, genetics and so on. In the US, in contrast, molecular biology had grown rapidly and had been incorporated into the more flexible structure of American higher education. Others had had similar concerns. Leo Szilard, an Hungarian émigré physicist-turned-biologist from

Washington DC, responded to the Cuban missile crisis in October 1962 by 'pack[ing] up his wife and belongings' and fleeing to Geneva. There he announced himself to Victor Weisskopf, the Director of the European nuclear physics laboratory CERN, with the words 'I am the first refugee from the Third World War.' Szilard then suggested that a European molecular biology laboratory should be founded on the model of CERN.

That December Kendrew and Jim Watson stopped off in Geneva on their way back from the Nobel ceremony in Stockholm and met Szilard and Weisskopf. As a result, Kendrew and some European colleagues convened a meeting at Ravello in Italy in September 1963. Max was a keen participant. The meeting rapidly developed two views. Kendrew championed the idea of a European Molecular Biology Laboratory, but a more modest proposal also emerged – endorsed enthusiastically by Max – for summer schools, visiting professorships and a fellowship scheme to enable European workers to gain experience in each other's labs. As Max later told the *New Scientist*, it was easier for a researcher in Genoa to get a fellowship to go to Berkeley than to collaborate with a colleague in Marseilles. In the event, the European Molecular Biology Organisation, established for the first time at this meeting, set up subcommittees to pursue both options. Without any apparent premeditation or fixing, Max was elected the first chairman of EMBO's Council. It was highly unusual for Max to take on this kind of administrative responsibility: according to Sydney Brenner 'he was a bit cornered but accepted it with good grace'.

Max threw himself into solving the immediate problem of how the organisation was to be funded. The lifeline came from the very wealthy Volkswagen Foundation in Germany: Max went to meet its officers in March 1964, and came away with an invitation to apply for funds. By December the Foundation had come up with 2.6 million Deutschmarks ($687,000 at the time). Another $10,000 a year came in from the Israeli government, whose scientists were anxious not to be left out, as did a sizeable one-off grant from the Swiss pharmaceutical company Interpharma.

At Max's side until his term as Chairman ended in 1969 was Raymond Appleyard, then head of biology at the European Commission's atomic energy community Euratom, who became EMBO's part-time Executive Secretary. Appleyard was immediately impressed with Max's skill as chairman of a council that, like the LMB, was full of prima donnas: it included five Nobel prizewinners. 'For my money, the key figure in the whole thing was Max,' he says now. During Max's chairmanship the fellowship scheme operated effectively and EMBO began a programme of training courses to rival the course on phage genetics at Cold Spring Harbor that had done so much to give American molecular biology the edge. Meanwhile, European governments took over the funding of the organisation, and they eventually accepted the need for a European Molecular Biology Laboratory: Kendrew achieved his ambition when the lab opened at Heidelberg, with him as Director, in 1978. Both EMBO and EMBL continue to flourish. Max remained a keen supporter and honoured visitor, presiding at the opening of EMBO's new office building in 2001 when he was eighty-seven.

Kendrew nominally remained head of Structural Studies at the LMB until 1975, but, according to Aaron Klug, he was rarely in Cambridge from the mid-1960s onwards and in practice Klug and Hugh Huxley ran the division. Max must have found Kendrew's enthusiasm for policymaking and administration incomprehensible, but publicly he was generous in recognising his talents: EMBL, he wrote in an obituary after Kendrew's death in 1997, 'stands as his monument'. Forever linked by their shared Nobel, the two men could not have been more different and had a relationship that was frequently difficult. However, although they were not friends, there is nothing in Max's correspondence to suggest that he did not hold Kendrew in high regard, and he was never less than loyal to him.

Another high profile departure came in 1976, when Francis Crick left for a year's sabbatical at the Salk Institute in La Jolla, Southern California. He had been a visiting fellow there since the early 1960s, and its President, Frederic de Hoffmann, 'went to great efforts to tempt [him] to stay on there'. Crick was already sixty, and the offer of a personal endowed chair – with no retirement age – did the trick: he remained at the Salk until his death in 2004. Max missed Crick terribly. He was unquestionably the towering intellect of the LMB, and had done much to attract highly able and imaginative colleagues. On a personal level, Crick had on more than one occasion put Max right on some theoretical aspect of his work and Max valued his judgement. Crick, for his part, respected Max for the body of work he had done and for his skilful management of the lab, though he clearly did not regard him as an intellectual equal: 'Max wasn't a particularly quick thinker. He was a plodder, but a very persistent plodder, and he had considerable insight as a result of his plodding. It didn't come out in flashes.' It was Max's insight and persistence that was eventually to lead to the full revelation of the mechanism of haemoglobin's action.

Despite their very different personalities the two men were good friends for the rest of their lives: after Crick's departure, each visited the other at home on transatlantic trips whenever it was possible, and Max sent Crick copies of many of his research papers. For the lab as a whole, Crick's going was certainly a loss, but given the general high standard of research across all the divisions, it did not materially affect the status of the institution.

Normal retirement age for the head of an MRC unit was sixty. Under this rule Max would be expected to bow out to tend his lawns and roses in May 1974. To him this was simply not an option. Not only was he midway through his 'haemoglobin battles' (see next chapter), in search of new forms of evidence to support his analysis of how the molecule worked, but also the lab as a whole was at a difficult moment in its history. With the election of the Conservative Party under Edward Heath in 1971, spending on science was under new scrutiny, and a committee under Lord Rothschild had decreed that the MRC should lose a quarter of its budget. The LMB accounted for five per cent of the MRC's total spending, and so there would be battles to be fought to defend its interests. Max was hardly going to leave his creation to be squabbled over by strangers.

There was much to fight for: the lab now employed 200 people. With the agreement of the governing board, Max applied to delay his retirement until May 1979. The Council agreed at once, no doubt relieved at being able to postpone the difficult question of the succession. A year later, the Council accepted the view that it was 'inconceivable' for the lab to close on Max's retirement – the normal fate of MRC units. At the same time, it asked the Oxford professor David Phillips to carry out a review of the whole field of molecular biology. Although he had never held a post there, Phillips knew the LMB very well. As head of the crystallography lab established by Lawrence Bragg at the Royal Institution, he had collaborated closely with Kendrew on myoglobin; as a part-time Reader at the RI, Max was also a frequent visitor there. On the face of it, Phillips might have been expected to give the LMB sympathetic treatment.

The main questions were: who should succeed Max; whether the existing administrative structure should continue; and what the scientific priorities should be in future. Max entered enthusiastically into discussions about his successor. There was no obvious candidate. It became more and more apparent to the MRC officials who had to wrestle with the problem that, for the role he had created, Max was effectively irreplaceable. After a visit in 1974 Dr Julie Neale wrote: 'I certainly find it difficult to conceive someone coming in with the abilities of Perutz and conducting with his skill the harmonious management of this group of really quite tricky distinguished senior scientists.'

The Phillips committee, following interviews with staff at the lab, picked up 'several hints that Dr Perutz might have played his directorial role with such skill as to leave many of his colleagues unaware of the real extent of directorial decision-making'. One member of the governing board told them that he was unaware that it had ever taken a decision, acting as a mere rubber stamp for the divisions. Opinion in the lab was sharply divided between those such as Kendrew who felt it needed a much stronger directorial hand, and others who liked Max's laissez-faire style. Interestingly, it was the members of the Association of Scientific, Technical, and Managerial Staffs (ASTMS), the trade union to which most of the technical staff belonged, that put forward one of the most powerful arguments for the status quo:

> We are very satisfied and believe it would be difficult to improve [the running of the lab]. There is a marked lack of unnecessary paperwork, and administration problems are usually dealt with promptly and efficiently. We find that senior members of staff are easily accessible ... the present method of control and administration while being informal is effective and should be retained in future.

In the event, the views of the staff, and the successes of the past, counted for less than what Phillips and his committee viewed as the challenges of the future. Their report recommended introducing a formal management structure with a director at the top, rather than a chairman, and the name that finally emerged for this new role was Sydney Brenner's. Max had recommended Brenner, though as chairman rather than director: he had reservations about giving one person authority over policy and spending. However, he was not able to counter the turning tide. In 1978, as rumours about the succession began to fly, he published a letter in *Nature* headed 'Perutz not retired':

> Contrary to widespread belief, I have not yet retired. I shall remain chairman of this laboratory until Dr Sydney Brenner takes over on 1 October 1979. The laboratory's governing board has kindly invited me to continue working here after that date ... I ... shall be pleased to receive graduate students and postdoctoral workers who want to come to work with me.

The Phillips review led to an immediate 25 per cent cut in budget. As long as money had been less of an issue, the lab had managed to live more or less within its means. However, this cut, coming at a time when the lab had grown to such an impressive size, was more than the system could cope with. Between 1977 and 1979, Brenner took over the financial side before he himself succeeded as director. At the same time, the MRC insisted that the LMB have a more senior full-time administrator on site, which had the effect of tying up much more of the director's time and attention and distancing him from his scientific colleagues.

Brenner's experience as director was not a particularly happy one, and he left prematurely in 1986, to be succeeded by Aaron Klug. Richard Henderson, who came to the LMB in 1966 as a PhD student, took over in 1996 and handed over to his former deputy Hugh Pelham in November 2006. All three of Max's immediate successors, therefore, began their association with the lab under his chairmanship, and all credit him with creating the conditions in which this remarkably successful institution grew and flourished. Today, the LMB houses an international community of 400 staff, students and technicians. Many features have endured from the early days – such as the ritual of coffee and tea in the canteen and the lab seminars each autumn – but the old idea that you could just hire bright people and give them the money to do what they liked has long gone. Today, although an MRC staff scientist has more freedom than a university researcher living from grant to grant, funding is tightly controlled and research productivity assessed through regular reviews.

There is a feeling that the glory days, when you couldn't throw a stick down an LMB corridor without hitting a Nobel prizewinner, are now long gone. However, if you plot Nobel prizes by decade – counting prizes rather than prizewinners, as several have been shared – you find one or two in each decade since the founding of the lab in 1962, and seven in all. In addition to those mentioned earlier, in 1997 John Walker won the prize for chemistry for unravelling the structure of the enzyme, ATP synthase, that catalyses production of the energy-carrying molecule adenosine triphosphate. Since 1962, the LMB alone has generated more Nobel laureates in the sciences than several sizeable developed countries including Canada, Australia and France. As I was writing this, Andrew Fire and Roger Kornberg, American scientists who each spent time at the LMB as post docs, won the 2006 prizes for medicine and chemistry respectively. This is a phenomenal record of consistent achievement over more than forty years for one small institution.

Max was there to enjoy each new success. Still fully engaged with research as he reached sixty-five, he persuaded the MRC to let him keep some lab and office space in the building, and he found ways and means to continue funding his work. A private benefactor, the Canadian millionaire Tom Usher, gave shares in his business to Max to be held in trust for research: this generated around two dozen fellowships that went to young people throughout the lab. Max successfully applied for grants from the US National

Science Foundation and National Institutes of Health, which enabled him to pay for the assistants he needed to continue his work. Awarded a cash prize by the newly-created international research organisation Forum Engelberg, he used it to endow a fellowship for young scientists. The first beneficiary was Marie-Alda Gilles-Gonzalez, a young Haitian-born black woman scientist then working in California. She was investigating a newly-discovered oxygen sensor protien in bacteria that she found to contain a haem group. Max's offer of a post-doctoral position at the LMB was a life-line, saving her, in her own words, from 'the humiliation of accepting a fellowship that would have tagged a label of "minority" to [her] vitae'. Moreover Max personally intervened with the Overseas Labour Section at the Home Office to ensure that her work permit was issued within days rather than months, and got her travel expenses paid in advance. He never forgot that Bragg's application for a Rockefeller grant in 1938 had saved his career, and he often took the opportunity to do the same for others who were just starting out. Gilles-Gonzalez now has her own laboratory at the University of Texas Southwestern Medical Centre.

The LMB invited tributes for a book honouring Max on his retirement, and Jim Watson's captured Max's character with remarkable perspicacity:

> You have what we always search out, an eternal, intense interest in science, coupled with that marvellous appreciation that you have been allowed to do exactly what you wanted to for virtually all your adult life. And of course, I remain so completely indebted to you for your never failing encouragement at a stage when I was clearly the most unqualified person ever expected to understand a rotation picture.

At the same time, the Royal Society awarded him its highest honour: the Copley Medal. It came with a £1,000 cheque, but it was the list of previous winners that caused Max the greatest gratification. They included Michael Faraday, Charles Darwin and Albert Einstein, as well as Dorothy Hodgkin and Francis Crick. The citation is perhaps the best testimony to Max's influence on the subject of molecular biology:

> In recognition of his distinguished contributions to molecular biology through his studies of the structure and biological activity of haemo-globin, the oxygen-carrying component of the blood . . . Under his leadership the Laboratory [of Molecular Biology] has been generally acknowledged as the world's leading centre of research in this subject.

10

The breathing molecule

Has science promised happiness? It has promised truth, and the question is whether one can ever make happiness from truth.

François Jacob, Le Souris, La Mouche et L'Homme

Being chairman of the LMB ought to have been a full-time job, but Max was no bureaucrat, and had never seen himself as principally an administrator. Once the lab was up and running smoothly, his priority was to sit down at his own bench and get on with his research. He astonished a visiting MRC officer in 1965 by claiming that he reckoned to spend 90 per cent of his time in the lab rather than the office, but most of his former colleagues corroborate the statement.

What was there still to do with haemoglobin? The whole philosophy of the Cambridge unit, and now its successor the LMB, was that knowing the structure of a molecule would tell you how it worked. From the double helix, for example, it was immediately obvious how the genetic material could make copies of itself so that characteristics could be carried down the generations. So far the same was not true of haemoglobin. Everyone who came to look at the chunky, black-and-white model of the low-resolution structure would immediately ask 'But what does it mean?', and to his chagrin Max would have to reply that he didn't know: it did not tell him how the molecule worked.

In the late nineteenth century, physiologists discovered something puzzling. If haemoglobin is in an environment where the concentration of oxygen (and therefore its pressure) is high, such as the lungs, then it remains fully oxygenated. According to the standard laws of physical chemistry, as the pressure falls, the oxygen content of haemoglobin should fall only slowly until, at very low pressure, it suddenly gives up all the rest of its oxygen – but this is not what happens. Instead, haemoglobin begins to give up oxygen in the tissues and organs where the oxygen pressure is still relatively high, then gives it up more and more readily as the pressure continues to fall. Conversely, it begins to pick up oxygen relatively slowly as the oxygen pressure rises, but then grabs it more and more avidly until it is fully oxygenated once more. If it did not do this, we could not survive: the oxygen in the air we breathe would simply not reach the tissues in time to stop us from suffocating.

The graph of this process is called the oxygen dissociation curve of haemoglobin, and its S-shape implies that the haem groups in each molecule act not as separate chemical entities but *cooperatively*. As each gains or loses an oxygen molecule, it changes the readiness of the next to do the same. This was the mechanism that Max's model had to explain: how was it that the haem groups cooperated, so that, as in the Biblical parable, more oxygen would be given to those molecules that had most?

The model made it clear that the four haem groups were almost as far apart from one another as it was possible to be; 'unaware of each other's existence', as Max put it in several papers. They clearly had no direct contacts. There must be some change, he reasoned, in the interaction of the globin chains that made it easier or more difficult for the haem groups to hold on to oxygen. After all, he had long known that crystals of oxy- and deoxyhaemoglobin had different forms. On reading Max's 1938 paper in *Nature*, Felix Haurowitz had written from Prague to tell him of his latest result. He had grown big, flat, dark red hexagonal crystals of haemoglobin without oxygen. While he was looking at them under the microscope, trapped under a thin glass cover slip, he saw them spontaneously change shape to the scarlet needles of oxyhaemoglobin: the change swept across from one side to the other as air seeped under one edge of the cover slip. If the crystal changed, then each molecule must also somehow change.

There were other differences between the oxy- and deoxy- forms. Christian Bohr, the Danish physiologist (and father of the more famous Niels, the nuclear physicist), had discovered as long ago as 1904 that as the blood becomes more acidic, haemoglobin gives up its oxygen more readily. In muscles and other hard-working tissues, high levels of carbon dioxide increase the acidity of the blood, forcing haemoglobin to release oxygen and so matching supply to demand. This so-called Bohr effect is one of the many unsung wonders of animal physiology that makes it possible for cheetahs to chase gazelles and sprinters to record sub-10-second times for the 100 metres. Any structure of haemoglobin would have to explain how it worked.

Another key piece of information went back to an observation made by Michael Faraday in 1845. He found that the magnetic properties of dried blood were different from those of two of its key components, iron and oxygen: in technical terms, blood was diamagnetic while iron and oxygen were paramagnetic. In any substance all the atoms that compose it generate tiny electrical currents as their electrons whirl in their orbits, which in turn make magnetic fields. The magnetic susceptibility of the whole substance is the sum total of all these. In the 1930s, the chemist Linus Pauling followed up Faraday's discovery by asking his student Charles Coryell to measure the magnetic susceptibility of haemoglobin with and without oxygen. Coryell found oxyhaemoglobin to be diamagnetic, and deoxyhaemoglobin to be paramagnetic, implying that there had been a shift in the orbits of some of the atoms. Pauling concluded that the shift was caused by the oxygen forming a chemical bond with the iron. When Max finally understood the haemoglobin mechanism, this turned out to be a crucial factor.

Before he could understand anything about how the molecule worked, he needed to find the structure of deoxyhaemoglobin, so that he could see how it differed from the oxygenated form. This was technically difficult because of the need to keep oxygen out of the crystal, and also because horse deoxyhaemoglobin grew in crystals of a form that could not easily be analysed using crystallography. In the mid-1950s, when members of Max's group had made preliminary studies of haemoglobins from a wide range of species, the Oxford biochemist Margaret Jope had sent him a crystal of human deoxyhaemoglobin that did have the right characteristics for study. So, when Hilary Muirhead came to work with him in 1959, he suggested that she take this as her PhD topic. They solved the problem of excluding oxygen by mounting the crystal in a sealed box filled with nitrogen – one of many technical fixes created by the workshop staff at the unit. Using the methods that had been applied so successfully to horse oxyhaemoglobin, Muirhead solved the structure of human deoxyhaemo-globin in three dimensions at 5.5 Å resolution by October 1962. Computer-based methods and better data collection had made it possible for a protein structure to be solved within the usual three-year time frame for a PhD thesis, in striking contrast to the twenty-two years it had taken Max to produce the first model of haemoglobin. By the time he won his Nobel prize, the work for which he was honoured was already within the compe-tence of an able graduate student.

Although the two molecules came from different species, the arrange-ment of the four chains in oxy- and deoxyhaemoglobin, as far as could be judged at this level of resolution, seemed to be the same. There was a difference, however. The two beta chains were over 7 Å further apart from one another in the deoxy than in the oxy form. There were only two possible explanations: that it was a species difference between horses and humans, or that it represented a structural change between oxygenated and deoxygenated haemoglobin. Other researchers had by this time deciphered the amino acid sequence of human haemoglobin, and partially finished the horse sequence as well. There were differences, but none that could account for a change in structure. Nevertheless, Max could not be absolutely sure that the presence or absence of oxygen caused the change: at the time they had neither horse deoxyhaemoglobin nor human oxyhaemoglobin for comparison.

Muirhead's structure came out just before the announcement of the Nobel prize in October 1962. It meant that Max was able to include it in his Nobel lecture in December of that year, though he felt the need to apologise for the many questions that still remained unanswered: 'Please forgive me for presenting, on such a great occasion, results which are still in the making, but the glaring sunlight of certain knowledge is dull and one feels most exhilarated by the twilight and expectancy of the dawn.'

While some remained sceptical about his findings, Max found enthusiastic support in an unexpected quarter. His discovery that protein molecules could change their shape caused excitement among a group of French scientists who were tackling a fundamental question: how does the living cell make just enough of the proteins it needs and no more? They knew that some of the end products on the assembly line from genes to amino acids could turn off their own production. Taking their cue from engineering, the French scientists dubbed this process 'feedback inhibition': as Max drily remarked, this term meant 'nothing more complicated than that you stop being hungry when you have had enough to eat'. But how did the inhibition work? It might turn off the gene itself, but in some cases it seemed that the inhibitor was turning off one of the enzymes – the catalysts that drove the various reactions in the pathway. The obvious way to do this would be to occupy the active site on the enzyme – but this did not seem to be what was happening.

In 1959 a new graduate student of Jacques Monod's in Paris, Jean-Pierre Changeux, was looking at feedback inhibition of an enzyme from the bacterium E. coli. He found that the amino acid that turned off this enzyme did not occupy the active site. How could an inhibitor binding at one site make the enzyme stop acting at another site? Monod came up with the idea that inhibitory molecules could cause a structural change in the enzyme that rendered it unable to catalyse its usual reaction. He invented the term 'allosteric' (meaning 'other-shaped') to describe such a process. No one at this stage had solved the three-dimensional structure of an enzyme, let alone identified its binding sites, so he had no experimental evidence.

In August 1963 Hilary Muirhead and Max jointly published a long paper in *Nature* on their deoxyhaemoglobin results. Max had already begun to correspond with Monod about this; Monod had also heard him speak at a conference in the US, just after the Nobel prize announcement, and had invited him to lecture at the Institut Pasteur. Max set off for Paris in October 1963, delighted to have this new finding to add to his already impressive body of work on protein structure. But if Max was pleased, Monod was ecstatic: the finding was the first direct evidence that proteins could change their structure. Not a man given to traditional scientific reticence and circumspection, he announced to his colleagues that he had discovered 'the second secret of life' – the first being the double helix of DNA. Haemoglobin was the first documented allosteric protein. Of course it was not an enzyme, but this did not bother Monod: jovially he dubbed it an 'honorary enzyme'. The point was that it demonstrated that proteins could move between two different states, and that their reactivity depended on the state they were in.

Together with Changeux and another colleague, the Rome-based American Jeffries Wyman, Monod published a classic paper in 1965 that marked the official launch of the concept of allostery. The so-called MWC (Monod-Wyman-Changeux) model defined two states of an allosteric protein: the tense, or 'T' state in which the molecule was clamped in a position that made it difficult or impossible for the substrate to bind, due to the action of other small molecules known as 'allosteric effectors'; and the relaxed, or 'R' state in which these constraints were released and the substrate could bind to the protein.

With his facility for adopting simplifying descriptions, Max announced that haemoglobin was not an oxygen tank, passively filling up and emptying, but a 'molecular lung' that actively opened and closed as it absorbed and released oxygen. He still did not know what triggered the change, or how the uptake of each oxygen molecule made it easier for the next. He could only assume that somehow the oxygen set off a series of shifts in the arrangement of the atoms in the molecule, starting near the iron atom. However, the resolutions of both Muirhead's structure and his earlier oxyhaemoglobin structure (strictly methaemoglobin, but the structures were the same at this level of analysis) were too low to see small shifts in the position of atoms. The only answer was to push on to a solution of the structure at a resolution high enough to reveal the positions of the individual atoms, as had been done for myoglobin.

Once again, the task was beset with frustrating delays. Collecting data at such a high resolution would be a very tall order, involving many thousands of reflections. Fortunately, the LMB had acquired a new instrument, called a three-circle diffractometer, that would measure the intensity of the reflections automatically, without the need for photographic film or poring over spots by eye. The instrument had been designed by Uli Arndt, an old friend and former Cavendish colleague of Max's, who was Bragg's chief instrument-maker at the RI. Max had been in frequent contact with him there, and in 1963, with Bragg's retirement imminent and Arndt's colleague David Phillips planning to go to Oxford, he invited him to join the MRC lab. Arndt was delighted to return to Cambridge, and remained at the LMB for the rest of his life, developing a series of devices to make the collection of crystallographic data easier.

Unfortunately, they did need skilled operation, and the American post doc to whom Max delegated the collection of his high-resolution data did not align the diffractometer correctly. Once again, precious time was lost (Max said 'some years') before anyone noticed the problem. After that, Max decided to collect the data himself. He began by designing a device to ensure the instrument was properly aligned, and had the lab engineer Frank Mallett rebuild the 'temperamental' electronics originally supplied by the commercial firm Ferranti. The task of analysing the data would be one for an experienced programmer, but Michael Rossmann had by this time started his new life at Purdue. David Blow had impressed upon Max the need to have a more mathematically-inclined member of his team, and in 1964 he had hired Joyce Cox (now Joyce Baldwin) for a project using the low-resolution data. She, with Hilary Muirhead, would now undertake the processing of the high-resolution data Max had collected. Max himself never really responded to the intellectual challenges and opportunities computing offered to crystallographers of a more mathematical bent: growing, mounting and collecting data from crystals always remained his forte:

Arndt and Mallett helped me to keep [the diffractometer] running day and night for 15 months while I measured the intensities of some 100,000 reflections between 5.5 and 2.8Å spacing from crystals of the native protein and three heavy atom derivatives . . . Hilary Muirhead and Joyce Cox processed the data and calculated the phase angles. The final electron density map was so beautiful that I soon forgot the tedium of data collection.

The beauty of an electron density map is very much in the eye of the beholder, but even the uninitiated can admire the model of haemoglobin that Max and Muirhead built, based on the atomic positions the map revealed (it is now in the Science Museum in South Kensington). Using the same technique that Kendrew had devised for myoglobin, they constructed a forest of metre-long vertical brass rods that provided the supports for the four haemoglobin chains. They assembled these from prefabricated wire 'amino acids', each bond length and angle precisely measured, on a scale of two centimetres to the angstrom. Suitably lit, the molecule hung suspended in space: the invisible was now not only visible, but large enough to walk round and examine from every angle. Such models are far more than the static endpoint of a crystallographic study: they provide a valuable source of reference whenever questions come up about how the molecule works in practice, and how it interacts with other molecules.

The atomic model of horse methaemoglobin was completed in 1968. The same year Max's mother died. She was eighty-three, and had been suffering from Parkinson's disease: when a summer cold turned into pneumonia, she was too weak to fight it. Max, despite the differences he had had with Dely over the years, had continued to show her the utmost duty and devotion. He saw her at least once a week, as long as he was at home: Gisela was if anything even more solicitous, and Vivien and Robin also played their parts. Max was at her bedside on the afternoon of her death (though Gisela reported that he had been reading a copy of *Nature*). How he felt at losing her is not recorded. As she had been ill for some years, he undoubtedly would have seen her death as a welcome release from suffering. His own bond of affection, forged in childhood, had been sorely tested by three decades of dependence. Now the burden of responsibility was lifted at last.

By 1970 Max had successfully obtained high-resolution structures of human and horse deoxyhaemoglobin. Only by comparing the fine detail of the oxy- (or met-) and deoxy-forms could Max finally have the answer to the question that had been in his mind all this time: how did the structure of haemoglobin make it such a superb oxygen carrier?

He focused his attention on the iron atoms: after all, it was the uptake of oxygen that seemed to trigger the switch from one structure to the other, and oxygen interacted with iron. Each iron atom sits in the centre of a disc of atoms (known as a porphyrin) made up of four smaller rings, each with a nitrogen atom pointing towards the centre. The porphyrin plus the iron atom is the haem group. The haem sits in a pocket formed on the outside of the globin chain, connected to it via bonds with an amino acid called a histidine. When Max and his colleagues came to study the position of the iron atom in deoxyhaemoglobin, they found that instead of sitting neatly within the four nitrogen atoms around the centre of the haem, like a planet orbited by its moons, it bulged out towards the histidine. They also looked at the tails of the alpha and beta chains, and found that, while in oxyhaemoglobin these could move freely, in deoxyhaemoglobin they were bent round and locked into place by bonds (that Max termed 'salt bridges') with nearby amino acids.

In order to understand what was happening to the iron atom, Max had to learn some unfamiliar chemistry. According to quantum theory, the same atom can exist in one of two different 'spin states', which are influenced by the bonds it makes with other atoms. Iron bound to oxygen is in a low spin state, while oxygen-free iron is in a high spin state. The difference in spin state accounts for the changes in magnetic properties found by Pauling and Coryell in the 1930s. In the 1950s, the Oxford chemist Bob Williams discovered, using spectroscopy, that spin state changes would affect the lengths of the bonds between iron and other atoms, as well as changing the atomic radius. Max heard Williams speak at a conference in Switzerland in 1961: in the course of his talk, Williams proposed that such changes might underlie cooperativity in haemoglobin through causing shifts in the protein chains. Max met him afterwards and asked him to explain the theory over dinner. A little later, Williams visited him in Cambridge and gave a lecture there on his findings, but Max appeared reluctant to accept the idea without crystallographic evidence. In the late 1960s, the American chemist J. Lynn Hoard, at Cornell University, made beautifully detailed crystallographic studies at high resolution of haem-like compounds and discovered that the spin state did indeed affect their conformation. High-spin iron atoms had a larger radius than low-spin atoms, which explains why the iron in deoxyhaemoglobin was squeezed out of the plane of its encircling rings: Hoard predicted that the difference between the two could act as a trigger.

Consulting Hoard's tables, Max realised that the difference between the two spin states would mean that in deoxyhaemoglobin the iron atom would bulge from the plane of the ring by 0.75 Å instead of 0.3 Å as in oxyhaemoglobin. Bob Williams had told him, furthermore, that theoretically the bond the iron atom made with the histidine on the globin chain would be stretched in the high spin (deoxy-) state by one or two tenths of an angstrom. 'I realised then,' he told the author Horace Judson later that year, 'that this difference of about half an angstrom would just about suffice to provide the long sought-for trigger.' (Half an angstrom. One-twentieth of a nanometre. One twenty-*millionth* of a millimetre. We need reminding, as Max looks at his metre-high models, that we are dealing with the very, very small – but on the scale of atoms, half an angstrom is a significant distance.) In a burst of concentrated activity, Max wrote a series of papers for *Nature*, the most important of which proposed a mechanism that explained the remarkable oxygen-carrying properties of haemoglobin.

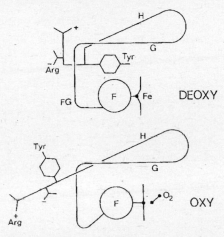

Max's diagram showing how the position of the iron atom (Fe) changes when oxygen (O_2) binds to it, leading to a relaxation of the globin chain

Thereafter known as the 'Perutz mechanism', it explained the molecule as a spring-loaded device, tripped from one state to the other by the seemingly weak trigger of an oxygen molecule attaching to the iron in one or more of the four subunits. For each iron atom that combined with oxygen, he suggested, tension at the haem would be released and one pair of salt bridges would break, successively releasing the constraints that prevented the remaining subunits from taking up oxygen. At some point on its journey through the high-oxygen environment of the lungs, the whole molecule would flip from the tense to the relaxed state, avid for oxygen and ready to make the circuit back to the muscles and organs with its precious load. The making and breaking of one particular pair of salt bridges near the end of the beta chains, he also argued, accounted for the Bohr effect by absorbing and releasing hydrogen ions.

In his more popular accounts, Max used a very Bragg-like metaphor to describe the surprising power of an infinitesimal change in the radius of the iron atoms: 'four fleas that make an elephant jump'. For him it was the triumphant conclusion to the quest which he had begun in 1937: finally he knew not only how a haemoglobin molecule was constructed, but how it worked.

The cooperative mechanism outlined here explains most of the important properties of haemoglobin, even though stereochemical details are still lacking. It is of a kind that could not have been guessed without knowing the structure of the two forms in atomic detail ... Even then it could be worked out only by combining the structural information with many other results.

It is remarkable that there should be such an exceedingly complex, subtle and elegant instrument of respiratory transport.

This paper was a milestone, perhaps even greater in its way than the work that had won him the Nobel prize. Aaron Klug, who was Director of the LMB from 1986–96, describes it as 'one of the most important [biological] mechanisms in the world'. 'Max was the first in history to allow us to look at a molecule as a machine – it is as big a step as understanding the planets,' says Guy Dodson, a crystallographer who solved the structure of insulin with Dorothy Hodgkin at Oxford, and subsequently worked on a number of molecules including haemoglobin in the lab he headed at York. Choosing a felicitous phrase for a dedicated striver for heights both literal and figurative, David Blow described the paper as 'the pinnacle of Perutz' achievements on haemoglobin'. His description of bonds made and broken, of components strained and relaxed, and of structures clicking from one conformation to another, turns this tiny molecule, out of reach of human vision and touch, into a mighty work of engineering. Jacques Monod once said that he arrived at his allosteric theory by imagining himself to be an allosteric protein. Max may not have gone so far, but he certainly must have imagined himself walking round a haemoglobin molecule, examining it from every angle and feeling its power.

Sir Lawrence Bragg, who remained Max's staunchest supporter, was understandably delighted to have lived to see his protégé's crowning success. After his operation for prostate cancer in 1962, he remained alert and active for a further nine years, even touring India to give a series of lectures. Less than a year after Max published his mechanism, Bragg went into hospital for a further operation, at the age of eighty-one, but this time did not pull through. With his death in July 1971, Max lost a true friend and the man he described as his 'scientific father'. They had been two of a kind in their delight in science for its own sake. Max wrote genuinely affectionate and admiring obituaries in *Nature* and *New Scientist*: in another, later article he wrote:

When reviewing scientific work I sometimes paraphrase people's papers, but when I tried to paraphrase Bragg's I always found that he had said it much better. Bragg's superb powers of combining simplicity with rigour, his enthusiasm, liveliness and charm, and his beautiful demonstrations conspired to make him one of the best lecturers on science that ever lived . . . [H]is approach to science was an artistic, imaginative one . . . Nowadays cynics want us to believe that scientists work only for fame and money, but Bragg slaved away at hard problems when he was a Nobel Laureate of comfortable means . . . So often men of genius are hellish to live with, but Bragg was a genial person whose creativity was sustained by a happy home life.

Max seemed to be writing the epitaph he would have liked for himself.

In 1971 Max was fifty-seven, theoretically due to retire as chairman of the LMB in three years' time. He would have been justified in thinking that with the oxygen-carrying mechanism of haemoglobin finally solved, he could turn his attention to new projects – but he was in for a shock.

The evidence for the stereochemistry of the cooperative mechanism seemed so obvious and convincing that I expected everybody to accept it, but it failed to convince even my closest colleagues, let alone others in the haemoglobin field.

As well as Bragg's, he had the backing of Crick and Monod – not exactly a lightweight crew – but that wasn't enough. The fact that in the wider haemoglobin literature his idea became 'controversial' he took as a personal affront, and so he set about chasing down every objection. It took him the best part of two decades, at the end of which, as he ruefully remarked, 'interest had long faded'. There were many skirmishes in what Max himself called 'the Haemoglobin Battles': his dogged determination to come out on top, and his unshakeable belief in the correctness of his position were absolutely characteristic.

In June 1971, Jim Watson invited Max to participate in a Cold Spring Harbor Symposium on the subject of protein structure, the first return to this subject since 1949 for the institution that had done so much to promote molecular biology. To his chagrin, Max's presentation on the stereochemistry of cooperative effects in haemoglobin did not provoke the universal acclaim he was clearly hoping for. One after another, participants walked up to him afterwards to raise objections.

After returning home, I complained to Bragg that I would have to spend years disproving these objections; he replied a few days before his death that I should not waste time with them, but forge ahead with new work.

In the event, new objections raised their heads like those of the Hydra as soon as I had decapitated the old.

Challenges like this are all part of the normal rough-and-tumble of science, a winnowing of ideas that ensures that only the most solid results endure. Max would not have been used to it, however: within the limits of the resolution, crystallographic studies give unequivocal results that are unlikely to provoke dissent. Now he had moved into territory occupied by biochemists to whom things look much less black and white: biochemistry studies dynamic processes, reactions and cycles and exchanges that take place ceaselessly in the living cell, rather than static structures. Max came to suspect that all his opponents acted out of bitterness that his results contradicted theirs, writing to Robin apropos of Einstein's failure to accept quantum mechanics:

> I have found so often that failure to discover the right solution turns scientists into cranks: that is to say they do not have the strength of character to admit that they were wrong or that someone else succeeded where they failed. Rather than admit it, they withdraw into life-long opposition to the new advance and so cut themselves off from the mainstream of scientific thought.

Max himself was certainly not going to withdraw. Throughout the 1970s he was driven more than ever by the desire to find answers. Vivien remembers that all his life, whatever he was doing – riding his bicycle, going for walks, dead-heading the roses – his mind would always be occupied by the latest problem. How else might he attack it? What might undermine his conclusions? Who might be able to help? His approach was that of Isaac Newton who, when asked how he made discoveries, answered: 'By always thinking about them. I keep the subject constantly before me and wait until the first dawnings open little by little into the full light.'

The outcome was a series of papers, using not only crystallography but also evidence from various different spectroscopic techniques, that between them clarified the way the haems interacted and documented the action of several 'allosteric effectors': small molecules that interacted with haemoglobin and made it more or less likely to switch between the T and R states. Robin Perutz is now a professor of chemistry at the University of York: he had many discussions with his father during this period and says that Max's example is one he has tried to follow in his own work: 'Max brought a chemical insight that was remarkable, together with a flexibility in his willingness to bring many techniques to bear on a single problem . . . He proved himself to be a finisher and willing to fight for his theories.'

Despite these advances, others who were less interested in the structure and more in the dynamic relationships among atoms in a molecule felt that his account was still too rigid to be plausible. Real atoms don't keep still: where any one of them decides to be at any time depends upon the sum of all the electrical charges in its vicinity as well as other factors such as temperature.

From the early 1970s, Robert Shulman and his colleagues at Bell Labs had been arguing that the tension at the haems that Max had proposed was unconvincing as a trigger mechanism. In 1976 Shulman told Max that he now had evidence from a new technique, X-ray absorption fine structure spectroscopy (EXAFS), that the distance between the iron atom and the nitrogens in the haem group changed too little to trigger the switch from T to R. Max decided that 'the only way to prove [Shulman] wrong was to repeat his work myself'. Max prevailed on an American friend to help him repeat the experiment (though the friend agreed rather reluctantly, and only if his name was kept off any publication that came out of it) at the Stanford Linear Accelerator in California. After all that, Max found that his EXAFS measurements were exactly the same as Shulman's. However, when he and his colleagues looked closely at what they had both measured – the distance from the iron atom to one of the nitrogen atoms in the surrounding ring – they realised that this distance alone, which represented one side of a triangle, could not determine the distance that was crucial to Max's theory: the displacement of the iron from the middle of the ring. This is something he could have worked out from looking at Shulman's data alone, without the need to rush off and repeat the experiment, but nevertheless he published his own results with a strong statement to the effect that Shulman's interpretation had been wrong.

The only way finally to resolve the question of how far the iron moved was to take the structural studies of haemoglobin to even higher resolution. Synchrotron radiation, a high-energy form of X-ray beam produced by an accelerator, made that possible. Beginning in 1981, Max undertook X-ray studies of his human deoxyhaemoglobin at the French synchrotron facility LURE, near Paris, and was able to raise the resolution of his images to 1.74 Å, high enough to 'see' individual atoms. Once again, he was doing the hard work of data collection with his own hands, assisted by his Israeli post doc Boaz Shaanan.

In exhausting 24-hour shifts, we took 211 1.1–1.5° oscillation photographs with the Arndt-Wonacott oscillation camera, which required changing triple film cassettes every 20 minutes. By day Shaanan and I did it together and during the nights we worked in alternate two-hour shifts . . . [after one long day] I was so tired that I could no longer perform the simple task of marking each pack of films at its four corners . . . and of labelling the films correctly. I was 67 by then and felt it.

The result confirmed that the iron atom was displaced in deoxyhaemoglobin, but by a rather smaller distance than Max had claimed in his 1970 paper – 0.4 Å rather than 0.75Å in the alpha haem. This brought it just within the range of possible errors in the EXAFS measurements, but was still enough, Max judged, to act as the initiator of the switch from one state to another. Hatchets were more or less buried when Max and Shulman published a joint paper in 1987, together with Max's colleague Giulio Fermi, reconciling the results from the two methods and acknowledging that Max's original criticism of Shulman's extrapolation was 'perhaps too harsh'.

Another dispute lasted even longer. In April 1980, the American biochemist Chien Ho, from Carnegie Mellon University in Pittsburgh, organised a symposium at Airlie House near Washington DC on the role of iron in proteins. Ho, who had previously been a co-author on some of Max's spectrosopy papers, convened the symposium in part to honour him on his retirement from the LMB. Shortly before it began, he told Max of new results (using a spectroscopic technique, nuclear magnetic resonance, that measures the environment of hydrogen atoms in a protein) that did not support his conclusions on the Bohr effect.

Max, John Kilmartin and several colleagues had published a paper as long ago as 1969 that attributed a large part of the Bohr effect to the making and breaking of a particular 'salt bridge'. This paper held a special place in Max's affection, coming even before his landmark 1970 paper on the switch between the oxy and deoxy structures. Now Ho had told him that the importance of this salt bridge depended on the experimental conditions, particularly the kinds of salts in the solution surrounding it. Max was not best pleased. He came to the meeting in combative mood: one participant remembers him standing throughout Ho's presentation of his work (he found sitting painful because of a back problem) and voicing challenging questions as soon as it was over. On his return to Cambridge, he plunged into a new round of experiments, once again linking up with specialists in the field of his challenger so that he could make his own measurements. In a typical comment to his sister Lotte, written four years after the Airlie House meeting, he wrote:

> I remonstrated that [Ho's] experiment must be wrong, because my mechanism is supported by overwhelming evidence, but he would not listen. Scientists at large have no judgement, with the result that my mechanism has now become 'controversial'. I therefore decided to repeat Chien Ho's experiment myself with the help of two young people in London ... We soon discovered that Chien Ho's work was nonsense, as I knew it must be, but it was essential to nail down the nature of his error.

Max was right that Ho had made a mistake: but it turned out that his own colleagues had made a similar error. Ho and his team eventually resolved that particular issue in 1991, but the dispute did not end there. Such scientific differences of opinion are not unusual: what was unusual was the nature of Max's response. On one occasion he published a paper refuting unpublished results that Ho had shown him: a breach of scientific etiquette, as it meant readers were unable to make up their own minds. Another time, he even wrote to the editor of a leading journal to ask him (unsuccessfully) to block publication of one of Ho's papers.

Ho remains saddened and baffled by the whole episode. As he wrote to a colleague,

> It is really difficult for me to understand how a scientific disagreement has turned into such an emotional issue for him. Looking back over the past 17 years, both Max and I have made a few conclusions, based on experimental results available at that time, which turned out not to be supported by subsequent findings. This is what science is supposed to be, but he took our disagreements very personally.

Several friends, including John Edsall and even his own son Robin, tried to persuade Max that he had misjudged Chien Ho, but without success. While both Shulman and Ho still differ from Max in the importance they accord to purely structural evidence, there is no evidence that their work on haemoglobin was anything other than a perfectly valid part of the collective effort to understand how this fascinating molecule works.

Max's desire to have the last word led to a bizarre development. In 1997, he published a volume of his selected papers, entitled *Science Is Not a Quiet Life*. A whole chapter is devoted to what he called 'the Haemoglobin

Battles'. Its introduction outlines the challenges that were posed to his model, and relates how he beat them down one by one and emerged victorious. The chapter includes more than one accusation of influence being brought to bear on junior scientists not to carry out experiments or publish results that might support him, several allegations of scientific incompetence and one of outright fraud. Some of those who might wish to tell their own side of the story have found themselves facing a peculiar difficulty: he has given most of them fictitious names. The true identities of 'Smith and Brown' (Ho and his colleague Irina Russu), 'James Lauterbrunner' (Robert Shulman) and one or two others emerge after only a few moments spent cross-checking this introduction with the published papers both included in the book and otherwise cited: yet Max offers no note of explanation, or even an admission that he has used pseudonyms.

Most of those obscured in this way he regarded as his 'scientific enemies', but others are friends such as the man he called 'Jacob Pinder' who with his female assistant 'Jane' helped with the EXAFS experiment (who cannot be identified from the literature as they did not co-author the subsequent paper). Some time before, in his correspondence with John Edsall at Harvard, Max referred to the difficulty of being frank in one's memoirs about scientific conflicts with others that had a 'strong personal element'. In a characteristically naïve way, he seems to have resolved the difficulty by not naming names. Quite apart from the fact that all but two of these fictitious characters can easily be identified from the published papers, it seems not to have occurred to him that to attack people in this veiled fashion effectively denied them any right of reply. 'I hate fights and would have been happier without them', Max concluded at the end of this chapter. He fought harder than anyone, however, and left some of his opponents bruised and embittered by the experience. Chien Ho, who surprised a lot of those who knew of their differences when he attended Max's memorial celebration, now says: 'Max was truly outstanding in establishing structural biology and in leading the MRC Laboratory of Molecular Biology . . . He was a very warm person, but with a very complex personality . . . It was difficult to work in his field (hemoglobin) if your results did not agree with his thinking.'

What is particularly sad is that the acrimony that Max felt towards his critics was really unnecessary. Their conflicting findings spurred him to devise a range of imaginative experiments, boldly deploying a wider range of techniques than he had previously, that defined for all time the nature of haemoglobin's stereochemical mechanism. Although he

Max and his colleagues outside the Hut, late 1950s

(*Below*) X-ray diffraction photograph of oxyhaemoglobin

(*Below right*) John Kendrew's 'visceral' model of myoglobin

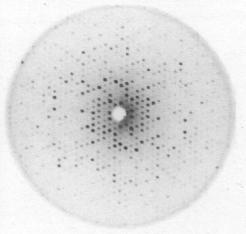

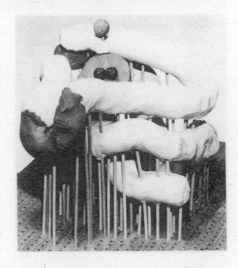

Christmas card from Gisela and Max to Gisela's brother and sister-in-law, 1959, with photo of the first haemoglobin model: 'this is our Happy Christmas'

Max describes his model to Raymond Baxter while John Kendrew looks on, in the *Eye on Science* programme 'Shapes of Life' televised by the BBC, 1960

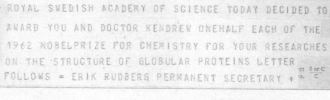

Telegram from Stockholm, 1 November 1962

Max, Gisela and John Kendrew celebrate the announcement of the Nobel prize

Nobelists gather in Stockholm before the ceremony in December 1962: Maurice Wilkins, John Steinbeck, John Kendrew, Max, Francis Crick and James Watson

Nobel medal

Gisela and Max at the Nobel ball

Max explains the structure of DNA to the Queen at the official opening of the LMB in May 1962

The governing board of the LMB in the mid 1960s: John Kendrew and Francis Crick (standing), Hugh Huxley, Max, Fred Sanger and Sydney Brenner (seated)

The newly-built LMB

Sanger, Brenner and Max celebrate Sanger's second Nobel in 1980

Max with his first high-resolution model of haemoglobin, 1968

After he retired as Chairman, Max kept an office at the LMB, the wall hung with portraits of his scientific heroes

Don Abraham, who worked with Max to find a treatment for sickle cell disease, with Max in the LMB canteen

Kiyoshi Nagai, the first to make genetically-engineered haemoglobin, with Max's model

Max's poor health often improved when he was able to relax in sunny locations in the US

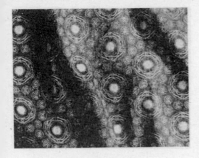

Designs for a dress fabric and a plate based on Max's first Patterson maps (bottom panel), made for the Festival of Britain in 1951

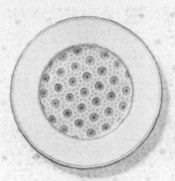

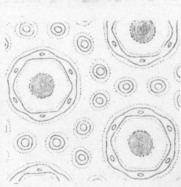

(*Above right*) Max greets Pope John Paul II in 1983: he was elected a member of the Pontifical Academy in 1981

Asked to nominate his favourite painting for a National Gallery newsletter, Max chose *The Agony in the Garden* from the school of El Greco

MAX PERUTZ · LECTURE THEATRE 2002

In science truth always wins MFP

Max gave enthusiastic and inventive public lectures – here he holds a model of a protein chain – until well into his eighties. (*Inset*) The LMB named its new lecture theatre after Max in 2002: it bears a plaque carrying one of his favourite sayings eventually came to acknowledge that some of his initial ideas were over-simplifications, in its essentials his mechanism was largely vindicated. With Guy Dodson, Max published his final haemoglobin review paper in 1997 (he was eighty-three at the time), assessing his mechanism in the light of all the work done since 1970. In Dodson's view, it stands up magnificently.

What Max said was that the T-R transition occurred when oxygen bound to the deoxy haem: each binding event put a strain on the iron stereo-chemistry and particularly on the iron-histidine bond ... so that the tension destabilises the T state. This is beautifully true for the alpha subunit; it is more complicated in the case of the beta subunit, but still true.

Perutz's Mechanism of Haem-Haem Interaction

Cartoon drawn by Max's colleague Kiyoshi Nagai, illustrating the shift between 'tense' and 'relaxed' states of haemoglobin

Max himself had to concede that his mechanism did not hold all the answers. The review concludes with a niggling list of observations on the molecule that still can't be explained: in particular, some of his own students and post docs found that there was no straightforward structural explana-tion for the precise relationship between the uptake of oxygen and the release of hydrogen ions when salt bridges broke. Nevertheless the 'Perutz mechanism' or 'MWC-Perutz model' is a fixture of the scientific textbooks, thanks to his formidable determination and willingness to engage on all fronts. Ken Holmes, who came to the LMB with Aaron Klug in 1962 and left in 1968 to found a biophysics laboratory at the Max Planck Institute for Medical Research in Heidelberg, is unequivocal in his admiration for Max's bold stance: 'One of Max's great offers to the field was that you can use this kind of mechanical thinking in protein structure in a predictive and interesting way. And he was proved dead right, in spite of lots of people trying to prove he was wrong.'

Health and disease

*Men that look no further than their outsides, think health an
appurtenance unto life, and quarrel with their constitutions for
being ill; but I, that have examined the parts of man, and know
upon what tender filaments that fabric hangs, do wonder that we
are not always ill; and considering the 1000 doors that lead to
death do thank my God that we can die but once.*

Sir Thomas Browne in *Religio Medici*, 1643

Ill-health was both an impediment and a spur to Max's research. His own
poor state of health frequently interfered with his capacity to work: at the
same time, he held fast to the vision that molecular biology would one
day help to provide an explanation of human vulnerability to disease and
open up new routes to treatment.

The notion that diseases can be inherited is an old one, older than the
understanding of genetic inheritance that began to develop with the
nineteenth-century monk Gregor Mendel's studies of pea plants. There are
rabbinical texts dating back to the second century AD that show that their
writers understood that a tendency to bleed excessively – the condition
later named as haemophilia – might run in families, and that therefore
boys in affected families might be excused circumcision. However, it was
not until the middle of the twentieth century that anyone made a link
between an inherited illness and a faulty protein molecule.

In 1949 Linus Pauling published a paper entitled 'Sickle cell anemia
– a molecular disease'. Sickle cell anaemia, or sickle cell disease, is an
inherited condition affecting mainly people of African origin, in which
the flexible, disc-shaped red blood cells roll up, as they lose oxygen, into
rigid tubes or sickle shapes that block the fine capillaries in the circulatory system. Sufferers experience painful 'crises' at regular intervals and die
young because these blockages gradually starve their organs of oxygen. At
that time, most people thought the sickling must be caused by something
outside the cells. Pauling and his colleagues discovered that there was a

small difference between the electrical charges on molecules of normal haemoglobin and those of haemoglobin from sickle cell patients. They made the bold and startling claim that a tiny change in the structure of a protein molecule could be all that was necessary to cause a serious disease. Moreover, they proposed that the sickling occurred because a change on the surface of the molecule caused it to form a weak bond with the molecule next door, so that eventually 'large aggregates' of molecules would be formed in the cell, distorting its shape. Pauling's hypothesis has proved to be entirely correct.

Max's friend Fritz Eirich told him about the paper, and sent him a sample of sickle cell haemoglobin, while another contact in the US sent him some whole sickle cells. Working with Murdoch Mitchison from the zoology department, Max showed that in the absence of oxygen the cells refracted light under a polarising microscope in exactly the same way as deoxy-haemoglobin crystals. They concluded that the sickle cell haemoglobin was crystallising in the cells as it lost oxygen, and returning to solution when oxygen was present.

They quickly published their result in *Nature*. To Max's chagrin, Pauling reacted not with pleasure that his work had been confirmed, but with irritation: he felt that Max's paper might lead readers to think it was all Max's own idea. Max felt obliged to write a conciliatory letter:

> I am very disappointed that you should have been annoyed with our paper, particularly because all the new experimental evidence we report seems to fit in so beautifully with the basic ideas set out in your first paper . . . I am sorry that this misunderstanding between us should have arisen, particularly as I have spent much effort trying to convert unbelievers to your scheme.

As he would be two years later with the alpha helix, Max had been inspired by Pauling's creative thinking to conduct a simple experiment that brilliantly illustrated the correctness of the older man's ideas. As a consequence, he developed a new interest in abnormal haemoglobins, both for what they could tell him about disease, and for what they revealed about the workings of the normal molecule. His early explorations in this area helped to usher in the era of 'molecular pathology', the study of the origins and mechanism of disease at the level of DNA and proteins.

Once again there was an early setback. The year after his *Nature* paper came out, a group of scientists he was visiting in Paris disputed his finding on the grounds that they could not obtain a crystalline X-ray diffraction pattern from sickle cells. Max later related that he had 'left the laboratory in shame' at having published a finding that could not be substantiated. Some years later, a young medical researcher, Tony Allison, came to spend a year at the Cambridge lab. Allison had been the first to demonstrate that the reason the sickle cell gene persisted in African populations was that children who had only one copy of the faulty gene, and therefore a mixture of normal and sickle cells, were much more resistant to the malaria parasite than those who had entirely normal blood cells. Now he wanted to study sickle cells using X-ray methods. But he too failed to get convincing diffraction patterns.

Allison's brief sojourn in the lab had a very important consequence, however. After the biochemist Vernon Ingram had succeeded in making mercury derivatives of haemoglobin for Max, he was pursuing studies of the molecule by using enzymes to break it up into smaller pieces, the method that Fred Sanger had developed for sequencing proteins. Nothing of great significance emerged, but when Allison departed, leaving his samples of sickle cell haemoglobin behind, Max and Francis Crick suggested that Ingram might try using his methods to compare normal and sickle cell haemoglobin. Rather than trying to sequence the whole protein, which would have taken years, he broke each chain into just a couple of dozen pieces each. Spreading the pieces out according to their electrical charges on a piece of damp filter paper provided a 'fingerprint' for each haemoglobin sample – and the fingerprints immediately showed that the difference between the two came down to a single amino acid. Max was delighted. At a time when his own work on the haemoglobin structure was going frustratingly slowly, he was able to report a major triumph for the lab to the Rockefeller Foundation: 'This is the first discovery of a specific chemical difference between two proteins of the same species differing by the mutation of a single gene. This discovery has opened up an interesting new field on the border line of protein chemistry and genetics . . .'

When Max had first arrived in Cambridge in 1936, the Secretary of the University's Board of Research Studies introduced him to a young German doctor, Hermann Lehmann, over lunch in Christ's College. In 1940 the two of them shared the iniquities of internment as 'enemy aliens' at Huyton, though Lehmann was not deported to Canada and his friends obtained his release by October of that year. Lehmann joined the Royal Army Medical Corps as a pathologist, and was posted to India. After demobilisation, he took up a medical research post in Uganda, specialising in malnutrition and anaemia. There he first came across sickle cell anaemia and some of the other diseases caused by abnormal haemoglobins that were to make his subsequent career.

By the time Vernon Ingram discovered the amino acid substitution in sickle cell haemoglobin in 1956, Lehmann had returned to England and was a senior lecturer in chemical pathology at St Bartholomew's hospital in London. Introduced by Max, the two quickly began to collaborate, Lehmann first supplying sickle cells from different patients to confirm the first result, then offering more types of abnormal haemoglobin – which he avidly collected on trips to the ends of the earth – for the same kind of analysis. When Ingram moved to MIT, he continued the work there, but the renewed link between the MRC Unit and Lehmann's lab was not broken. Lehmann came to Cambridge as Professor of Clinical Biochemistry in 1963. Five years later, Max and Hilary Muirhead built the first atomic model of haemoglobin, and it became possible to see in three dimensions how one or another amino acid substitution would affect the way the molecule worked.

Max invited Lehmann over one day – he worked in Addenbrooke's Hospital, just across the way from the LMB – to have a chat about possible links between disease symptoms and the positions of the 'mistakes' in the variants of haemoglobin he had collected, of which there were now more than a hundred:

> My day with Lehmann became one of the most interesting and satisfying in my scientific career. One by one the positions of the substitutions explained the abnormal properties and clinical symptoms. It was the first time the causes of human disease could be seen at the atomic level.

The two of them published a table giving the amino acid substitutions, abnormal properties and clinical symptoms of thirty-four variants of haemoglobin, explaining each in terms of the structural alterations the rogue amino acid would cause. That 1968 publication in *Nature* was a landmark in the field: according to the Oxford haematologist Sir David Weatherall, 'very few new concepts have been established which are not to be found in this paper'. It was really the curtain-raiser for the era of molecular medicine, which today represents a substantial proportion of all medical research.

Using a new and very sensitive instrument for measuring oxygen affinity, invented by Kiyohiro Imai and built for Lehmann by his colleague, Hideki Morimoto, they went on to study the mutations in more detail. They found that alterations to amino acids on the outside of the molecule had little or no effect (with the exception of the sickle cell mutation). For those on the inside of the molecule, at the interfaces between subunits or around the haems, it was a different story: these could destabilise the molecule in either its oxygenated or deoxygenated form, so that it would no longer function effectively as an oxygen carrier. Human carriers of mutations causing these changes would be mildly to severely or even fatally ill, depending on which mutation they suffered from and whether it affected one or both copies of a gene.

Even at this very early stage in the history of molecular pathology, Max believed it was important not to oversell the significance of his findings for medical treatment.

> The data presented here do not hold out any hope that the lesions in mutant protein molecules could be repaired directly; repair at the genetic level . . . is imaginable but still Utopian . . . At present the only hope lies in preventing the conception of homozygotes [individuals receiving the same mutation from both parents] or even of heterozygotes carrying strongly pathogenic mutations.

Genetic testing and counselling for couples who are at risk of carrying a genetic disease remain the first line of defence for preventing ill-health among future generations. Gene therapy, though many have cited it as the ultimate goal of research into such diseases, has proved to be extremely difficult to implement in practice. Repairing genes, however, is not the only possible route to treating a molecular disease. To a chemist, a faulty protein is a chemical problem that might have a chemical solution.

With sickle cell haemoglobin, it was still not clear by the end of the 1960s how Pauling's proposed aggregation of molecules worked. The breakthrough came in 1972, when Beatrice Magdoff-Fairchild at Columbia University, who had trained in crystallography with Lindo Patterson in the 1930s, obtained the first successful X-ray diffraction pattern from crystals of sickle cell haemoglobin. It turned out that neither Tony Allison nor the French workers had been careful enough to exclude oxygen, which made the molecules go back into solution, and Max's 1950 paper had been right after all.

Max immediately started a collaboration with Magdoff-Fairchild's laboratory, and rushed out a paper on the structure of sickle cell haemoglobin. The Columbia group had found that the molecules aggregated into long fibres. Max announced that these consisted of six strands of individual molecules, each like a string of beads, winding slowly round a hollow core. In fact he was too hasty: by the end of the 1970s, it was clear that there were fourteen strands per fibre, and no hollow centre. Max found it hard to believe that nature would use such an unlikely number, but he eventually had to accept it in the face of overwhelming evidence.

In 1971 President Richard Nixon, partly in an effort to improve his standing with black voters, announced to Congress that sickle cell anaemia would now be a 'priority disease'. In the years that followed, and in the light of the advances made by Max and others, funds for research began to flow. One of those charged with distributing this money was Alan Schechter of the National Institutes of Health in Bethesda, Maryland. He dates his interest in haemoglobin to the lectures Max and John Kendrew gave in New York in 1962, just before the Nobel ceremony: then, a few years later, he saw one of Max's first stereo projections of the high-resolution structure. Max's paper on the structure of sickle cell fibres started him thinking that it might be possible to design a drug that would stop molecules of sickle cell haemoglobin from sticking together: 'Even though the structure was wrong it revived the field . . . The sickle cell money came along in part because of that work. People began to feel that maybe something could be done.'

At Schechter's invitation, Max became an adviser to the US-funded programme, a role that kept him up to speed with the latest research from around the world. In his own lab, Schechter soon found a peptide, or small protein fragment, that would block sickle cell haemoglobin from crys-tallising in a test tube. However, there were all sorts of problems with making it work in a living body: getting it to pass through the red cell membrane was one, and another was the enormous quantity that would be needed – at least one peptide for every one of the 1.5 million mol-ecules of haemoglobin in each red cell, if not many times more – in order to compete strongly enough for binding sites to block the crystallisation. 'In research you work at proof of principle,' says Schechter. 'If you always think about the problems ahead, you never make the first step on the journey.'

Others were also making their first steps in the same direction. At the same 1980 Airlie House conference in Washington where Max began his 'battles' with Chien Ho over the Bohr effect, he met Donald Abraham, a chemist from the University of Pittsburgh. In the early 1960s, Abraham had worked with another distinguished Viennese-born chemist, Alfred Burger, who had been one of the founders of the field of medicinal chem-istry in the US. From that time, when the first protein structures began to be solved, he had made it his ambition to design drugs based on what was known of the structures of target molecules. It was an entirely new idea at the time. Drug development had typically involved a large degree of hit-and-miss serendipity, perhaps most famously when Alexander Fleming discovered penicillin through failing to wash up his laboratory glassware. From the mid-70s, Abraham had made sickle cell haemoglobin the target of his research, as haemoglobin was the only protein that had both a well-characterised structure and a known link to disease. Unable to persuade the National Institutes of Health to fund him, he had obtained a small grant from one of Pittsburgh's star baseball players, Willie Stargell, who had organised fellow black players to raise funds for sickle cell research. However, Abraham was isolated and making little progress until he booked himself a last-minute place at the Airlie House symposium.

Buttonholing Max between sessions, he enthused to him about the possibilities he saw for designing antisickling drugs. Within a few minutes, Max had invited him to come to Cambridge and collaborate with him. Abraham was ecstatic.

> My spirits rose to the heavens. I was invited to the best laboratory in the world, not with recommendation letters or CV, but on an idea. This I learned was a hallmark of Max's genius. Good ideas, whether his or others, were treasured. Formalities were not to stand in the way . . .

Between 1980 and 1988 Abraham made sixteen visits to Cambridge. The routine was that he and his colleagues in Pittsburgh would make trial compounds and test their antisickling properties; he would then fly to Cambridge with crystals of haemoglobin with the agent attached; Max would collect X-ray data on them and Abraham would run the programs that analysed the results. These studies of how small molecules bind to proteins were another pioneering step taken in Max's lab that is now fundamental to the search for new drugs.

Unfortunately, however – in the vast quantities that would have to be delivered to the red cells – all the agents they tried were too toxic to use in humans. Abraham decided to take another route: to start with drugs that had already been approved as safe for other human diseases and see if any of them had antisickling properties. The most promising was clofibrate (prescribed to reduce levels of various fats in the blood), but, though it had a moderate effect on pure sickle cells, too much of it got mopped up by serum albumin, which Abraham describes as the 'garbage truck' of the blood that goes round picking up anything that might be toxic.

In the course of their studies, Abraham, Max and a French colleague, Claude Poyart, discovered that clofibrate and a related drug, bezafibrate, lowered the oxygen affinity of haemoglobin so that it gave up more oxygen than it would normally. Poyart pointed out that this might be useful in other areas of medicine, such as improving the delivery of oxygen to damaged heart muscle after a heart attack: but bezafibrate had the same problem as clofibrate, in that serum albumin grabbed it before it could get into the red cells.

Despite the setbacks, Max was invigorated by this new line of research, and Pittsburgh became one of the regular stops on his (at least) annual trips to the US. To his great satisfaction, the University of Pittsburgh appointed him to a visiting lectureship, from 1983 to 1986, that substantially enhanced his income for a commitment of two or three seminars per year. But, although scientifically much of this work was proving very interesting, the prospect of a clinically useful drug seemed as far off as ever. In 1988, the School of Pharmacy at Pittsburgh closed Abraham's Department of Medicinal Chemistry, and he moved to head a new lab at Virginia Commonwealth University in Richmond, Virginia. He could now do his own crystallography and no longer needed to go to Cambridge. Almost immediately, he discovered two new agents that seemed to have the advantages of clofibrate and bezafibrate, but did not get mopped up by serum albumin. Encouraged by the president of his university, in the mid-1990s Abraham set up a company, Allos Therapeutics, to develop his new agents. Soon afterwards he received a visit from Max, who wrote home excitedly to Robin:

> Don Abraham, my long-time collaborator, has finally found a compound that lowers the oxygen affinity of haemoglobin even in the presence of serum albumin. True to present fashion, he has founded a company designed to exploit this compound commercially. It improves delivery of oxygen to tumours which in turn increases their sensitivity to radiation, so that smaller doses can be used . . . I am delighted that the work which Abraham and I began in 1980 is finally giving medically useful results.

These things take time, but by 2006 one of Abraham's drugs was showing promise for women with breast cancer who were receiving radiotherapy for secondary brain tumours. Two of his other inventions have been taken up by drug companies as part of the slow haul towards a new treatment for sickle cell disease, and one is receiving support from the rare diseases initiative of the National Institutes of Health. Abraham is a man of boundless enthusiasm who has doggedly pursued his dream of designing a drug based on understanding the structure of its target molecule, but he states unequivocally that without Max's collaboration he would never have made the progress he has. One of the most important lessons Abraham says he learned from Max was not to give up on a line of research because he couldn't do the experiments himself, but to find someone who could and start a collaboration. Apart from their shared scientific enthusiasms, the two men formed a genuine friendship based on a common generosity of spirit.

With advances in technology and computing, the number of known protein structures has now reached tens of thousands: accordingly the field of molecular medicine has also expanded exponentially. The number of approved drugs based on structural knowledge is growing much more slowly, but it includes Tamiflu for influenza, herceptin for some forms of breast cancer and some of the anti-HIV drugs. While drugs for blood disorders based on knowledge of the haemoglobin structure remain elusive, Alan Schechter believes Max's work on abnormal haemoglobins underlies advances in many other fields, having provided an independent justification that the crystal structure of a protein was an accurate guide to how the molecule behaved in the living cell.

Max himself, while never over-hyping the promise of a molecular approach to disease, remained optimistic that it would pay off in the long term. Asked to lecture to the Royal College of Physicians, he wrote a talk entitled 'Determining the atomic structure of living matter: what use to medicine?' His research for the lecture, ranging far beyond his own work with blood diseases, eventually grew into a short, introductory book entitled *Protein Structure: New Approaches to Disease and Therapy*, which he published in 1992. He dedicated the book to Sir Harold Himsworth, 'whose foresight and courage led him to support my colleagues and my early work on protein structures when there was only the faintest hope of it ever benefitting medicine.'

The work Max undertook on mutant forms of haemoglobin also led him to renew his interest in the differences between haemoglobin from different species. The haemoglobin of almost all vertebrates looks superficially identical, with its two pairs of linked alpha and beta chains around a central water-filled cavity. Yet the sequence of amino acids varies quite widely between one species and another. Although haemoglobins of all these species bind oxygen in a cooperative fashion, they vary in how strongly other constituents of blood affect their affinity for oxygen. These variations relate to the animals' different lifestyles: those that live at high altitudes, for example, have haemoglobins with unusually high oxygen affinities so that they can survive in the thin mountain air. By the early 1980s, Max had atomic-level structures of both human and horse haemoglobin, and could see that, despite the fact that forty-two out of 287 amino acids differed between the two, the molecule worked in exactly the same way. Meanwhile, at the Max Planck Institute in Munich, the biochemist Gerhard Braunitzer, who had been the first to decipher the amino acid sequence of human haemoglobin, had gone on to sequence a veritable zoo of different species, including fish, crocodiles, geese and llamas.

Haemoglobin was then the only protein for which a wealth of information on species differences could be related to its three-dimensional structure. It provided a perfect natural experiment for anyone interested in the way that species adapt to their environments right down to the level of molecules, and therefore for the process of evolution by natural selection. Charles Darwin had proposed in 1860 that variations arising by chance in a population would be 'selected' if they increased an individual's chances of surviving and – most importantly – breeding. His theory of evolution argues that this process, continuing over millions of years, was all that was needed to get from the earliest, single-celled, ocean-dwelling life forms to the diversity of species on earth today. He did not know of Gregor Mendel's work on inheritance in pea plants and so offered no mechanism for this variation and selection, but a series of classic studies in genetics during the twentieth century, followed by the unravelling of the genetic code, showed exactly how a chance mistake in the copying of DNA could lead to altered proteins that might have good, bad or neutral consequences.

It was an unmissable opportunity for Max: in 1983 he sat down and wrote a long and detailed review that revealed the wonderful diversity of the living world from the perspective of the haemoglobin molecule. He began with the primitive lamprey and proceeded via fish and crocodiles to the amazing bar-headed goose, which has a special adaptation to enable it to migrate over the tops of the Himalayas, and on to the mammals. Constantly referring to his model of human haemoglobin, he picked out the amino acid substitutions that he believed could account for the special abilities of all these other species. He was aiming higher than a mere catalogue: plunging into one of the more lively debates in evolutionary theory, he gave it as his view that most species differences at the molecular level were neutral in evolutionary terms, mere accidents that had been passed on because they didn't do any harm. Just a few, in contrast, changed the function of the molecule in a way that provided a selective advantage: a single change in llama haemoglobin was enough to enable it to take advantage of meadows at higher altitudes as its oxygen affinity increased. (On discovering that elephants have the same modification, even though they don't live on mountain tops, Max speculated that 'this might have helped the elephants who carried the supplies for Hannibal's army across the Alps in 218 BC'.)

A rather acid anonymous referee's report, accusing Max of 'dabbling' in an area not his own and of being 'naïve' in his conclusion that the haemoglobin structure was largely conserved throughout evolution, was enough to get this paper rejected by the widely-read *Science* magazine, though he had no difficulty getting it published in two more specialist journals. To prove his point he needed to harness another new technology. Protein science was being revolutionised by using recombinant DNA – genetic engineering – to put amino acids into particular positions at will. The trick was to take the DNA sequence that coded for a particular protein, alter it in a predetermined way and insert it into a bacterium such as *Escherichia coli* that would then make millions of molecules of the altered protein, so that its properties could be studied. Max wanted to use this method to help him distinguish between species differences that had just arisen by chance and had no effect on the molecule's function, and those that offered some selective advantage.

He had just received the funding from his Canadian benefactor Tom Usher that would allow him to hire a specialist to do the work. Almost ten years earlier, Hideki Morimoto and Kiyohiro Imai, co-inventors of the device Max and Lehmann had used to measure oxygen affinity in mutant haemoglobins, had sent their PhD student Kiyoshi Nagai to work at the LMB. Nagai had spent eighteen months working with John Kilmartin to make chemically modified haemoglobins to test Max's model of the mechanism of the 'breathing molecule'. For him it had been a dream come true: 'I thought what we were doing in Japan was very interesting, but students there normally work in one lab and don't know anything else, whereas [at the LMB] everyone talks freely about their research. Though I came to work on haemoglobin I could learn a lot of other things, which fascinated me.' Max thought Nagai was the best person to undertake new work on genetically-engineered proteins and, as soon as he had the funds, wrote and asked him if he would like to come back. Nagai accepted at once.

The project did not get off to a good start: *E. coli* can be fussy about the proteins it is prepared to make, and for two years Nagai could get no more than the smallest traces of beta globin – the beta chain of haemoglobin without its haem. At that point, with no guarantee that *E. coli* ever would produce large amounts of human protein, he had to decide whether to go back to his job in Japan, or take a gamble and stay. He stayed, and the gamble paid off. In the third year, with the help of a Danish post doc, Christian Thøgersen, he devised an ingenious method of getting *E. coli* to make any protein he wanted, including of course the engineered forms of beta globin. Eventually, it became possible not only to make both globin chains in *E. coli*, but to assemble them with the haems into haemoglobin: colonies of bacteria would turn red, just like red blood cells.

The way was now open to test amino acid substitutions in human haemoglobin one by one to see which were critical for adaptation to different environments. Max's view that most were evolutionarily conservative, with no effect on the molecule, was largely vindicated, but he also had to accept that some of his predictions turned out to be wrong. One of the most curious adaptations is found in fish, and is called the Root effect. It is a form of very extreme Bohr effect that uses high acidity to force oxygen out of the blood and into the swim bladder, so that the fish can control its buoyancy. Max had predicted that a single specific amino acid substitution would be enough to account for this. Five years after Nagai started working on the problem, he went to Paris to make the very sensitive oxygen affinity measurements in Claude Poyart's lab.

Max always had very simple ideas, and sometimes he was right and sometimes wrong . . . I first did a control with the native protein, then tried this mutant. Max was wrong! This mutation was not sufficient to create the Root effect in human haemoglobin. I phoned him from Paris and he was disappointed, but was happy for me that I had made the first man-made mutant of haemoglobin.

Later, a student of Nagai's, Noboru Komiyama, spent some time trying to establish exactly which of the 110 differences between human and crocodile haemoglobin lowers the oxygen affinity so that crocodiles, which cannot store oxygen in their muscles as whales do, can extract more oxygen from haemoglobin and stay under water for up to an hour. Max had predicted that as few as three key substitutions would account for the difference. Life turned out to be not so simple: not only did Komiyama show that Max had picked the wrong ones, but that human haemoglobin needed another nine substitutions to work as well as a crocodile's. Nagai recalls that Max's disappointment when he was proved wrong was always tempered by pleasure that the problem was solved.

One of the most interesting spin-offs of Nagai's work with engineered haemoglobins was the possibility of making blood substitutes. Donated blood for transfusions in operations or after accidents carries a risk of potential infection, for example by HIV. However, purified haemoglobin is no use as a substitute: the molecule works efficiently as an oxygen carrier only when it is in a red blood cell. Much of its structure, as Max discovered, is designed to respond to other agents in the cell, such as chloride or hydrogen ions, by giving up oxygen more rapidly and completely than it would otherwise. Left to itself, haemoglobin clings on to its oxygen too tightly. Worse still, the four-subunit molecules tend to break into two halves that damage the kidneys.

Working with the American pharmaceutical company Somatogen, Nagai made a haemoglobin called rHb1.1 that avoided both these problems. Unfortunately, it proved to have potentially toxic side-effects: research continues to find a safer second-generation version, but Nagai is no longer involved. He found it 'interesting to see how the other world worked', but the commercial aspects of the research did not appeal to him. Max himself never collaborated with an industrial partner, although ironically it had been a consultancy post he was offered with Schering in 1969 that first forced the MRC to develop a policy for such partnerships. Max's 'consultancy' consisted of visiting the company headquarters in Germany each year and giving a few lectures on his work: the arrangement continued until the company terminated it in 1978. He also sat on the advisory board of the company founded by Don Abraham from 1994.

It was obvious to the governing board of the LMB that in Kiyoshi Nagai, Max had found a formidable asset to the lab. One of Sydney Brenner's last acts before he resigned as director was to offer him a permanent job. Today, Nagai is joint head of the Structural Studies Division. By 1990 he had personally stopped working on haemoglobin and moved on to other fields, but still saw a lot of Max. Although in his 'retirement' Max had a small office, he did not have his own lab: if he wanted to do an experiment he would come to Nagai's and ask if he might have a 'square foot' of bench space among the students and post docs. The two of them would take walks together in the nearby Wandlebury country park, and Max enjoyed chatting to Kiyoshi's children who often came to the lab on Saturday mornings. Reading Max's 1970 paper on the haemoglobin mechanism had first inspired Nagai to think of becoming a molecular biologist, and Max's affection and encouragement continued to inspire him throughout his career.

Max continued to engage in debates about evolution. He was present in the audience when the philosopher Sir Karl Popper gave the first Medawar lecture at the Royal Society in 1986, taking evolution as his theme. Max knew Popper slightly, as a fellow Viennese emigré, and shared his anti-deterministic, anti-relativistic views: he certainly supported Popper's dismissal of the claims of either Freudianism or Marxism to be scientific. Charles Darwin, on the other hand, was one of Max's greatest scientific heroes. When Popper launched into a critique of Darwin's theory of evolution by natural selection, asserting in passing that 'biochemistry cannot be reduced to chemistry', Max could not contain himself: he vigorously challenged Popper from the floor to explain on what grounds he thought this to be true. Popper gave him the 'magisterial reply' that he 'would see the answer if he thought about it for an evening'. Max did so and, having failed to think of a single example of a biochemical function that did not work as well in a test tube as it did in a living creature, he sat down and wrote a refutation of Popper's argument that was published in the *New Scientist* a week or two later.

It drew heavily on the knowledge he had collected about species differences in haemoglobin. Popper had rejected what he called 'passive' Darwinism, the idea that evolution was driven by random mutations that just happened to make animals more successful, in favour of an alternative model in which the individual actively sought out advantageous environments. Max found this implausible: how could the llama, for example, seek out the higher pastures *before* it had fortuitously acquired the mutation that would allow it to survive there? The article also allowed Max to air his dissent from Popper's most influential idea: that scientific method invariably involves an initial hypothesis followed by experiments to test its validity. Indeed this notion of 'falsifiability' – that a scientific idea must generate experiments that could in theory prove that it is wrong – has come to be seen as the very essence of the scientific method. Max conceded that much of science worked in this way, but not all: 'In practice, scientific advances often originate from observations, made either by accident or by design, without any hypothesis or paradigm in mind.'

The example he gave was from astrophysics, but his own field of X-ray crystallography had led him to this view. He had begun his research career with no idea at all of what a haemoglobin molecule looked like: as new methods and technologies developed he eventually found the answer. Crystallography could unambiguously reveal the structures of simple salts such as sodium chloride in a manner that could never be falsified. While a structure as complex as that of haemoglobin might subsequently be made more precise, its ability to explain the effects of human mutations and species differences tended to confirm its essential correctness.

Max's interest in the haemoglobins of different species led him, as he entered his ninth decade, to embark on a wholly new area of medical research. Friends and colleagues had wondered why he had pinned his whole career on a single molecule, rather than increasing his chances of success by developing some new projects. Now he was to surprise them all.

Musing about the differences between the oxygen affinities of one species and another, he began to look more closely at the haemoglobin of a rather unpleasant intestinal worm called *Ascaris*. Almost seventy years previously, Max's mentor David Keilin had discovered that *Ascaris* haemoglobin gave up its oxygen with a reluctance unmatched anywhere in the animal kingdom, an adaptation to living in the oxygen-poor environment of another animal's guts. To understand what feature of the molecule gave it this property was a formidable problem: it had eight subunits instead of the usual four, and no one had troubled to unravel its amino acid sequence. Max couldn't do this himself, so he resorted to his usual method of finding someone who could. A lab in Antwerp specialised in the haemoglobins of invertebrates, and Max found a student there, Ivo de Baere, who could do the sequencing.

Using the genetic engineering technique pioneered by Kiyoshi Nagai, they showed that a rearrangement of the amino acids near the haem meant that there was an extra bond holding on to the oxygen molecule. Substitute one *Ascaris* amino acid with the one found in most vertebrates, and the high oxygen affinity dropped back to normal. Case solved. But there was something else odd about *Ascaris* haemoglobin. At one end of the protein chain there was a sequence of amino acids, glutamine-glutamine-histidine-lysine, repeated four times. Max saw that strands of this sequence lined up side by side would bond together into sheets, because of their complementary electrical charges. He hadn't seen this formation in proteins before, and dubbed the sequence a 'polar zipper'. With his Belgian colleagues and Rodger Staden, the LMB's pioneer in the computerisation of DNA sequences, Max searched the international protein databases to see if any more polar zippers were out there. He found several, often consisting exclusively of long runs of glutamines. Max built a model to show how such glutamine repeats might aggregate into sheets, sent off a short paper and thought that was 'the end of the story'.

In March 1993, a remarkable scientific odyssey ended when an international team identified the genetic mutation that causes Huntington's disease, an extremely distressing inherited degenerative brain disease that usually begins in middle age with abnormal movements and quickly progresses to dementia. It had taken fourteen years and had required the cooperation of the people of fishing villages along the shores of Lake Maracaibo in Venezuela where the disease was unusually prevalent. As Max was reading about the discovery in the journal *Cell* he nearly jumped out of his seat. The only difference between the normal protein made by the newly-discovered gene and the protein from families with the disease was the number of glutamines in a repeat sequence: fewer than thirty-five in healthy people, from 40 to over 100 in those with disease. The larger the number of glutamines, the more severe was the disease and the earlier in life it struck.

Intrigued, Max got colleagues at LMB to make some synthetic protein fragments that consisted almost entirely of glutamines. As his model had predicted, they zipped themselves together to form strongly-bonded sheets. Max wrote up the results and suggested that in brain diseases such as Huntington's, proteins with long glutamine sequences were either clumping together in the nerve cells or affecting the activity of other essential proteins by sticking to them. As a complete newcomer to the field, he circulated his paper to anyone he could find who was working on the disease. One of these was Nancy Wexler, President of the Hereditary Disease Foundation in the US, who had led the search for the Huntington's disease gene in Venezuela. For her it was personal: her mother had died of the disease, and she and her sister each had a 50 per cent chance of inheriting it. When Max asked her 'how this work might be pushed further to gain an insight into the molecular mechanism of these neurodegenerative diseases' she welcomed him with open arms. After meeting her for the first time in Washington in 1995, Max wrote to Robin: 'Nancy Wexler, the remarkable woman . . . whose initiative and drive led to the discovery of the gene . . . is outgoing, emotional, bright, knowledgeable, warm-hearted and dynamic; my only worry is that she expects too much from me.'

By this time, Max and three of his colleagues in Cambridge had discovered that engineering glutamine repeats into a small protein molecule did indeed make it aggregate into clumps – but, frustratingly, pathologists who looked at the damaged nerve cells in the brains of Huntington's patients could see no evidence of any abnormal deposits of protein. That all changed two years later, when Gillian Bates at King's College in London put part of the human Huntington's disease gene, with a very long glutamine repeat sequence, into laboratory mice. Not only did the mice develop symptoms similar to those of human patients; when pathologist Stephen Davies at University College London looked at their brain cells, he found deposits of protein had built up inside the cell nuclei. Once they knew where to look, he and others found similar clumps in the nuclei of human brain cells from patients with Huntington's and a variety of other brain diseases. The fact that there were deposits of protein in these cells did not *prove* that they caused the disease, but *if* they did, then at last there was a possible target for a drug.

Max was delighted that his hunch had got the Huntington's researchers to 'think molecular' and had kick-started such a fruitful line of enquiry. He continued to attend the meetings organised by Nancy Wexler until 2000: she was always effusively grateful for his presence, writing to him that:

> Your logic, knowledge and keen insight has graced these meetings and made them so much better. Your insistence that people speak English and not jargon saved many participants there from not understanding – they were too shy to ask but you were never afraid to be bold and daring and imaginative and candid.

As late as July 2001, aged eighty-seven, Max published a paper in *Nature*, co-authored with Alan Windle who was a Cambridge professor of materials science, offering a perfectly valid theoretical explanation for the relationship between the length of the repeat sequences, the age of onset of the disease and the death of nerve cells. Although there was still no unequivocal proof, he himself did not doubt that it was the accumulation of protein inside cells that destroyed brains and killed people.

For Max, the goal was now to solve the structure of the fibrous clumps that formed in this and other diseases. Tough deposits of abnormal protein, called amyloid, also form in the brain in Alzheimer's disease: although this disease is not caused by glutamine repeats, the deposits are made up of fragments of a normal protein forty-two amino acids long. Max wondered if there was a common mechanism. Enlisting a set of colleagues both from LMB and further afield, he amassed a collection of data on abnormal protein fibres from both X-ray crystallography and electron microscopy. The data were suggestive, but not definitive, and so he had to fall back on the approach he, Bragg and others had taken in the earliest days of protein crystallography: to imagine a structure that would give the observed results. He concluded that in each case the fibres consisted of water-filled cylinders made up of slowly spiralling strands of protein. About twenty amino acids were enough to make one complete turn. Max argued that such a structure would not be very stable, but that one that completed a second turn – taking around forty amino acids – would be stable enough to set off a process of aggregation. The best hope for therapies, he concluded, was preventing the growth of these structures.

Max worked frenetically on the problem through the autumn of 2001. Vivien remembers how excited he was by it all:

> He could hardly sleep because he would always find some flaw in his ideas and then lie awake trying to think a way round the flaw. He would then be cock-a-hoop because he'd convinced himself that his ideas could

stand up. Uncharacteristically he was trying to deduce the structure of amyloid theoretically and he sat at home building little models and trying to think out how it might work.

That December Max learned that his own fragile body had finally fallen victim to an invader he could not repel: a rare and incurable cancer. Wanting to leave a definitive statement, he rushed to complete two papers before going into hospital. As they were not quite finished on the day he was due for admission, just before Christmas 2001, he and his colleague Alan Fersht went to see his consultant in the morning and persuaded him that the LMB was part of Addenbrooke's. That gave an extra half day for him and Fersht to finalise the papers before despatching them to the *Proceedings of the National Academy of Sciences*. He then walked back to the hospital for his operation.

He drew a stronger conclusion in these papers than was perhaps justified by the evidence, claiming unequivocally in one title that 'Amyloid fibres are water-filled nanotubes'. Aaron Klug strongly advised Robin Perutz not to let Max publish, or at least to make him change the title, fearing that his reputation would be damaged if he published something that turned out to be wrong. Robin in turn argued that even if it was wrong 'it might start something'. It did indeed turn out to be wrong, but meanwhile others were starting to look for the right structure. In 2005, David Eisenberg of the University of California at Los Angeles reported an X-ray crystallographic study showing that amyloid fibres derived from a form of yeast were made up of tightly-zipped protein segments, with no water in between. These structures have subsequently also been found for 40-amino-acid repeat units in amyloid filaments. Today, the goal in Huntington's research is to find an agent that stops the abnormal protein from clumping together, just as Max had urged. Klug now concedes the truth of the philosopher A.N. Whitehead's dictum, 'It is more important that an idea is fruitful than that it is correct.'

Max made one other foray into the battle against disease that is less widely known as it resulted in no published papers of his own. HIV/Aids had emerged as a new and terrible threat to human health in the early 1980s. In 1983, the MRC launched its Aids Directed Programme, a funding initiative designed to encourage British research into vaccines and drugs against the human immunodeficiency virus. In 1987, its steering committee divided into two parts: one would devote its energies to funding research into vaccines, while the other would concentrate on antiviral drugs. The head of the MRC, Sir James Gowans, invited Max to chair the second of these, with Raymond Dwek, now Professor of Biochemistry at Oxford University, as its only other member. Max was initially reluctant to accept the post:

> I suggested other people who might be more suitable, but failed to persuade him to another choice. His main reason: that I have no need to get any money out of his special AIDS fund for myself. I have mixed feelings about the job: I know too little virology; the problem is appallingly difficult; we would have to do something highly original if we are not to duplicate the American effort.

But having accepted, Max threw himself into the task. He and Dwek travelled round the country, visiting laboratories and trying to identify leads, as well as arranging for facilities to be set up to test hopeful candidates. Although at that time the structure of the virus's proteins was still unknown, scientists had discovered that its surface had many sugars attached. Dwek credits Max with spotting a class of compounds called imino sugars, derived from natural plant extracts by scientists at Kew Gardens, as potential antivirals. Max went to Chicago with Dwek to encourage the pharmaceutical company GD Searle to investigate further, and they got as far as a clinical trial. Unfortunately, such drugs turned out to have unacceptable side-effects in the doses that would be needed against HIV. (However, Dwek's colleagues in Oxford have gone on to develop them successfully as a treatment for a rare metabolic condition called Gaucher's disease, in which a fatty substance accumulates in the organs.)

Max's deeply-felt concern about this modern plague emerges from a series of letters he wrote to newspapers after a Channel 4 *Dispatches* programme in June 1990 gave massive exposure to the theory of Dr Peter Duesberg that HIV did not, after all, cause Aids. As the programme had failed to do, Max was able to provide figures for the increasing death rate among British haemophiliacs during the 1980s, after they had been exposed to contaminated blood: this was one of the strongest pieces of epidemiological evidence linking HIV to Aids. His letter to *The Times* expressed his belief that to show such a programme, when there was a risk that it might increase the number of people infected, was a 'terrible crime': 'I cannot understand the insensitivity and thoughtlessness of the people who produced this programme . . . Have they ever seen an Aids patient? It is an appalling disease, one that causes terrible, long drawn-out suffering before it kills.'

Max himself had indeed seen an Aids patient: his own nephew, Jim, then in his forties and the only son of his brother Franz. Franz, with his wife Senta, had moved back to Vienna from America during the 1980s: after Senta's death in 1985, he had made a happy second marriage to Gerda Steyrer, one of the family of girls who had grown up with the Perutz children in the 1920s and 1930s. Jim, who was homosexual, had stayed on in New York: with awful inevitability he became one of the victims as the city's gay community began to be ravaged by the disease. Max consulted the network of Aids researchers that he knew and, using his contacts, supplied Jim with a treatment that he hoped would boost his immune system, but his case was already hopeless. On a visit in 1991, Max reported:

> Jim, looking more haggard and pock-marked . . . than last September and limping on his sore feet, was there to meet me . . . We had tea together, first talked about Jim's troubles and then tried strenuously to talk about other subjects to take our minds away from the question that overshadowed everything else: 'How long will he live?' . . . He is a tragic sight.

Jim died the following year.

Fighting ill-health was a natural preoccupation for Max: he had been doing it all his life, on his own account. Having begun as a sickly child, he probably owed his survival into adulthood to his parents' wealthy circumstances and solicitous care. Recent memories of him are invariably coloured by his obvious preoccupation with his physical well-being, perhaps not surprising in one of his advanced years, but unusual in the degree to which he enlisted others in its cause.

After the 1950s, when Werner Jacobson's insight unmasked his intolerance of wheat, the change in his diet improved his health considerably. The threat of illness, however, remained a source of intense anxiety to him and his immediate family. Even colds would lay him low: in 1960 he wrote to Gisela: 'I wish I were not such a glasshouse plant, always catching something. I am so cheerful when I am well.'

Because he absorbed food poorly through his damaged gut, he was constantly at risk of dietary deficiencies, and to make matters worse the range of foods he could eat safely progressively narrowed to little more than meat and fish, mild cheese, boiled potatoes, a limited range of vegetables, and bananas (but only if overripe to the point of blackness). To compensate, from the late 1950s onwards he injected himself with vitamin supplements, which he commissioned American colleagues to buy for him (along with other remedies unavailable in the UK).

During the 1970s, he developed a pain in his hip that made walking, cycling and sitting at a desk difficult, but apparently disappeared when skiing or mountain walking. Later his back trouble worsened and he had to give up skiing, cycling and the tougher mountain walks. More dramatically, he lost his voice for a period of several months. The first time this happened was after giving a lecture in the Cambridge chemistry department in February 1977 shortly after recovering from laryngitis; it became much worse when he visited some of his chain-smoking Italian colleagues later that year. He was reduced to communicating with family and colleagues by scribbling notes on paper. Doctors could find no explanation for his symptoms, which led to the cancellation of a number of high-profile engagements: both Sydney Brenner and the Oxford haematologist David Weatherall remember stepping in at less than a day's notice for public lectures at the Royal Institution. More in hope than expectation, his doctor referred him to a speech therapist, who set about teaching him to speak all over again: he was delighted with the exercises she set him and eventually the problem disappeared.

He never gave up on the health service – his voluminous medical notes show that he regularly demanded yet another line of investigation from his doctors. At the same time, and in parallel with whatever his doctors recommended, he would often (not always with their approval or knowledge) try anything else he heard of on his travels abroad or through his reading of the scientific literature. During one month in 1977, he reported a whole series of self-diagnoses and self-treatments to Robin:

My throat is so sore that I can hardly speak ... [Werner] Jacobson and I came to the conclusion that lack of Ca[lcium] was most likely, since none of the food I eat contains much Ca ... So I started to take some Ca gluconate which has made me feel better. [He had also arranged with a lecturer in chemical biochemistry to do tests for various deficiencies 'which the textbooks suggest'.] ... I hope that the concerted efforts of my various friends will lead to a diagnosis.

I cannot do spectrophotometry any more, because the traces of O_3 [ozone] from the fan motor ... make my throat and mouth so sore. ... The new Vitamin C from America arrived a week ago and has made me feel much better and fitter all round. It has also improved my hip a lot already. So perhaps in due course it may also help to cure my throat.

The *E. coli* [a preparation of what we would now call 'friendly bacteria', supplied by a visiting French colleague] have had a marvellous effect in

relieving my digestive troubles, and have also relieved much of the sore-
ness in my mouth, but the throat remains terribly sensitive.

During this period he was also seeing the Cambridge ear, nose and throat
specialist, starting the speech therapy that would eventually restore his
voice and consulting an orthopaedic surgeon about his back.

Soon afterwards, like many of his fellow Nobelists, Max received a request
from the 'Repository for Germinal Choice', inviting him to consider
becoming a sperm donor. Anything that smacked of eugenics was anathema
to Max, but he also took the opportunity to disabuse the sender of the
idea that he was some kind of super-being:

> Let me tell you that I am small, bald, short-sighted and cross-eyed, that
> my testes have been exposed to X-rays these 44 years . . . and that I am
> plagued by multiple allergies and crippled by back trouble. This shows
> that the winning of the Nobel Prize does not necessarily go with other
> desirable genetic traits.

Nothing made a definitive difference to his overall health until the late
1970s when Max's daughter Vivien, whose own health had not been good
since her late teens, was referred to Dr John Mansfield's private allergy
clinic in Surrey. She benefited so much from the treatment that she urged
Max to try it himself. Mansfield put him through a course of desensitisa-
tion to the less severe allergens that affected him, expanding the range of
foods he could eat. Some years later a joyful letter to Lotte, written from
Switzerland where he was giving a series of twelve lectures in a fortnight,
gives some idea of the difference it made:

> My walking was excellent: after my daily training in Zurich I managed
> to climb steeply up 1000 [metres] in $2^{1}/_{2}$ hours which is no slower than
> in younger days, and also to run down in an hour without pain. The
> sun shone every day and I felt fantastically happy.

It was just bad luck that later the same year he went down with shingles,
an appallingly painful condition that he could relieve only by listening to
recordings of Beethoven piano sonatas. When Lotte complained that he
had not written, he replied that after three weeks without sleep he was
too 'ill, depressed and worried about myself' to do so, and that, although
the shingles was now better, he was depressed again because his infant
grandchildren Timothy and Marion had given him colds: 'This sort of thing
has happened so often at Christmas that I have come to dread it.'

Unlike his mentor David Keilin, whose chronic asthma made him reluctant to leave the security of Cambridge, Max's difficulties did not stop him from maintaining an exhausting schedule of international visits. He or Gisela simply sent his hosts in advance a typed list of acceptable foods (he had versions in at least five languages). Smoke and candles were taboo, and hotels – and the LMB's maintenance staff – had to be briefed that he could not tolerate the smell of most cleaning products. Old paper, which harbours mould spores, turned out to be another threat, causing Max to turn the offending books out of the sitting room at home, and develop a cavalier attitude to the retention or otherwise of many of the papers in his office. As for travelling, he hated long flights, trusted the food and hygiene only in Western Europe and North America, and so declined all invitations to visit more exotic locations, other than Hawaii where he went with Gisela in 1979: 'I have been swimming every day . . . I am pleased that American hygiene has made this heavenly place accessible to me, while other tropical countries are closed to me . . .'

Every one of Max's colleagues has stories to tell about the completely unselfconscious manner in which he went about ensuring his elaborate needs were met – and about their own efforts, sometimes comically unsuccessful, to accommodate him. One tells of spending half a day finding an acceptably candle-free restaurant in Stockholm for a party in Max's honour, only for Max to send a message saying he was too tired to eat out and would stay in his hotel. Another remembers watching in astonishment as Max cut a wedge with his penknife from each of the varieties of apple on offer in a Washington gourmet food store, tasting each before putting them back, and buying just a few of the one he found acceptable. The need to find ripe bananas has exercised many: Don Abraham was stumped one February in Pittsburgh, so put a green one in the oven to blacken it artificially – but he was caught out; Max could tell by the taste. An Italian colleague reports that he had carefully served Max bottled mineral water at dinner, thinking that, unlike his hosts, he would not trust the Roman water supply. The following morning Max declared himself too ill to give that day's lecture in Italian to an audience of schoolchildren, saying that calcium in the water had caused heart palpitations.

From the late 1970s onwards, Max's back problems meant he could stand or lie down, but not sit comfortably. Many a young researcher was discomfited as this legendary scientist, having introduced the lecture, proceeded to lie at full length at the front of the auditorium while the speaker attempted to gain the audience's attention. (This posture once provoked Sydney Brenner to growl, 'Any questions from the *floor?*') He was famous for travelling with an assortment of special cushions to make sitting on planes, buses and trains more comfortable. Once, he had organised two coaches to bring guests from Cambridge for a soirée at the Royal Society in London and, in one of those little mishaps that beset him throughout his life, both coaches left for the return journey while Max was still inside the building, making sure there were no stragglers. As Uli Arndt remembered:

> He came running back and found both coaches leaving. Everyone thought he was in the other coach. He had no money on him, with a gong round his neck, in tails, standing there: fortunately someone gave him a lift back to Cambridge. I was rather blamed for that: he said 'You should have seen that my cushion was still on the coach!'

Most who knew him were tolerant of what they called Max's 'funny little ways'. In his own eyes there was nothing eccentric about doing whatever was necessary to preserve his health, but others saw it differently. Under the heading 'The Nobel art of eccentricity', the *Observer* relished the chance to congratulate Max for 'maintaining a standard of eccentricity that goes above and beyond the call of duty'. It reported how, as guest of honour at the 1993 International Congress of Genetics in Birmingham, seated high on the podium in front of the assembled dignitaries, he had taken out and read his papers, removed his jacket and jumper and finally eaten his snacks, while a series of civic speakers droned on. It is hard to avoid the conclusion that Max sometimes used his 'eccentricity' as a conscious rebuke to those he thought were wasting his time. On one occasion, he was invited to lecture on the medical implications of molecular biology at the Italian National Research Council in Rome, at a symposium on Medicine and Morals:

> When I got there . . . I was collared by a man from the Vatican Radio . . . and then by the Italian Television on the dangers of molecular biology. He was disconcerted when I replied that I could not think of any. To their further embarrassment I then proceeded to eat my usual cheese and banana to stoke up for the lecture ahead . . .

The multiple symptoms and elaborate precautions, combined with a formidable output of scientific work and a busy travel schedule, inevitably raised doubts in many minds that Max's health was really as fragile as he thought. Even Aaron Klug, one of his closest friends and confidants, admits that 'we used to wonder how real all these things were'. He sees Max's preoccupation with his health as the necessary flaw that so often seems to be part of the make-up of highly creative people. There seems little doubt that Max's health was more than usually precarious, his coeliac disease making him vulnerable to debilitating stomach upsets, low blood sugar and various dietary deficiencies. But it is also true that few people worry about their health more than he did, that worrying made him feel worse, and that stress often made him more likely to succumb to symptoms that disappeared when he was cheerful. Throughout his life, his letters often mentioned depression or anxiety and physical ill-health in the same breath, and cheerfulness and well-being, not surprisingly, also went hand in hand.

The extended trips he made to the US almost every year, especially if they included California, seemed to have an almost magically restorative effect. Gisela was moved to remark, having received a cheerful missive during a coast-to-coast trip that Max made in 1973, that he seemed to be more relaxed when he was away from home. Max's reply seems frankly defensive:

> I think I get tense only when preparing for a trip, but there are many weeks when I work happily with little tension. There is always a bit more to do than I can manage and the choice of priorities sometimes generates stress or makes me try to do more than I can stand. I don't think it is true that I am more relaxed here than at home. In fact the timing of lecture preparations, personal chores and engagements with people can be tricky.

His own letters rather give the lie to this. At home, especially during the long, dark Cambridge winters when colds threatened, he kept Gisela and Vivien, who had moved home to live with her parents, in a constant state of anxiety about his health. Vivien now says she never believed he would live as long as he did, having expected from the 1980s onwards that an infection would one day carry him off. With another trip to America scheduled in February 1976, which included speaking at a seventy-fifth birthday celebration for Linus Pauling, Max was feeling unwell as usual and thought of cancelling. Eventually, colleagues and family persuaded him that he should go, that he would enjoy it. It was the right decision:

> If ever I get ill again as I did last winter, I shan't wait for months before coming out here, where all my troubles have disappeared in less than 2 weeks. Even my crippling sciatica has vanished . . . It is marvellous not to go to bed and wake up in pain, not to live in dread of the next tummy attack and sore throat, and to have regained my stamina. So I am becoming my cheerful and enterprising old self again.

In conclusion one can only say that health and happiness were perhaps even more inextricably entangled in Max than is usual. One wonders if this might be an unconscious reaction to his experience of illness in early childhood, to which his dazzling mother had responded with alarm but also with devoted attention – an idea I am sure Max would have roundly rejected as Freudian nonsense. In fairness to him, I should point out that he was almost as concerned about the ailments of his friends and family as he was about his own, and if he could use his knowledge and contacts to get them better treatment, he never failed to do so.

In his eighty-fifth year, Max began to suffer chest pains. In September 1998 he underwent a coronary bypass operation, and soon afterwards had a pacemaker fitted. This development, understandably enough, engendered a new set of worries: he would anxiously phone his cardiologist Andrew Grace to describe any symptoms he thought might be untoward, though they would usually disappear if Grace told him they were nothing to worry about. Having a heart condition slowed Max down very little in practice. His output of published work, now mostly review essays but also including some scientific papers, barely declined. He continued to make daily appearances in the lab, and to travel in the US and Europe. In November 2000, after giving a series of talks in Philadelphia and Washington and visiting friends in New York, he wrote to his granddaughter Marion that he had felt 'on top of the world' during his trip and had returned 'very cheerful and not a bit tired'.

Sadly it was now Gisela's health that was causing him most concern. While she had had her own career in counselling, she had rarely joined Max on his overseas trips, but during the 1990s she began to do so more often: her brother Steffen and his family lived near Washington DC, and most of Max's friends were her friends too. Unfortunately, though, she was too ill to accompany him in 2000, and during the following year he was distressed to see her become ever more frail. It fell to Vivien – since 1991 living in a new house built in the grounds of her parents' home – to provide support to both her parents. Despite her frailty, however, Gisela was to survive Max by almost four years.

In the end it may have been the mountains he loved so much that killed Max. In early 2001, a wart appeared on his nose that the dermatologist pronounced harmless. The following autumn, he developed a small tumour on his cheek that grew rapidly but at first seemed to respond to radiotherapy. However, by the time the tumour had been identified as a rare Merkel cell cancer, it had spread uncontrollably to other parts of his body. By December he knew he had only weeks to live. No one knows exactly what causes this form of cancer, but exposure to sunlight is thought to play a role. Max's early letters often express his joy at feeling the sun 'burning his face' as he climbs into the mountains.

For one whose fear even of the common cold had been legendary, Max faced the prospect of death with a degree of serenity that was much more than resignation. Far from taking to his bed, he became suffused with a new urgency. There were things to be done. First, he must complete and submit his two papers on the structure of amyloid protein. That was done by Christmas. At the same time, with the old-fashioned consideration that had always governed his relationships with friends and colleagues, he must make his goodbyes. (A typical example of this consideration was the warmly sympathetic letter of condolence he wrote to the widow of his French colleague Claude Poyart towards the end of November. He did not mention his own illness.) With Vivien's help, he first of all planned to hold a party, perhaps in February, as an early celebration of his and Gisela's diamond wedding anniversary, but as his prognosis worsened they dropped the idea, and instead he sent off written farewells that also encouraged friends to visit him. A remarkable series of letters took much the same form:

> I have some good and some bad news; the good news is that I have solved the structure of amyloid fibres . . . I have been thinking of nothing else for the last 3 months, but it was finished last Monday and I was able to send off two papers to PNAS [*The Proceedings of the National Academy of Sciences*] . . . I am enormously pleased to have done this . . . The bad news is that I have a rare cancer that started with a lump in my cheek and all attempts to contain it have failed . . . I have had 65 years of fantastically productive research and a happy marriage, delightful children and grandchildren surrounding me with affection and finally I have enjoyed the friendship of so many wonderful people . . . I should have liked a few more years, but am lucky to have had so much.

In response people came in droves, and found him in remarkable spirits. Marie-Alda Gilles-Gonzalez, who credited Max with saving her research career, was one of those who felt she had to see him one last time, and in January 2001 she and her husband flew over from Columbus, Ohio.

> The pilgrimage would be sad, I thought . . . Instead I arrived in Cambridge to find Max busy, cheerful and enjoying a social life that would have exhausted most healthy people half his age. The first day I tried to see him, he was busy with a tea party. He began the next by dictating to

his secretary for several hours, after which I managed to visit him. Seeing Max again turned out to be a very happy occasion ... [we] even had some great laughs together.

Other, more formal letters ('It seems that my days are numbered ...') were dictated and typed, the recipients including the Queen, the President of the Royal Society, the Chancellor of the Pontifical Academy and the former Prime Minister Margaret Thatcher, in which Max elegantly summarised his lifetime's interaction with the organisation or person concerned. When I went to see him in mid-January 2002, he was in hospital having a blood transfusion: the cancer in his liver had made him jaundiced. He lay on the bed, frail and yellowed but smiling: he gleefully showed me the letter he had just drafted to Margaret Thatcher, pointing out, among other things, that he had been right and she wrong about Robert Mugabe. He knew he was facing death, and had no expectation of an afterlife, yet he exuded the grace and tranquillity of a man confident in a life well lived. It was an unequivocally uplifting encounter.

Max Perutz died in hospital on 6 February 2002.

12

Truth always wins

Liberalism remains the basis of all essential decencies: scepticism, curiosity, love of the individual and the personal, the concept of an inner conscience.

Malcolm Bradbury

Max delighted in the written word. Like his mentor W.L. Bragg, he kept a commonplace book in which he wrote down quotations that particularly appealed to him. His model was the poet Milton, who kept such a book 'in a search for truths, moral, political and economic, with which he might serve England, mankind, and God'. When he published his own collection, Max denied having such a lofty purpose, but added that 'many [of my quotations] have become my guiding mottos'. Often they are taken from the works of fellow scientists, such as Peter Medawar, André Lwoff and François Jacob, but others testify to the breadth of Max's reading: there are selections from Pope, Vasari, Goethe, Shakespeare, George Eliot and Boris Pasternak. He particularly enjoyed a line from the memoirs of the ballerina Margot Fonteyn: 'The important thing I have learned over the years is the difference between taking one's work seriously and taking oneself seriously. The first is imperative and the second disastrous.'

The very act of collecting and using such quotations suggests that Max certainly took his own writing seriously. In his later years, he willingly took on the role of wise commentator on science and the wider world, and expected that people would pay attention. His articles reflected above all his engagement not just with science and the scientific community, but with humanity as a whole: his examination of science always took place through the lens of its capacity to contribute to human understanding and human well-being. He has left a legacy in his less specialist publications that deserves recognition along with his scientific achievements.

His sense of himself as a writer certainly dated from his student days in Vienna, if not earlier. He wrote a good letter, and he knew it: though he recognised that he was inferior to one of Vienna's greatest authors, it was audacious for the 22-year-old chemistry student to make the comparison at all:

> A volume of the early letters of Hugo van Hofmannsthal has come out ... Even by the most self-indulgent comparison, his letters are better. He writes so naturally and in such an unforced way just what he thinks. I can also do that if necessary. But there is a difference between what I myself think and what occurs to Hofmannsthal on the spur of the moment. What a pity!

He had earlier hoped that an article he had written about his expedition to Jan Mayen would be published, though his mother thought that it was 'too boring and dry for anyone to accept it'. He persisted, making the acquaintance of an American writer who thought he could turn it into 'a witty and very long piece of nonsense' that might be accepted by *National Geographic* magazine. The piece never seems to have appeared anywhere, and Max's literary efforts were confined to his letter writing for the next decade.

His 1943 trip to the US indirectly kick-started his career as a writer of popular articles on science. Soon after his return he completed one entitled 'Proteins: the machines of life', which appeared in the July 1944 issue of the American Association for the Advancement of Science's *Scientific Monthly*. Max may have begun work on the article while kicking his heels in Washington at the end of September 1943, when he discovered the delights of the Science Library in the Library of Congress. He begins by complimenting his American readers on their proficiency with machines:

> The stranger is led to the conviction that Americans are born with an understanding of machines: indeed, it would not have surprised me in the least to see a baby driving its own perambulator.
>
> However, much as I admired the ingenuity of American machines, I could not help thinking how very simple even the most complex man-made machines are if we compare them with the machines that make man or any other living organisms.

He goes on to deploy, with tremendous assurance, the full range of devices that science writers have always used to make science palatable to the uninitiated. He draws analogies, he gives historical vignettes, he uses examples from everyday life such as the silkworm's cocoon and the permanent wave. He does not forget that his readers will have little comprehension of how very small are the protein molecules he describes:

The number of protein molecules of the smallest size that could be accommodated in the volume of a pin's head is about the same as the number of pins' heads that could be packed into a sphere of the diameter of La Guardia Airfield in New York.

He uses just three case studies, so as not to overload his readers: the role of globulins in immunity to infection, the oxygen-carrying properties of haemoglobin and the catalytic action of enzymes. Throughout, he returns to the metaphor of his title: that proteins are 'the machines of the living organism'. This was all written long before he or anyone else had conceived the idea that proteins could move between different forms. Proteins meet his definition of a machine by virtue of their capacity to do work. It was an insight that most of his colleagues would have taken for granted, but which was not widely current in the public at large (nor is it now).

Encouraged by his success in having the article accepted, Max also submitted 'an improved version' to the British monthly *Discovery*. Interestingly, this version, to which he made many corrections at David Keilin's suggestion, lacks much of the liveliness of the original. A new introduction plays to the wartime British public's concern with protein in the diet. The machine analogy has almost completely disappeared: the title is simply 'Proteins'. Max even introduces the erroneous idea that proteins may be important in heredity; although this was then a widely held belief, he did not include it in the American version.

While there is no evidence that the motive for these popular articles was mercenary, he was surprised to discover that they could be very lucrative: he told Lotte that his article would bring in 'about as much as a month's salary'. Thereafter, he was on the lookout for more opportunities to enhance his income in this way, as well as to hone his writing skills and to bring his research to a wider audience. He first broadcast on BBC radio in 1948: a ten-minute talk on proteins, which he was careful to ensure that his parents-in-law in Switzerland had heard.

As you can well imagine I approached this task with some trepidation. To develop a worth while idea in ten minutes without using any technical jargon beyond the absolutely necessary and without dropping into a string of meaningless platitudes, is not easy in any subject. Atomic physics is fundamentally simple; the entities with which it deals any intelligent schoolboy can understand. But these large biological molecules, their chemistry, structure and function merely seem dull and senselessly complicated to the intelligent layman. Hence my stress on their fundamental biological significance rather than on what they are. I am glad if you think I succeeded in making the talk interesting . . .

Other broadcasts followed, including one on his glacier research. He did not dash off these popular works, which others might have seen as a distraction from the serious business of science, but took time to get them right: he told the Peisers in the same letter that every piece of writing he published 'is generally twice rewritten'. This preoccupation with presentation, with making an impact, pervades his popular work and even, as he wrote to Lotte in 1950, informed the way he wrote his scientific papers:

I have become more interested in writing as such; the presentation of a scientific discovery is, or at least it should be, a work of art. Scientific papers ought to be written so that they grip the interested reader, to be so clear that you don't have to read each sentence twice, and to explain to the reader not only what you have done but also why.

Later that same year he received a commission that gave him a chance to exercise to the full his talent for presenting complex ideas in a widely accessible form:

I have . . . been given a task so fascinating that I could not face refusing it, however pressed I am. There is to be a grand Science Exhibition as part of the Festival of Britain 1951, where the present state of fundamental knowledge is to be represented in popular form. The climax of this exhibition is to be a hall dealing with 'Growing Points of Science': one of these is to be on the 'Nature of the Universe' and the other on the 'Problem of Life'. I have been asked to do the second of these and to present the Problem in its chemical and physical aspects . . . Isn't this a wonderful opportunity for applying one's creative imagination? It's a nice subject to think about night and day.

The Festival of Britain, which opened on 3 May 1951, was designed to mark the centenary of the Great Exhibition of 1851, and to be a celebration of British art, science and industrial design. It was pervaded by a mood of optimism, sorely needed after the long years of war and post-war privations, with science especially being promoted as the source of future prosperity. Surviving visitors remember the astonishing new buildings on London's South Bank: the Skylon, the Dome of Discovery and the Royal Festival Hall (only the last has survived) – but science also had another site, close to the Science Museum in South Kensington, where Max's exhibit would be on display.

He was to use 'any means of expressions and display, static or dynamic, and have a staff of designers and craftsmen to put my ideas into practical shape'. The fee for the twelve-page script, 'Molecular aspects of living processes', was £25, which made the task well-paid as well as fascinating. He based it around five questions, starting from the premise that the living cell was 'a kind of clockwork of great complexity': What are the cogs of the clockwork? What drives them and how do they work? What coordinates their function? How does the clockwork reproduce itself? And why does it die? I have not been able to discover the fate of the exhibition materials themselves, or indeed any images of them, though the script remains in the National Archives.

Several other crystallographers also contributed to the Festival. Kathleen Lonsdale, one of the first women to be elected to the Royal Society, and Helen Megaw, a physicist from Cambridge who worked with Bernal, had long seen the potential of crystallographic patterns as inspirations for designers. They worked with a wide range of designers and manufacturers in the Festival Pattern Group to produce fabrics, lamps and other decorative items. Designers transformed Max's and Dorothy Hodgkin's Patterson maps of haemoglobin and insulin into curtain material, silk ties, gift wrap, fake leather upholstery, a dinner plate and even (in the case of insulin) a lace bridal veil. Similar patterns adorned the furniture, floor and wall coverings in some of the Festival restaurants: Gisela subsequently had a dress made up in the 'haemoglobin' pattern. The mood of creativity and collaboration that made the Festival an inspiration to so many visitors genuinely reflected the energy and new ideas that characterised much of 1950s Britain.

It became a secondary ambition of Max's to become an author as well as a scientist, bringing to the scientifically uneducated some idea of the whole nature of the enterprise:

> I have long been toying with the idea of writing a book about research, conveying the fun and excitement of it, the triumphs and the disappointments and to destroy the popular misconception that research is always done with some utilitarian object in view (to cure diseases, to invent things, to predict what is going to happen) rather than for its own sake, to discover the strange workings of a wonderful world.

While there seemed at the time to be no demand for such a book, Max was courted by a number of publishers to write a textbook on molecular structure, a prospect he did not particularly relish, though he thought he might take it on 'partly to fill a gap that needs filling, and partly to earn money and repay my large debts'. In the event, a salary increase at the end of 1950 and the accelerating pace of his research persuaded him to put other activities on the back burner for a while, and the book was never written.

Max's first book emerged over a decade later from an enterprise he undertook with quite a different purpose. In 1960 he received an invitation to deliver the prestigious Weizmann Lectures at the Weizmann Institute in Israel the following year. It was just after his successful solution of haemoglobin at low resolution, and Kendrew's of myoglobin, and at the point when Crick and others were tantalisingly close to cracking the three-letter DNA code: he was therefore to speak on the structure and function of proteins and nucleic acids. Max was intrigued by the idea of visiting Israel, where he had a number of friends and colleagues who were former refugees. However, he was determined that any visit to the Middle East should take in the Arab perspective as well. The lecture fee seemed sufficiently generous to allow him to take Gisela, Vivien (then sixteen), Robin (eleven) and his mother for an unforgettable trip, visiting Cyprus, Lebanon and Jordan before arriving in Israel.

Max was very far from being a Zionist – indeed, he showed considerable ambivalence about his own identity as a Jew – and this trip confirmed him in an implacable opposition to Israeli policy towards the Arabs. On his arrival in Beirut, where he was to lecture at the American University, he discovered to his mortification that his host and hostess, whom he found had 'warmth, tact, and a natural dignity, without any of the ingratiating Eastern manner which I expected to find in this part of the world' were refugees from Palestine, even though they were Christian Arabs.

> This put our minds into a state of divided loyalty from which we have not recovered. Every time we think of them, and of the fearful refugee camps which we were to see later on, we feel that the State of Israel is a terrible wrong. And then we think of the stories we heard from the Jews to whom Israel was the haven from the concentration camps, and of their fabulous achievements, and we feel if only this could have been accomplished without another, and equally great, tragedy.

The Weizmann Institute in Rehovot, where Max gave his three lectures, was founded in 1934 as the Daniel Sieff Institute by Israel Sieff, one of the founders of Marks & Spencer. In 1949 its name was changed in honour of Chaim Weizmann, chemist and first President of the State of Israel. Despite Max's misgivings about Zionism, he remained an admirer of Weizmann, writing later, 'I wish that the liberal ideals for which he stood would still guide Israel's rulers today.' He was struck by the contrast between the well-funded labs, libraries and lecture rooms at the Hebrew University in Jerusalem and the Haifa Technion, and the struggles of the American and French universities in Beirut to provide high quality education with very limited resources. He was deeply sceptical that the powerful military presence in Israel was, as he was assured, purely defensive: 'In fact the government's attitude strikes the observer as ultra-nationalist and aggressive. They would certainly try to push the Israeli frontiers back to the Jordan and the Suez Canal if they got half a chance.'

That chance came in 1967, the occupation of the West Bank, Gaza and other areas confirming Max in his opposition to Israeli policy. It was more than thirty years before he returned to the country, for a symposium at the Weizmann in honour of Ephraim Katzir's eightieth birthday in 1996. Katzir (formerly Katchalsky) was the former head of biophysics at the Weizmann who was elected Israel's fourth president in 1973: Max had got to know him in the early days of EMBO and liked and admired him.

On his return from his 1961 visit, he sat down to write his first book, based on the lectures. *Proteins and Nucleic Acids* appeared in 1962 and constituted an introduction to molecular biology that at the time had no competitors. It was not in any sense a popular account, assuming at least undergraduate chemistry, but its message was in essence a simple one: 'The entire edifice of molecular biology is based on the hypothesis of the sequence of bases in a nucleic acid chain forming a code which determines the sequence of amino acids in a polypeptide chain.' The book was partly superseded three years later by James Watson's textbook *The Molecular Biology of the Gene*: by the time this came out in 1965 the details of the genetic code had been almost entirely cleared up and molecular biology had begun to move into the mainstream of biology teaching. Today, Max's book is an interesting historical snapshot, while Watson's became a standard text that has remained in print ever since (now in its fifth edition).

While Max enjoyed writing and thought himself a good teacher, his work in the lab always meant more to him and the labour involved in producing a textbook never appealed. His daughter remembers him being unhappy and frustrated all the time he was preparing the Weizmann book for publication, though he grudgingly accepted that both the wider audience, and the further fee, would be welcome. Having just bought a detached house with a large garden in Sedley Taylor Road, the Perutz family were as hard up as ever. In the summer of 1962 Max was taking his customary break in the mountains, walking with colleagues after a conference in Munich, and missing Gisela:

It is absurd that you should not be here. I hope that my book will at least bring in enough money to enable us to go together to the mountains again, rather than having my expenditure strictly limited to whatever I manage to earn through my lectures abroad . . . My companions enjoy the flowers and the scenery but it does not make their blood flow faster, and I wish I could share with you my delight and thrill with every view and flower and pretty old farmhouse. I can see . . . how much less beauty means to them than to you and me . . . We miss it more when it is absent, yet our life is so much richer through that extra sense we are lucky to possess.

Max's Nobel prize, announced in October the same year, both put an end to his financial worries and gave him a platform that would enable him to reach a wide audience. As long as he was still Chairman of the LMB and engaged in developing his model of the haemoglobin mechanism, however, he confined his popular authorship to writing up his own work in science magazines such as *New Scientist* and *Scientific American*. It was in the title of a 1971 *New Scientist* article that he first coined the brilliantly descriptive phrase 'the molecular lung' to describe haemoglobin's active engagement with oxygen molecules. Proud of his skill as a literary stylist even in scientific writing, Max was irritated by over-intrusive editing. On receiving an extensively rewritten proof for one of his *Scientific American* articles, he found it drab as a textbook and drained of all personal feeling. He sat on it until the deadline approached, then wrote to say the editor could publish that version if he liked, but not with Max's name on it. His original text was reinstated.

The whole question of how a scientist might write about his work for a wide audience exploded in the spring of 1968 with the publication of James Watson's *The Double Helix*. Along with everyone else who was involved in the story of the DNA discovery in 1953, Max had received a copy of the draft manuscript in 1966, then provisionally (and to Crick's fury) entitled *Base Pairs*. He had sent almost by return a signed release form agreeing that it did not libel him in any way. Though he protested strongly about Watson's depiction of other characters in the story, particularly Lawrence Bragg and Rosalind Franklin, there is no evidence that he objected to the book in principle, and when it was eventually printed he told Watson that he was 'glad to have it'. Crick and Wilkins, however, vehemently opposed publication, and persuaded Harvard University Press not to put its name to the book, but Watson quickly found another publisher and it duly appeared under the title that has become an iconic phrase.

Not until *The Double Helix* was in print did Max begin seriously to worry that the passage in which he gives the MRC report on the work of the King's group to Watson and Crick showed him in rather a bad light. He realised something was amiss when he heard from Harold Himsworth that the historian Robert Olby, who was working on an academic account of the DNA discovery, had asked the MRC for a copy of the report, so that he could see whether or not it had been marked 'Confidential'; his fears were increased when Hans Krebs almost cut him dead at a Royal Society meeting in June. Worse still, reviews of the book appeared that highlighted the episode: one in *Scientific American* by the French Nobel prizewinner André Lwoff said that Max's action 'might . . . be considered a breach of faith'. He decided that he must clear his name, and began to draft letters to both *Science* (which had published an equally scathing review by Erwin Chargaff) and *Scientific American*, explaining the circumstances under which he gave the report to his colleagues. The points he wished to stress were that the report was not confidential, that it contained nothing that either Watson or Crick could not have gleaned already from conversations or public presentations by the King's researchers, and that therefore it was not particularly useful in finding the structure.

For over a year the draft letters circulated round the scientific community, and the correspondence fills a large file. As well as Crick, Watson, Bragg, Himsworth and Wilkins, Max consulted Wilkins's former boss Randall and other eminent scientists who were not directly involved, such as Hans Krebs and John Edsall. Most thought he was right to set the record straight, though Himsworth feared that he would make matters worse by reopening the issue after such a long interval, and that as an interested party he would not be able to convince a public already 'immunised' by Watson's picture of cut-throat scientific competition. Randall insisted that even though the MRC report was not marked 'confidential', he had never imagined that it would go beyond the committee. Watson, who agreed to write a letter for publication alongside Max's, baldly stated that Max's claim that the King's researchers had shared their data with Cambridge was 'polite nonsense', and that the contents of the report had in fact been very useful to Francis.

Letters from Max, from Wilkins and from Watson finally appeared in *Science* in the last week of June 1969, a year and four months after the publication of the book. Despite an appeal from Crick to make it much briefer, Max's letter ran to about 1,500 words, not including two extracts from the original report that he appended. In it he regretted that he had failed to ask Watson to change the manuscript, and on the facts of the issue itself conceded that:

> [A]s a matter of courtesy, I should have asked Randall for permission to show [the report] to Watson and Crick, but in 1953 I was inexperienced and casual in administrative matters and, since the report was not confidential, I saw no reason for withholding it.

In his own letter, Watson apologised that by omitting to say that the report was not confidential he had allowed reviewers to 'badly misconstrue' Max's actions (he made this point clear in subsequent editions). At the same time, he added further points in support of his original account, including a detailed explanation of the importance of the report to Crick's contribution to the solution, and a slightly less blunt rejection of Max's suggestion that the King's researchers were 'generally open with all their data'.

The rest of Max's letter does rather smack of protesting too much, exactly as Himsworth had feared: it did nothing to mollify Randall. Wilkins wrote privately to Max that he 'respected its tone', adding:

> I think you will agree that during the period described by [Watson], *none* of us behaved quite impeccably . . . I think everyone is now beginning to admit that frank disclosures of how research is done are, in spite of their interest and entertainment value, not, on balance, a good thing.

While Max was dealing with the fallout from *The Double Helix*, he met Horace Freeland Judson, a writer with *Time* magazine who was reporting from Europe and had become interested in the DNA story. As Judson talked to Max, he conceived the idea of a book about molecular biology for a general readership, which eventually, after ten years of research, appeared in 1979 as *The Eighth Day of Creation*. Judson included long, verbatim extracts from his transcripts, so that the stories of the discoveries of DNA and protein structures emerge from the mouths of the protagonists.

Max was initially wary – in fact, he tried to put Judson off by asking for a donation of fifty guineas to the lab's recreation fund – but was immediately persuaded by the writer's enthusiasm and willingness to learn. Between 1968 and 1977, Judson made frequent and extensive visits to Cambridge, even moving there to live for some years, and recorded long conversations with Max about his life and work. Judson encouraged him to talk not only about what he had done, but how he felt as he recalled the details of the momentous occasions when he knew his research had taken an important step forward. Max had always liked to tell stories, and he warmed to the task. Judson won him over completely when he seized the opportunity to interview him on the train journey from London to Cambridge after a symposium in honour of Lawrence Bragg's eightieth birthday.

> He plans to write a book about the development of molecular biology and I think wants me in it to give it a human and personal flavour. Also my way of explaining things and making them interesting has impressed him. He has recordings of my 'explanations' on tape to which he has listened several times, as he likes them so much.

Max was one of those to whom Judson gave drafts of the finished work. One of his key recommendations was that the author cut a lot of the stuff about 'how haemoglobin was not solved', but Judson's aim was to tell the story as it happened, not a sanitised version of how the scientists might have liked it to. His seventy-page chapter on Max and the solution of protein structures is entitled, quoting from its subject: 'As usual I was driven on by wild expectations'. It does include many of Max's false steps and dead ends, but overall presents a picture that clearly mirrored Max's own sense of himself:

> The totem of his intelligence was neither the tiger nor the shark but the elephant: will and a colossal attention span. Behind the industry and an unflagging, paradoxical urgency . . . looms a brooding weightiness of mind.

From this point on, Max's own writing acquired a new, more personal perspective. On his retirement from the chairmanship of the LMB in 1979, the popular magazine *New Scientist* invited him to write an autobiographical account of the founding of the lab. The article that appeared in January 1980 is chatty, anecdotal, humorous and self-deprecating, a style he had developed in his personal letters, but not previously in his published articles: 'Bragg sounded the University about a job for me but received no encouragement because I was a chemist working in the Physics Department on what was in fact a biological problem. What could one do with such a misfit?'

Perhaps emboldened by the success of this autobiographical approach, Max began to compose a long essay, 'Enemy Alien', about his wartime experiences. It was a personal and passionate account, beginning from the moment the policeman arrived to arrest him and recounting the bizarre history of Project Habbakuk. In 1981 he submitted it to the *New Yorker*, in the hope that an American publication would gain him a wide readership. They accepted the piece at once, although publication was 'incredibly slow': it finally appeared in August 1985 under the title 'That was the War: Enemy Alien'.

Max was almost overwhelmed by the response. Of the deluge of letters he received, two touched him particularly. The first was from a retired Canadian civil servant called Jack Pickersgill, who had been an assistant to the Prime Minister, Mackenzie King. In his article Max had mentioned that the release of internees in Canada had accelerated after Ruth Draper, a performer of comic monologues who was then extremely popular on both sides of the Atlantic, appealed to King for the release of an Italian schoolboy she had known since his birth. Pickersgill was puzzled that this episode had not been mentioned in King's diaries, so he went to the government archives, found all the correspondence between Draper, King and various obstructive bureaucrats, and sent copies to Max. This gesture began a friendship that continued for many years: the two men regularly exchanged letters, books and opinions, and to their mutual delight met when Max visited Ottawa in 1987.

A second contact was even more poignant. Through the *New Yorker* a letter came from Valerie Finnis, the wife of Sir David Douglas Scott (father of the young officer Merlin Scott who had alerted the government to the plight of the Italian refugees on the *Dunera*), to say that Max's tribute to his son had moved him to tears. This led to an invitation to Max and Gisela to visit the Scotts. Sir David, a cousin of the Duke of Buccleuch, lived in the Dower House on the estate of Boughton House in Northamptonshire. Max described this palace to his sister in wonder as 'like a French chateau of the 16th century'. He found Sir David, who was ninety-eight and totally blind:

> A shrunken image of the ideal English gentleman, decent, handsome, well mannered, tactful, considerate, erudite in the classics and history, and utterly devoted to the English countryside and to his own home and garden where he had lived since 1902 ... We felt that we had witnessed the approaching death of a gracious way of life and seen a home that is beautiful without being showy, encumbered with all the bric à brac of past generations, which piety had preserved for a future that no longer exists.

Both Sir David and Lady Scott had wondered 'what a molecule is': puzzled by Max's explanation, they asked him to write it in the visitors' book. Lady Scott, who was a professional gardener and plantswoman, pressed numerous plants and cut flowers into Gisela's hands as they left. It was a sudden and unexpected glimpse into another world that made a powerful impression on Max and Gisela, both displaced from their original homes and families.

The *New Yorker* article, though written with a light touch and a humour that forty years' distance had made possible, was also a paean to the value of human freedom. No political demagogue, if Max had a creed it was that people should live in peace, health and freedom and that science should be used to promote that purpose. In addition to a personal and autobiographical slant, a second thread that developed in his writing after his retirement was a passionate advocacy of the benefits of science to society. During the 1950s, as epitomised by the Festival of Britain, science and technology had flourished in a mood of optimism and excitement. By the 1970s a backlash was in full swing. Many trace its origin to the publication of Rachel Carson's book *Silent Spring*, which in 1962 alerted the world to the impact of DDT on wildlife. Campaigning against nuclear weapons, the realisation that 'energy too cheap to meter' was a fantasy, and the accident at the Three Mile Island nuclear power station in 1979 led to a wider anti-nuclear movement. The invention of recombinant DNA technology in the early 1970s caused anxiety inside and outside the scientific community about the social and ethical implications of genetic engineering. A number of commentators published books during the 1970s that together amounted to an 'anti-science movement', arguing that science was out of control, inhumane, dominated by a technocratic elite and likely to cause more problems than it solved.

Such views caused Max much pain. Although he did not often engage in face-to-face debates, he once wrote triumphantly to Robin that, during a meeting of the Society for Social Responsibility in Science on genetic manipulation, he had forced another speaker to withdraw an accusation of irresponsibility against Jim Watson, in front of a mainly hostile audience. He was sympathetic to the views of scientists who had genuine fears on the potential dangers of recombinant DNA research: of the others he dismissed one as an 'unpleasant and dangerous demagogue' and another as a 'childlike hippy and back-to-nature type', saying that they were 'against genetic manipulation for motives other than scientific ones'.

So the opposition is very heterogeneous, ranging from the sane to the lunatic fringe and from the genuinely concerned to the political exploiters. . . . I do not believe that dangers will arise if the safeguards recommended by the appropriate bodies here and in the US are observed, and rather wish that more publicity were given to the dangers brought about by the indiscriminate use of antibiotics.

With hindsight his assessment seems to have been spot on. While Max was only too aware that science could be used for ill ends – he was himself a vocal opponent of nuclear weapons, or indeed the use of any weapons – he regarded the practice of science as an overwhelmingly benign and humanising influence. Around 1980, he was approached by one of his Italian colleagues and asked to speak on 'The impact of molecular biology on society'. He thought perhaps it was too soon for it to have had any impact, but instead agreed to speak on 'The impact of science on society'. He gave the Schiaparelli Lecture in Venice in September that year, and covered the same material soon afterwards in a *New Scientist* article, 'Why we need science'. Anxious to show himself in tune with the times, he began with a feminist point:

> You need go back only to your grandmother's early days to realise that Adam's eating of the apple of knowledge has been of great benefit to Eve. Remember the beginning of *Anna Karenina* . . . Dolly, the princess, is only 33, and already the mother of five living, and two dead children. Her many pregnancies have left her faded and plain, making the prince lose interest in her . . . For working-class girls who were unmarriageable because they lacked a dowry, domestic service was the only outlet. Women's liberation could not have succeeded if science had not provided contraception and household technology.

He goes on briskly to dispose of the myth of an idyllic past without science, adding a jibe at modern 'Arcadian seekers who make a cult of health foods, frequent the herbalists, dress in romantic rustic floral prints, buy pinewood furniture for their suburban cottages, or turn to organic farming'. Having seized our attention with this polemical introduction (at the risk of alienating the owners of floral dresses and pine furniture – probably many of his *New Scientist* readers), he accepts that science can bring risks as well as benefits. He then gets down to his true task, which is to address the question of how risks and benefits are to be balanced in meeting the health, food and energy needs of the world population. The reader comes away from this article with a clear sense of Max's personal philosophy, which one might characterise as compassionate pragmatism.

For him, the millions dead from malaria make the risk to the environment from DDT worth taking; organic farming is too unproductive and carries too high a risk of famine through sudden blights to be a secure way of feeding a growing population; and nuclear energy must be accepted as part of the mix if energy supply is to meet the world's needs. (Much later, on learning more about the problems of decommissioning, he expressed some reservations on this point.) His final paragraphs enter the realm of politics: here he draws heavily on Karl Popper's book *The Open Society*, concluding that no 'laws of history' have stood up to scrutiny and that therefore 'the future depends only on ourselves and not on any historical necessity'. Appearing only weeks after a quarter of a million people protested in Hyde Park against the siting of American nuclear missiles in Britain, his *New Scientist* article argued on both rational and compassionate grounds against nuclear war:

> There is no war that would make the world safe for either capitalism or communism, or for any militant creed or race . . . Militarists and extremists in both [the Soviet Union and the US] are trying to shift their strategic policies from that of deterring a nuclear war to that of preparing to fight one. Such a war would destroy us and all civilisation. The most urgent task of scientists everywhere is to expose the folly of the military planners on both sides in order to prevent that catastrophe.

Max continued to develop this article over the years, eventually publishing it in book form in 1989 as *Is Science Necessary?* The book also included 'Enemy Alien' and a number of the book reviews on science and scientists he had begun to publish on both sides of the Atlantic. Throughout the early 1980s, there had been occasional pieces in the *London Review of Books* and the *Times Literary Supplement*, beginning with a review of Peter Medawar's *Advice to a Young Scientist*. Max's review is full of droll anecdotes from his own life and experience. On Medawar's advice not to fear plagiarists but to keep one's door open Max comments: 'I am suspicious of scientists who tell me that others have pinched their ideas: far from preventing people from stealing it, I have always had to ram any new idea of mine down their throats.'

In 1985 Robert Silvers of the *New York Review of Books* asked him to review a book called *Solid Clues: Quantum Physics, Molecular Biology and the Future of Science* by Gerald Feinberg. Max wrote to Lotte that he was 'very pleased to be asked, as the NY Review is far more widely read than the London one', but thought the book so bad he declined. To his surprise, Silvers asked him to go ahead anyway, as the pithy reasons behind Max's refusal had so amused him. The book made a number of predictions about such things as the autonomy of computers and the availability of off-the-shelf organs for transplant, and Max set himself the task of investigating the likelihood of each of these. Although a champion of science, Max was always realistic about its limitations and irritated by extravagant claims.

> Feinberg sees the future mainly as a collection of technological fixes in the United States, but he does not consider how science might be used to eliminate poverty, ignorance and disease in the rest of the world. This surely is our greatest challenge . . . [His] glib forecasts . . . are linear extrapolations of current progress, but carried into the clouds of science fiction. I believe that scientists writing for the general public should keep their feet on the ground, since otherwise they destroy credibility. Besides, just because the human mind is not like a computer, past progress has rarely been linear, and the greatest advances, like Puck, have popped out of unexpected corners.

Silvers subsequently sent him many commissions and warmly encouraged his writing. Max met him in 1995 and was greatly impressed.

> He is [in his office] day and night, yet is the reverse of a neurotic workaholic, an expansive, jovial, relaxed man of 50 or so who knows and remembers everything he has ever published and gives his authors no end of encouragement. The best thing is that he likes my writing.

To Max's great delight, the series of essays won him the 1997 Lewis Thomas Prize for 'the scientist as poet', awarded by Rockefeller University. Today it is possible to find former colleagues who grumble about Max's presumption in writing on subjects that were often outside his own field: but this is a frequent complaint that scientists make against one another, and underestimates the scarcity of authors who can write knowledgeably and entertainingly about science in literary journals. Others of Max's scientific acquaintances were delighted and surprised by this second career: most had not previously realised how widely read and cultured he was, though Graeme Mitchison, a mathematician and musician who joined the LMB to work with Crick and Brenner, often took the opportunity to talk to Max about literature and music.

Max was not content merely to read the book in question. He approached each review as a new piece of research, consulting sources that the author might have missed and always adding his own personal knowledge. He often remarked that the nice thing about living in Cambridge was that, if you were uncertain about any field, a world expert would not be too far away: and he had no hesitation about consulting such experts whether he knew them or not. Many of his reviews were written during summer holidays in the Dolomites, during which Vivien and Robin were pressed for their 'uninhibited criticism' of his manuscripts before final revision.

Max too could be uninhibited in his criticism. His review of a reassessment of the career of Louis Pasteur by a respected historian of science, Gerald Geison, was written in 'a rage of indignation' against what he saw as an unjustified attack on 'one of the greatest benefactors of mankind':

> Toppling great men from their pedestals, sometimes on the slenderest of evidence, has become a fashionable and lucrative industry, and a safe one, since they cannot sue because they are dead. Geison is in good company, but he, rather than Pasteur, seems to me guilty of unethical and unsavoury conduct when he burrows through Pasteur's notebooks for scraps of supposed wrongdoing . . .

Geison was the first to make a detailed study of Pasteur's extensive private notebooks, where he turned up some discrepancies with his public statements. He was careful to acknowledge that history's judgement of the man as 'one of the greatest scientists who ever lived' was fully justified; however, he felt he had enough evidence to 'deconstruct' the Pasteur myth of the 'selfless seeker after truth' and render him in the less heroic colours more typical of contemporary biography. Max deplored his 'relativist' approach: the idea that science was in any way qualified by its cultural context was anathema to him, and he was happy to fire a salvo in the 'culture wars' then sweeping American campuses. Max was right that nothing Geison had discovered undermined the importance of Pasteur's discoveries in many areas of science, but the book does give interesting insights into the working methods of a man whom Max himself admitted was 'domineering, intolerant, pugnacious and . . . a hypochondriac'.

Max was not wholly blind to the faults of scientists. His review of a life of Fritz Haber, the German Jewish chemist who won the Nobel prize for his method of making ammonia and developed his country's use of poison gas in the First World War, damns him for his inhumanity in developing such a terrible weapon. Having decided he was a bad person, he could not bring himself to admit him to the first rank of scientists: 'any number of talented chemists could, and no doubt would, have done the same work before very long.'

The physicist Erwin Schroedinger is another whose work comes in for a harsh assessment at Max's hands. In 1943 Schroedinger gave a series of public lectures at the University of Dublin entitled 'What is Life?', which were subsequently published as a book. Several physicists, including Wilkins and Crick, have written that the book was influential in persuading them to look for the physical mechanisms underlying biological processes. Accordingly, the editors of a 1987 volume celebrating Schroedinger's centenary invited Max to contribute a chapter on the book's significance in the history of molecular biology. Max devoted most of his chapter to an obscure but extremely important paper published in German in 1935, from which Schroedinger had (with due acknowledgement) derived many of his ideas. Of Schroedinger's book itself, Max wrote that: 'What was true in his book was not original, and most of what was original was known not to be true even when the book was written.' This statement, for which Max found much of his evidence in a careful study by historian of science Edward Yoxen, is not in dispute. Some have felt, however, that in such a celebratory volume Max might have made more of the fact that, regardless of whether his ideas were true or original, Schroedinger's book was inspirational to others and therefore deserves its place in the history of the subject. In treating him so dismissively, was he influenced by his feelings about Schroedinger himself? Viennese by birth (and not Jewish), Schroedinger had fled his post in Berlin on Hitler's accession, but returned to Austria in 1936. He then tried (without success) to make some accommodation with the Nazis after the *Anschluss*, writing a grovelling letter to Hitler that was publicised in *Nature*. None of this would have endeared him to Max, but his chapter was scrupulously objective and gave no hint that personal animosity might have coloured his assessment of the book's importance.

In his introduction to *Is Science Necessary?* Max wrote:

> In science, as in other fields of endeavour, one finds saints and charlatans, warriors and monks, geniuses and cranks, tyrants and slaves, benefactors and misers, but there is one quality that the best of them have in common, one that they share with great writers, musicians and artists: creativity ... Imagination comes first both in artistic and scientific creation – which makes for one culture rather than two – but while the artist is confined only by the prescriptions imposed by himself and the culture surrounding him, the scientist has Nature and his critical colleagues always looking over his shoulder ...

After reviewing so many biographies, what kept Max from writing his own? He certainly had the opportunity. A friend sent his *New Yorker* article 'Enemy Alien' to the Sloan Foundation in New York, which published a series of scientific autobiographies, and they sent Max a contract by return, together with an offer of a $50,000 grant while he wrote the book. He did not immediately turn the invitation down, but, he told Lotte, he 'decided not to sign the contract until [he had] actually written a chapter'. He never did, later telling Lotte that he was not prepared to give up scientific work to make time for the writing, and that anyway he 'found other scientists' lives more absorbing than my own'. He told Vivien that he would write his autobiography when he could no longer do research – but that day never came.

Unlike John Kendrew, who kept a meticulously ordered and comprehensive archive that went back to his prep school days, Max made little effort to conserve his papers and correspondence, other than letters from close friends such as Dorothy Hodgkin, Lawrence Bragg and John Edsall. When Jeannine Alton of Oxford's Contemporary Scientific Archives Centre came to see him in 1984 about cataloguing his archives, Max gave her the impression there was nothing worth having apart from his correspondence with Bragg, Hodgkin and Evelyn Machin.

In early 1999, an American cognitive psychologist from CalTech, Al Seckel, appeared in the lab and offered him $65,000 for his notebooks, the saved sets of correspondence and some other miscellaneous items. Seckel was, he said, buying material relevant to the history of molecular biology on behalf of a publisher, Jeremy Norman of San Francisco, who wished to compile an archive of the subject and donate it to a university. Max was astonished that someone should offer so much money for something he regarded as worthless and, after checking that the MRC had no objection, accepted the offer without hesitation, as did others in his field. Belatedly, the Wellcome Library and the Royal Society became anxious about so much of Britain's scientific heritage being held overseas in private hands. When Jim Watson alerted them that Seckel was on the point of buying Francis Crick's papers for Norman, the Wellcome Trust stepped in and bought them at a cost of £1.8 million. Norman then sent the whole of his molecular biology collection for sale at Christie's in New York. In 2005 Craig Venter, founder of the private genome sequencing company Celera Genomics, bought the archive for his Venter Institute in Rockville, Maryland, and there a small collection of Max's papers – though rather more than he had acknowledged to Alton – remains to this day.

In the late 1990s, Max published two more books that had strongly auto-biographical overtones. The first, *Science Is Not A Quiet Life: Unravelling the Atomic Mechanism of Haemoglobin*, contained his selected scientific papers on haemoglobin, as well as sections on polar zippers and on glaciers. They are gathered into chapters with very brief introductions that enable Max to construct a narrative of his decades-long effort towards the haemoglobin structure and mechanism. At about the same time, he put together another collection entitled *I Wish I'd Made You Angry Earlier: Essays on Science, Scientists and Humanity*, echoing Lawrence Bragg's wearily ironic response to Max's explanation for his sudden insight about the alpha helix structure. Much of the book examines the nature of scientific discovery, but it also goes further than the previous collection, *Is Science Necessary?*, in exploring the relationship between science and scientists on the one hand, and humanity as a whole on the other. A section on 'Ploughshares into swords' looks at military technology and scientists caught up in war: another, 'Rights and Wrongs', includes essays on human rights, contraception and nuclear energy. Reading this collection gives a clear picture not only of his views on science, but his political and religious standpoints, his moral outlook and even his taste in music, art and literature.

Max's published writings were merely the visible product of his unflagging engagement with ideas, and his ever-widening acquaintance with like-minded people inside and outside biology. In 1981 he received an invitation to join the Pontifical Academy of Sciences. This body consists of eminent scientists from all over the world who are appointed (generally) without regard to religious belief to discuss scientific developments and produce reports for the Pope. It has ancient origins (Galileo was a founder member of the original academy), but was refounded in 1936 with a strengthened brief to promote scientific progress. Max first encountered it in 1961, when he was invited to contribute to a study week on large biological molecules: a good illustration that the Academy kept well abreast of all the latest developments.

The telegram announcing Max's appointment as an Academician reached him on 13 May 1981. That same day, a Turkish fanatic fired three bullets into Pope John Paul II as he was blessing the crowds in St Peter's Square. Max was as shocked as anyone.

> My pleasure and gratitude at the honour which his holiness has bestowed on me were at once overwhelmed by horror at the ghastly deed against the gentlest and most peace-loving of men. I had been deeply moved by the Pope's speech against violence in Ireland ... please convey to his holiness my sympathy in his suffering.

He was soon offered a challenging task: to join the Academy's President, the Brazilian Nobel prizewinner Carlos Chagas, in presenting an appeal from the Pope to the Queen of England asking for action on the proliferation of nuclear weapons. Max accepted at once. He related the story of this 'great adventure' at a meeting held in memory of Chagas, who died in 2000, in February 2001.

I felt a bit of a Charlie Chaplin about this, an immigrant to Britain of Jewish extraction acting as a messenger from the Pope to the Queen of England . . . [T]he Foreign Office decided that it was a political message and that it should be delivered to the Prime Minister . . . So Chagas and Hermann Brück [the astronomer with whom Max had been interned] . . . and I made an appointment with Margaret Thatcher at Number 10 Downing Street. She received us . . . dressed to the nines and with not a single hair out of place . . . she read [the message] and she said: 'I shall decide, I shall tell the Queen what to reply.' We were amused by this outcome, although a little saddened because we realised that we could not get any further.

On attending his first plenary session of the Academy, Max poured out his impressions to Lotte:

The meetings take place in a beautiful Renaissance villa in the Vatican Garden, over which the Dome of St Peter's towers like the Matterhorn over Zermatt . . . In [the Pope's] speech he made great play of the scientists' duty to discover and publish the truth. I used this cue in my own lecture to the Academy for a plea to the church to recognise the appalling prospects of famines with which the world will be faced unless underdeveloped countries take vigorous measures for birth control . . . The interior of St Peter's was brightly lit for the occasion. Just as when I first saw it 31 years ago, I thought that it expressed the worldly grandeur of the Popes rather than the teachings of Christ. They emerge solely, but movingly from Michelangelo's Pietà.

Max decided to make the population question the main issue he would address as an Academician. He genuinely believed that Catholic opposition to contraception could be reversed if enough facts were amassed of the threat that unlimited reproduction posed to the health of women and children, and to human survival if population exceeded resources. He persistently proposed 'Population and Resources' as a topic for a workshop and finally, almost ten years after his election, his persistence was rewarded. Preparation for the workshop proved to be a huge task. Finding people prepared to serve on the programme committee or to speak at the workshop – and getting them accepted by the Papal authorities – was far from straightforward: by July 1991 (the workshop was due to take place in November that year) the programme was in its nineteenth draft. Max himself was due to chair a session on Population, Health and Human Potential.

In the end he didn't go. As had happened so often in his life, with a big event in which he had a huge personal stake in prospect, he felt too ill to participate. Nevertheless, he was assured the workshop was a success. From his personal point of view it was a failure, however. The Pope continued to denounce international initiatives on population control, such as the United Nations International Conference on Population and Development that took place in Cairo in 1994. The book containing the proceedings of his workshop finally appeared in 1996: it was very carefully worded, making no strong statements about contraception or abortion, but eloquently describing the health and resource implications of having too many people. To date there is no evidence that Catholic teaching on the subject has changed as a result. The whole exercise was an interesting example of the mutual incomprehension of rationality and ideology.

Yet Max continued to express admiration for John Paul II's qualities, commenting in particular on his 'courage and determination in the face of severe physical handicaps' during a visit to Cuba in 1997. After a brief period of protest, he resumed active membership of the Academy: candidates he suggested for election included Te Tzu Chang, the Taiwan-based expert on rice genetics, and he successfully proposed Gillian Bates and Stephen Davies for the Pius XI medal for their work on Huntington's disease. He had just offered to speak on the subject 'In Science, Truth Always Wins' at the 2002 plenary meeting (he would have been eighty-eight), when he received his diagnosis of terminal cancer.

The Pontifical Academy gave Max a platform, sometimes an influential one, from which he could voice his concern that every human being should be able to live free of oppression and with adequate means to survive. His fundamental humanity was unambiguously the driver for his anxiety about overpopulation: unlike some other biologists of the twentieth century, he never for a moment entertained the idea of policies to 'improve' the genetic stock of mankind, whether voluntary or compulsory. (A recent biography shows that Francis Crick, in contrast, harboured startlingly illiberal eugenicist views: James Watson too has gone on record as saying that genes for low intelligence should be eliminated by prenatal screening.)

In 1993 Max became a founder member of the executive committee of the International Human Rights Network of Academies and Scholarly Societies, with fellow Nobelists the Swedish (US-based) neurophysiologist Torsten Wiesel and the French biochemist François Jacob, and the Dutch human rights lawyer Pieter van Dijk. With a secretariat based at the US National Academies in Washington, the organisation was to campaign on behalf of persecuted scientists and doctors worldwide. Most of the other representatives were nominated by the human rights committees of their national academies, but Max was embarrassed to have to confess at the inaugural meeting that the Royal Society did not have one (it still does not, though it has always participated in the Network's activities and recently hosted its annual meeting).

This was not a new interest for Max: as early as 1964 he had mobilised powerful scientific friends all over the world in support of two Italian colleagues who had been imprisoned for embezzlement, caught up in the chaotic cross-currents of Italian political and academic life. He had also joined an (ultimately successful) international campaign of scientists to protest to the Soviet authorities about the victimisation of Benjamin Levich, an internationally distinguished chemist, who in 1972 had applied to emigrate to Israel, and more recently had concerned himself with a worrying resurgence of anti-Semitism affecting scientists in the new Russia. Joining the committee of the Human Rights Network merely increased the amount of time he spent on this kind of activity. He formed a good working relationship with the executive director, Carol Corillon, writing warmly to Robin: 'The people who run this committee are idealists with a thoroughly practical bent and a shrewd understanding of human nature. It is a pleasure to work with them.'

He began to collect information on prisoners and victims of torture, protesting to governments and writing letters of support to the victims' families. At about the same time, he agreed to become a Patron of the Redress Trust, whose aims included obtaining reparation for victims of torture, writing to its founder Keith Carmichael: 'When I was a schoolboy I believed that torture had gone out with the Inquisition, and I find it hard to reconcile myself to the fact that the countries that do *not* practise it are in the minority today.' His friend the former Peterhouse Master John Meurig Thomas recalled that the idea of torture revolted Max to such a degree 'he could be seen to cringe when talking about it'. He spoke movingly at the launch of the Trust's 2001 report only months before the end of his life.

Already in his eighties, Max now set himself the task of getting to grips with human rights law, both national and international: among his causes was to get some 'humane modifications' made to the Asylum Bill then passing through Parliament. Most of his work took place well out of the public eye, but in 1995 he wrote a substantial paper for the Network's symposium in Amsterdam, entitled 'The Power of Scientists as Human Rights Activists'. The paper was subsequently published as 'By what right do we invoke human rights?', and included in his collection *I Wish I'd Made You Angry Earlier*. It was a succinct review of the history of Western thinking on human rights since the time of the Romans, but, just as Max's readers have been lulled by quotations from liberals through the ages, he hits them with a list of over thirty countries thought by the Redress Trust at that time (1995/6) to practise widespread, constant use of torture. He went on to deplore the 'fashionable notion ... that self-fulfilment ... should be man's or woman's ultimate aim', and called for a revival of the 'old fashioned virtues, love, loyalty, honesty, sense of duty, and compassion', along with human rights.

His paper began with a statement that, if not true in every case, certainly reflected his own motivation:

> Scientists the world over are united by a common purpose, ideally to discover Nature's secrets and put them to use for human benefit ... When a scientist who has committed no crime is imprisoned, we feel ... [that] he or she is one of our brothers or sisters, and we feel a duty to appeal for his or her release.

The notion of science as an international brotherhood without boundaries appealed very strongly to Max. He despised any form of racial prejudice. Once, after being stuck in a railway compartment from Paris to Brussels with two Walloon Belgian couples who spent the whole journey abusing the Flemish, he wrote despairingly to Robin, 'Why are men no better than rats who murder any that does not smell like their clan?' He told successive Presidents of the Royal Society that when asked whether he felt he belonged to Austria or to Britain or to the Jews, what he really wanted to say was that he 'belonged to the Royal Society and that was all [he] needed'. His experiences had given him a proper wariness of national or religious affiliations that might override a sense of common humanity: science, he believed, gave him an identity and a shared set of values that were proof against such a risk.

Whether or not Max's interventions made any difference, every letter he sent was heartfelt, and in an age when many shrug their shoulders at the world's injustices, he was never less than outraged, almost never 'too busy' to act, and never (despite all too much evidence to the contrary) in doubt that his voice would be heard. When British troops became engaged in the second Gulf war he wrote letters protesting about civilian casualties. 'Wonderful relief!' he wrote to Robin on hearing of the 1991 ceasefire in

Iraq – though his excited announcement of this news to a conference banquet in California had been greeted with 'lukewarm applause, after which people resumed their interrupted conversations. How heartless people are!'

Within days of 11 September 2001, with the dust still rising over Ground Zero, he wrote to the British Prime Minister Tony Blair to appeal to him not to respond with military force:

> I am alarmed by the American cries for vengeance and concerned that President Bush's retaliation will lead to the deaths of thousands more innocent people, driving us into a world of escalating terror and counter-terror. I do hope that you can use your restraining influence to prevent this happening.

At the same time he circulated the heads of all the Cambridge colleges with an anti-war statement drafted by Robert Silvers of the *New York Review of Books*, though it was overtaken by events and never sent to the editor of *The Times* as intended. He told the Chancellor of the Pontifical Academy that he did not have time to correct the proofs of his recent article, as he was engaged in 'trying ... to prevent a war'. It must have cut him to the quick, as US and British troops poured into Afghanistan the following month, that all the reason and compassion he could muster made no difference. He did not live to see the second invasion of Iraq.

If Max could suffer agonies at man's inhumanity to man, he was also highly susceptible to the beauty of human creativity. That science was a creative exercise like art or music he had no doubt, and the joy he felt at a new result was a visceral as much as an intellectual pleasure, but his upbringing and temperament also disposed him to seek out more obviously aesthetic delights. He was knowledgeable about most of the major art galleries of Europe and many of those in the US. His taste was conventional, with a preference for the art of the Renaissance and a surprisingly strong response to religious art. He saw no contradiction between his personal atheism and his respect for the beliefs of others, and was critical of the antireligious stance of Richard Dawkins:

> It is one thing for scientists to oppose creationism which is demonstrably false, but quite another to make pronouncements which offend people of religious faith – that is a form of tactlessness which merely brings science into disrepute. My view of religion and ethics is simple: even if we do not believe in God, we should try to live as though we did.

When London's National Gallery asked him to choose one picture from its collections and write about it, he chose El Greco's *The Agony in the Garden of Gethsemane*, which had made a powerful impression on him when he first saw it in 1936.

Seeing this wonderful picture again made me realise that it must have been the artist's ingenious abstraction, his omission of all irrelevant detail, his beautiful composition and colours, and, above all, Christ's superbly expressive face and figure that moved me so much as a young man. They still move me today.

On one of his visits to Italy, he went to see Leonardo's *Last Supper* at the convent of Santa Maria delle Grazie in Milan. Vivien was by this time a lecturer in art history and had warned him (this was before its recent restoration) that it was in a poor state.

Nothing that I had read had prepared me for its powerful impact. Standing near the back of the refectory Leonardo's perspective makes you feel that Jesus and the apostles are there with you, and this is enhanced by the natural, vivid, excited grouping of the apostles. None of the Giottos in Padua nor the Ravenna mosaics had that emotional impact on me; only the stern, over-life size mosaic of Jesus in the cathedral at Cafalu [Sicily], telling you to follow him.

Max also had a soft spot for the bright colours of Gauguin, and was thrilled to discover, on a scientific visit to Chicago in 1988, that its Institute of Art was hosting an exhibition of his paintings from galleries and private collections all over the world. With Gisela and Vivien's guidance, he certainly had the opportunity to develop a more educated eye for works of art, but he seems to have preferred to engage his heart rather than his head when contemplating works of human creativity. Commenting to Lotte on Vivien's book on Manet, which he was helping to 'polish', he remarked: 'I often wonder whether art historians don't try to put more meaning into pictures than was ever in the painter's mind.'

As with art, so Max enjoyed music for its consoling or uplifting qualities. He had a large collection of records, but it was only in his later years, as he took more and more opportunities to hear top class performers, that he came to appreciate the impact of a live concert:

I wish you had all been with me last night to hear Giulini conduct [Christoph Eschenbach playing Beethoven's *Emperor Concerto*] at the Los Angeles Music Center. . . Just the sound of the music was tremendous

. . . tears of emotion came to my eyes when I heard the familiar tunes rendered a hundred times more beautifully than on our record. It was as big a difference as between a photograph and reality.

He loved Bach, Mozart, Haydn, Schubert, Beethoven and the less challenging works by Brahms: Gisela regarded it as one of her greatest achievements to get him to listen to Prokofiev. He 'couldn't bear' Wagner. In art as well as science, Max hated to see his heroes shown to have personal failings. He was violently repelled by Peter Shaffer's play *Amadeus*, which portrayed Mozart as both a creative genius and a nincompoop with a scatological sense of humour: he, Gisela and Vivien walked out at the interval.

If you want to make money on the stage, write a play that represents St Francis as an obscene pervert . . . I thought the play presents a travesty of Mozart . . . We were disgusted by *Amadeus*, but the ignorant mob laps it up as gospel truth and loves it.

Perhaps he, like the character of Salieri in Shaffer's play, found it impossible to accept that divine gifts should have been given to a man of less than perfect virtue. Certainly he missed the point that this was precisely the dilemma that Shaffer sought to explore.

In his retirement, Max's membership of a variety of institutions provided opportunities to form new friendships with non-scientists, whom he could engage on topics in the humanities that most interested him. A notable example is the very warm and mutually respectful correspondence he struck up with the historian Hugh Trevor-Roper, Lord Dacre. After a long career as Regius Professor of Modern History at Oxford, Dacre accepted election as Master of Peterhouse, the Cambridge college where Max was an honorary fellow. The fellowship was dominated by a group of right-wing historians led by Maurice Cowling, who said they shared 'common prejudices . . . in favour of irony, geniality and malice as solvents of enthusiasm, virtue and political elevation'. Max called them the Gang of Four. Repelled by their cynicism, Dacre discovered in Max someone with a passionate interest in history as the lived experience of individuals rather than as the play of abstract social and political forces. Dacre showed a touching humility about his own ignorance of science, and urged Max to send him all his less technical publications: he pleased Max enormously by choosing *Is Science Necessary?* as his 'book of the year' in the *Daily Telegraph*. For his part, Max was agog to learn more of the time Dacre had spent in Germany immediately after the war, researching his book *The Last Days of Hitler*.

Dacre retired in 1987, soon after he and Max had begun to know each other better, and retreated to Oxford. Thereafter they exchanged a series of letters in which they guided each other's reading, shared views on a wide range of subjects in the arts and humanities, and compared notes on their reactions to the extraordinary events that unfolded in Eastern Europe as the decade ended. The day after the Czech government resigned in November 1989, Max wrote:

> I usually hate watching the news as it homes in on every place in the world where people are killing each other, but last night I was pleased that Gisela made me turn it on. The scene in Wenceslas Square was quite fantastic ... Not in my wildest fantasies would I have imagined that something like that would happen. I was in Prague in August 1968 ... two weeks after I left the terrible blow fell ... Events in Czechoslovakia touch me even more than those in Germany. The news from Prague and the beautiful weather put me in a euphoric mood this morning, but when I tried to convey my joy to some of the Fellows of Peterhouse I found that they could not have been less moved if last night's momentous events had taken place in Polynesia.

Max's dialogue with Dacre also gave him an opportunity to discuss frankly his feelings about his own origins and about modern Germany and Austria. This was something he had rarely done with colleagues, even those, such as Vernon Ingram and Michael Rossmann, who were also refugees from Central Europe. (One exception was Aaron Klug. Klug remembers Max suddenly saying out of the blue, as they looked at the giant statue of Einstein that sits outside the National Academy of Sciences in Washington DC, 'There he is, Einstein, a Jew, sitting there in front of the NAS!') Austria had always exerted a powerful pull, and he and Gisela went there for commemorations held in 1988 to mark the fiftieth anniversary of the *Anschluss*. Most outsiders felt that Austria had been too eager to adopt 'victim' status after the war, and slow to engage in '*Vergangenheitsbewältigung*' – coming to terms with a past in which many, possibly including the President, Kurt Waldheim, had embraced Nazism. Max was pleased to discover that this was not universally true: 'Gisela and I ... found our friends deeply divided about Waldheim. In general there was less complacency about the Nazi past and there were more reminders about the reality of Nazism than the English press reported.'

While the young Max had been virulently anti-German in the years after he fled his home, in his mature years he bore no grudge against the country. In 1987 he had been happy to accept membership of the Orden

Pour le Merite, the German equivalent of the Order of Merit. On reading an article of Dacre's about post-war Germany, he responded:

> I thoroughly agree with your views about Germany . . . I regard attribution of genetically-determined and inherited national character as humbug. I therefore disagree with the attitudes of some of my ex-German-Jewish colleagues who believed in collective guilt and swore never to set foot in Germany again.

Max in his eighties was a man supremely at ease with who he was, confident in his identity as a scientist and counting himself richly endowed with the cultural heritage of Europe. Once an outsider, he had established himself at the heart of both the international scientific community and the British intellectual elite. There could be no more eloquent testimonial to his 'national treasure' status than his selection, in his eighty-seventh year, as a guest on the classic BBC radio series *Desert Island Discs*. The format of the programme – courteous but probing questions from Sue Lawley on the subject's life, interspersed with eight pieces of music that he would like to have if stranded on a desert island – suited Max's storytelling style perfectly. He felt sufficiently relaxed in Lawley's hands to deal frankly with some of the more difficult episodes in his scientific career, mentioning his early despair that he would ever solve the problem of protein structure that so many others regarded as impossible. He also confessed to a sense of inferiority in the Cavendish days that lasted until he won the Nobel prize.

> I was a chemist but worked in a department of physics on a biological problem. There were all these clever physicists who knew more mathematics than I did. I always felt very small in relation to them and not all that sure of myself. But somehow the Nobel prize told me I was probably quite good at research and that really boosted my determination to carry on.

Max put a lot of thought into his musical choices. He decided not to pick his all-time favourites, but to choose passages that resonated with particular episodes in his life. Listeners were moved by his choice of the Act II Finale from Beethoven's *Fidelio*, in which Don Fernando comes to release the prisoners with the words 'A brother has come to seek his brothers.' He said it reminded him of the opening of the concentration camps at the end of the war: Lawley had to press him gently to tell his own story of rejection and incarceration in those dark years. For the lighter moments in his life he chose Figaro's Act I aria from *The Barber of Seville*, Jessye Norman singing 'He's got the whole world in his hands', and the

Pangloss aria from Leonard Bernstein's *Candide*, remarking that Voltaire's satire was very apt for the events of the twentieth century. But if he could take only one piece, it would have to be the third movement of Beethoven's piano sonata in E major played by Vladimir Ashkenazi, the 'consoling' music that had got him through so many periods of ill-health and misfortune.

Finally Lawley asked him what luxury he would take to the island, where the solitude would make any scientific activity 'hopeless'.

'A pair of skis', he replied.

'You never know – it might snow.'

Select Bibliography

Three of Max's own collections of articles have provided a starting point for much of the writing of this book:

Max Perutz, *Science is Not A Quiet Life: Unravelling the Atomic Mechanism of Haemoglobin*, Imperial College Press/World Scientific, 1997 (abbreviated in the Notes as *SINAQL*).

Max Perutz, *I Wish I'd Made You Angry Earlier: Essays on Science, Scientists and Humanity*, Cold Spring Harbor Laboratory Press, expanded edition 2003 (abbreviated in the Notes as *IWIMYAE*).

Max Perutz, *Is Science Necessary? Essays on Science and Scientists*, Oxford University Press, 1991.

Other books that have provided essential background information are:

Andrew Brown, *J.D. Bernal: The Sage of Science*, Oxford University Press, 2005.

Soraya de Chadarevian, *Designs for Life: Molecular Biology after World War II*, Cambridge University Press, 2002.

Francis Crick, *What Mad Pursuit: A Personal View of Scientific Discovery*, Basic Books, 1988.

Thomas Hager, *Force of Nature: The Life of Linus Pauling*, Simon & Schuster, 1995.

Richard E. Dickerson, *Present at the Flood: How Structural Molecular Biology Came About*, Sinauer, 2005.

Graeme K. Hunter, *Light is a Messenger: The Life and Science of William Lawrence Bragg*, Oxford University Press, 2004.

Horace Freeland Judson, *The Eighth Day of Creation: Makers of the Revolution in Biology*, Jonathan Cape, 1979; paperback reprint Penguin, 1995.

Brenda Maddox, *Rosalind Franklin: The Dark Lady of DNA*, HarperCollins, 2002.

James D. Watson, *The Double Helix*, critical edition edited by Gunther S. Stent, Weidenfeld & Nicolson, 1981.

A note on original sources

I have drawn on a number of archival resources, which are listed in the Acknowledgements and cited in the notes. Max's own papers were never formally collected and archived. Family letters (designated 'private papers' in the notes) remain in the hands of his family. Other material that was either in his home or his office at the LMB at the time of his death is to be deposited at the library of Churchill College in Cambridge (designated Perutz papers, Churchill College, Cambridge in the notes). A small collection of correspondence and notebooks is held at the Venter Institute, Rockville, Maryland, (designated Perutz papers, Venter Institute).

Letters derived from all these sources are to be published shortly: Vivien Perutz (ed.), *Selected Letters of Max Perutz*, Cold Spring Harbor Laboratory Press, forthcoming.

In 2001 the Vega Science Trust recorded eight hours of video interviews with Max about his life and work. Extracts are available to view on the web at www.vega.org.uk/video/programme/1

Notes

Chapter 1 Scenes from a Vienna childhood

1 '[E]ven if there had been no Hitler': Max Perutz, Foreword to *Hitler's Gift: Scientists who Fled Nazi Germany* by Jane Medawar and David Pyke, Piatkus, 2000.

2 Hugo ... had served his apprenticeship in Manchester: Tape 1, C464/22, British Library Sound Archive.

   [T]hey founded Brüder Perutz: Felix Perutz, *100 Jahre Bruder Perutz 1862–1962*, Volkswirtschaftliche Verlags-Gesellschaft, Vienna.

3 Dely's father Ferdinand Goldschmidt had been married twice: interview with Alice Frank, Max's first cousin.

4 [T]he Empire was a political mosaic: Barbara Jelavich, *Modern Austria: Empire and Republic* Cambridge University Press, 1987.

5 '[H]e once told my sister': Max Perutz, unpublished, undated handwritten memoir, private papers.

   [A] contemporary of Max's remembers: Fritz Eirich, interview with the author.

   Dely took the family to live in Reichenau: Max Perutz, unpublished, undated handwritten memoir, family papers.

   [A]s memorably described: Eric Hobsbawm, *Interesting Times: A Twentieth Century Life*, Allen Lane, 2002.

   [T]he Hungarian and Czech factories of Brüder Perutz supplied fabrics: Felix Perutz, *100 Jahre Bruder Perutz 1862–1962*, Volkswirtschaftliche Verlags-Gesellschaft, Vienna.

6 Felix converted to Catholicism: Marion Turnovszky, interview with the author, June 2003.

   Lotte gave her religion as '*Konfessionslos*': Nationalen S.S. 1932, O-P, Vienna University Archives.

   [H]e 'became a very devout little boy': Tape 1, C464/22, British Library Sound Archive.

   Although Jews constituted no more than 12 per cent: Steven Beller, *Vienna and the Jews 1867–1938*, Cambridge University Press, 1989.

7 [T]he conversation was mostly about servants: Alice Frank, interview with the author, October 2003.

7   One of the few possessions she took with her: Vivien Perutz, interview with
    the author.
    'My parents had wide interests which included almost everything except
    science': Max Perutz, unpublished, undated handwritten memoir, family
    papers.
    His earliest memory: ibid.
8   'She wasn't very bright': Tape 1, C464/22, British Library Sound Archive.
    'If I had been more with my parents': ibid.
9   'When I woke up at six in the morning,': Max Perutz, unpublished, undated
    handwritten memoir, family papers.
10 She had first noticed: Eve Machin, unpublished memoir.
    'I went with him on sedate outings': ibid.
    Knöpfelmacher advised them 'not to coddle me': Max Perutz, unpublished,
    undated handwritten memoir, family papers.
11 The Theresianum had been founded: Eugen Guglia, *Das Theresianum in
    Wien*, Bohlau Verlag, 1996.
    'There were barons, counts and princes': Max Perutz, unpublished, undated
    handwritten memoir, family papers.
    [T]hey boarders tended to make friends: Gottfried Peloschek, interview with
    the author, June 2003.
    'If my mother had been with me always': Tape 1, C464/22, British Library
    Sound Archive.
12 He learned Latin and Greek: Gottfried Peloschek, interview with the author,
    June 2003. Max told his daughter that he had never studied Greek, but
    then he remembered very little about his schooldays.
13 'It was difficult to see the delicate little boy': Eve Machin, unpublished
    memoir.
    'During this stay his parents gave a ball': ibid.
14 'The chemistry teacher also taught': Tape 1, C464/22, British Library Sound
    Archive.
    'In old Austria advancement used to depend': Max Perutz, unpublished,
    undated handwritten memoir, family papers.
15 'Very happily for me': Eve Machin, unpublished memoir.
    'We laughed as we parted': ibid.
16 'Half an hour ago I was told': Max to Evelyn Baxter, 1 November 1933
    (translated from the German).
    'He said that there's no point': Max to Evelyn Baxter, 5 August 1932 (trans-
    lated from the German).
    Fritz Eirich ... successfully talked Hugo Perutz round: Fritz Eirich, inter-
    view with the author, May 2004.

17 'I who had never worked in my life': Max to Evelyn Baxter, 7 December
    1932 (translated from the German).

    The 759 pages of Karl Hoffman's *Inorganic Chemistry*: M.F. Perutz, 'Linus
    Pauling (1901–1994)', *Nature Structural Biology*, vol. 1, 1994.

18 'I imagined that the atmosphere': Max to Evelyn Baxter, 7 December 1932
    (translated from the German).

    'It was horrible, but I did it': Gretl Petziwal (née Schloegl), interview with
    the author, June 2003.

    'Originally I wasn't in love with her': Max to Evelyn Baxter, 8 December
    1933 (translated from the German).

    'Now I stand here poor idiot and am a function of her moods': I am grateful
    to Dr Ronald Gray for pointing out that this is a quotation from Goethe's
    Faust.

    'And depending on whether I am seeing her today, whether she's been with
    me or with my friends, I'm either very happy or sad to death': Dr Gray
    points out that this is a quotation from Goethe's *Egmont*.

19 'Arriving by rocket on the moon': Max to Evelyn Baxter, 20 August 1933
    (translated from the German).

    'You can't imagine how the Hitler psychosis': Max to Evelyn Baxter,
    27 April 1933 (translated from the German).

    [H]e was 'not terribly interested in politics': eg Tape 1, C464/22, British
    Library Sound Archive.

20 'My parents are terribly upset': Max to Evelyn Baxter, 27 April 1933 (trans-
    lated from the German).

    'Actually the government here': Max to Evelyn Baxter, 24 February 1934
    (translated from the German).

21 Up to a third of the students were Jewish: Steven Beller, *Vienna and the
    Jews 1867–1938*, Cambridge University Press, 1989.

    'Nazi thugs permanently patrolled': Interview given by Max for Imperial
    War Museum oral history project on Britain's Response to the Refugee
    Problem, 24 May 1980, IWM Collection.

    'At the University there are similar incidents': Max to Evelyn Baxter,
    25 October 1932 (translated from the German).

    [H]is family had made plans: Max to Evelyn Baxter, 5 May 1933 (trans-
    lated from the German).

22 'My father's business is going very badly': Max to Evelyn Baxter, 8 December
    1933 (translated from the German).

23 'In chemistry you simply get to know': Max to Evelyn Baxter, 31 May 1933
    (translated from the German).

    'Peter Wooster complained': Fritz Eirich, interview with the author, May 2004.

24 'I would like to ask you to-day': H. Mark to J.D. Bernal, 11 October 1935, J. 178, Cambridge University Library MS. Add. 8287 John Desmond Bernal Papers.

Eirich took the line: Fritz Eirich, interview with the author, May 2004.

Chapter 2 'It was Cambridge that made me'

26 'In 1936 I left my hometown of Vienna': Max Perutz, 'How the Secret of Life was Discovered', *IWIMYAE*, pp. 197–205.

'Sage', John Desmond Bernal: See Andrew Brown, *J.D. Bernal: The Sage of Science*, Oxford University Press, 2005.

27 'He was the most incredibly magnetic': Max Perutz, tape 3, C464/22, British Library Sound Archive.

While he would not commit himself to funding an institute: See Gary Werskey, *The Visible College*, Free Association Books, 1988.

28 'Why water boils at 100°': Max Perutz to Gerald Holton, 9 July 1996, copy in Perutz papers, Churchill College, Cambridge.

29 Bragg was born in Adelaide: See Graeme K. Hunter, *Light is a Messenger: the Life and Science of William Lawrence Bragg*, Oxford University Press, 2004.

32 [T]heir 1934 paper on pepsin: J.D. Bernal and D.M. Crowfoot, 'X-ray photographs of crystalline pepsin', *Nature* 133, 794–5, 1934.

Crowfoot published X-ray photographs: see Georgina Ferry, *Dorothy Hodgkin: A Life*, Granta Books, 1999.

'It took me some time to realise': Max Perutz, Tape 3, C464/22, British Library Sound Archive.

33 CSAWG had been formed: Gary Werskey, *The Visible College*, Free Association Books, 1988.

'I realise that this application': J.D. Bernal to Burkhill, 7 October 1936, J. 178, Cambridge University Library MS Add. 8287, John Desmond Bernal Papers.

35 Max spent his time 'sitting at this beastly tube': Max Perutz, Tape 3, C464/22, British Library Sound Archive.

'I must have received a large dose of radiation': Max Perutz, Tape 3, C464/22, British Library Sound Archive.

Max wrote up his preliminary findings: M.F. Perutz, '"Iron-rhodonite" (from slag) and pyroxmangite and their relation to rhodonite', *Mineralogical Magazine* 24, 573–6, 1937.

'I wanted to work like Bernal': Max Perutz, Tape 3, C464/22, British Library Sound Archive.

36 One of the group, John Constant: Interview with John Constant, 3 September 2003.

36 Carter told Max of a discovery: Max Perutz, 'My first great discovery', unpublished lecture given at Peterhouse, Cambridge, 21 September 1994.

37 Max found that the 'complex external structure': M.F. Perutz, 'Radioactive nodules from Devonshire, England', *Mineralogische und Petrographische Mitteilungen* 51, 141–61, 1939.

38 Sir Frank Smith ... left Bernal in no doubt: Sir Frank Smith to Bernal 15 October 1937, J. 178, Cambridge University Library MS. Add. 8287, John Desmond Bernal Papers.

    'I feel sure that the paper is not one for the Royal': Bragg to Bernal, 26 May 1938, Box 77L, Bragg archive, RI.

    [H]e eventually succeeded in getting the paper published: M.F. Perutz, 'Radioactive nodules from Devonshire, England', *Mineralogische und Petrographische Mitteilungen* 51, 141–61, 1939.

39 In July 1937 he went home to Austria: Max Perutz, Tape 3, C464/22, British Library Sound Archive.

40 [H]is cousin Gina: Daughter of his father's brother Robert Perutz.

    'It has emerged during my stay in Cambridge: Max to Felix Haurowitz, 1 August 1937, Haurowitz Archive, Lilly Library, Indiana University, Bloomington, Indiana.

    Felix Haurowitz was Max's senior: Frank W. Putnam, 'Felix Haurowitz March 1, 1896-December 2, 1987', *Biographical Memoirs of the National Academy of Sciences* 64, 134–63, 1994.

42 'We were refreshingly vulgar:' Max to Evelyn, 6 September 1937.

    'We had only 4 hours to spend together': Max to Evelyn 6 September 1937.

43 'My fellow students regarded me': M.F. Perutz, 'The Haemoglobin Molecule', *Scientific American* 211, 64–76, 1964.

    Adair was 'a terribly modest man': Max Perutz, Tape 4, C464/22, British Library Sound Archive.

    [H]e arranged for Gowland Hopkins's daughter: Max Perutz, 'Origins of Molecular Biology', *New Scientist*, 31 January 1980, 326–9.

    'He never washed any of his glassware': Max Perutz, 'Keilin and the Molteno', in *IWIMYAE*, p. 375.

44 'The damn crystals are not growing': Max to Evelyn, 26 September 1937.

    'I am supposed to do another work': Max to Evelyn, 6 November 1937.

    'Just imagine me standing': ibid.

    'In spite of all the compliments': ibid.

45 'I started a love affair': Max to Evelyn, 26 September 1937.

    'I suppose Max and I had a basis': Anne Corden, letter to the author.

    'It would have been quite easy': Max to Evelyn, 6 November 1937.

46 'There is always plenty of excitement': ibid.

47 'I found three good haemoglobin crystals': Max to J.D. Bernal, undated, must be November/December 1937, J. 178, Cambridge University Library MS. Add. 8287, John Desmond Bernal Papers.

After Dorothy Crowfoot had developed her first insulin photograph: Georgina Ferry, *Dorothy Hodgkin: A Life*, Granta Books, 1999, p. 110.

Hodgkin later wrote of them: Dorothy Hodgkin, 'John Desmond Bernal 10 May 1901–15 September 1971', *Biographical Memoirs of the Fellows of the Royal Society* 26, 17–84, 1980.

48 Max did admit to showing them: Max Perutz, Tape 4, C464/22, British Library Sound Archive.

'This was the first structural information': Max Perutz, SINAQL, p. 35.

49 The paper describing their results: J.D. Bernal, I. Fankuchen and M.F. Perutz, 'An X-ray study of chymotrypsin and haemoglobin', *Nature* 141, 523–4, 1938.

'There are certain events, like the death of a loved person': Max Perutz, 'Origins of Molecular Biology', *New Scientist*, 31 January 1980, 326–9.

'I feel as if something had died within me': Max to Evelyn, 19 June 1938.

50 'I must apologise for disturbing your "peaceful" existence': Max to J.D. Bernal, 26 April 1938, J. 178, Cambridge University Library MS. Add. 8287, John Desmond Bernal Papers.

'Firstly my chances of getting a grant': ibid.

51 'That my own brother should have become': Max to Evelyn, 19 June 1938.

53 'Our fancy was fired': W.L. Bragg quoted in Horace Judson, *The Eighth Day of Creation*, Jonathan Cape, 1979.

He had a preliminary meeting in London: Walter Tisdale, officer's diary, 23 November 1938, folder 561, box 43, series 401, Record Group 1.1, Rockefeller Foundation Archives, Rockefeller Archive Center, Sleepy Hollow, New York (hereinafter designated RAC).

'[E]ventually this scientific center of the British Empire': ibid.

'In assuming the professorship at Cambridge': Discussion on Grant in Aid no. 38163, 3 January 1939, folder 561, box 43, series 401, RG 1.1, Rockefeller Foundation Archives, RAC.

Chapter 3 'The most dangerous characters of all'

55 'In College ... they admired what Hitler had done': Interview given by Max for Imperial War Museum oral history project on Britain's Response to the Refugee Problem, IWM Collection.

56 [Ewald's] mother, wife and children all came with him: Max Perutz, unpublished, undated ms of talk, private papers.

'Ewald's upright and loveable character': ibid.

56 Max told a story of how he raised the money: Unpublished memoir dictated January 2002, private papers.

57 There are some inconsistencies: In an interview given to the National Life Story Collection of the British Library (tape 6, C464/22, British Library Sound Archive) only a few months before he wrote this account, Max said that Seligman had refused his request and that he couldn't remember how he borrowed the money. Many later letters refer to a long-term debt of £500 to Heini Granichstaedten, who had come to the UK and worked in Edinburgh before moving to the US after the war. Max and his brother and sister jointly repaid this loan over several years. There are no references to repayment of a loan to Seligman.

The couple ... spent the rest of their lives in ... 'hideous lodgings': Unpublished memoir dictated January 2002, Perutz papers, Churchill College, Cambridge.

58 In May 1939, for example, he wrote to Bragg: Max Perutz to W.L. Bragg, 31 May 1939, 78A/47, Bragg Archive, Royal Institution.

[W]hat ensued has been exhaustively researched: Peter and Leni Gillman, *Collar the Lot!*, Quartet, 1982.

59 'His book transformed the chemical flatland': M.F. Perutz, 'Linus Pauling (1901–1994)', *Nature Structural Biology* 1, 667–71, 1994.

60 Max successfully submitted his thesis: Max Perutz, 'The Crystal Structure of Horse Methaemoglobin', Cambridge University, copy in MRC LMB archives.

The policeman said it would be 'just for a few days': Max Perutz, 'That Was the War: Enemy Alien', *New Yorker*, August 1985, republished in *IWIMYAE*, pp. 73–106.

61 [T]hey were 'herded into a huge empty shed': Max Perutz, 'That Was the War: Enemy Alien', *New Yorker*, August 1985, republished in *IWIMYAE*, pp. 73–106.

'The duration of our internment': Max to his parents, 20 May 1940, private papers.

62 'If the internment were not': Max to his parents, 21 May 1940, private papers.

'Yesterday ... we arrived at a new camp': Max to Evelyn, 21 May 1940, private papers.

'Then a German had been a German': Max Perutz, 'That Was the War: Enemy Alien', *New Yorker*, August 1985, republished in *IWIMYAE*, pp. 73–106.

63 'I am very well, sleep well, eat enough': Max to his parents, 24 May 1940, private papers.

63 '[T]wo bright German medical researchers': Max Perutz, 'That Was the War: Enemy Alien', *New Yorker*, August 1985, republished in *IWIMYAE*, pp. 73–106.

64 'For Heaven's sake shake my professors': Max to his parents, 17 June 1940, private papers.

'[O]ught to abide by the decision of the Government': W.L. Bragg to Walter Tisdale, 17 July 1940, folder 562, box 43, series 401, RG 1.1, Rockefeller Foundation Archives, RAC.

65 'Never in my life was I so helpless': Dely Perutz to Max, 26 June 1940, private papers.

66 'To this revolting scene': Max Perutz, 'That Was the War: Enemy Alien', *New Yorker*, August 1985, republished in *IWIMYAE*, pp. 73–106.

'We must be glad if Max is safe': Dely Perutz to Hugo Perutz, 6 July 1940, private papers.

'We cannot carry on the work here without him': W.L. Bragg to Walter Tisdale, 17 July, folder 562, box 43, series 401, RG 1.1, Rockefeller Foundation Archives, RAC.

Bernal wrote to Linus Pauling: J.D. Bernal to Linus Pauling, 25 July 1940, Ava Helen and Linus Pauling Papers, Oregon State University Special collections, box 01.304.1.

68 'To have been arrested, interned and deported': Max Perutz, 'That Was the War: Enemy Alien', *New Yorker*, August 1985, republished in *IWIMYAE*, pp. 73–106.

69 Max ... found [Fuchs] 'very distant': Max to Lord Dacre, Dacre papers.

70 'Now things do not look quite as black': Max to Anne Hartridge, 14 August 1940, private papers.

'Most important is to get father released': Max to Lotte Perutz, 23 August 1940, private papers.

'He seems keen to return to England': W.L. Bragg to Dely Perutz, 30 August 1940, private papers.

71 '[T]hat the xray photos were amazing': Dely Perutz to Hugo Perutz, undated but inferred to be 13 August 1940, private papers.

Testimonials from researchers in the US: See folder 562, box 43, series 401, RG 1.1, Rockefeller Foundation Archives, RAC.

'There has been a serious misunderstanding': Max to David Keilin, 8 September, private papers.

72 'Competent for our release is solely the Home Office': Max to Lotte, 22 September 1940, private papers.

72 'I should like to swim across the ocean': Max to Anne Hartridge, 8 October 1940.

73 '[T]he angry inmates spent their time . . . "locked in futile arguments"': Max Perutz, 'That Was the War: Enemy Alien', *New Yorker*, August 1985, republished in *IWIMYAE*, pp. 73–106.

'My release order arrived': Max to Lotte, 29 November 1940, private papers.

74 'I slept in a warm cabin': Max Perutz, 'That Was the War: Enemy Alien', *New Yorker*, August 1985, republished in *IWIMYAE*, pp. 73–106.

'In the camps, concerts were often the only diversions': Max to Franz Perutz, 21 January 1941, letter intercepted by the Home Office censor, HO 214/4, National Archives.

75 'He was smiling and sunburnt': Letter to author from Anne Corden, 2006.

'I had spent days and nights': Max to Evelyn Machin, 29 January 1941, private papers.

Chapter 4 Home and homeland

76 Bragg seemed genuinely delighted to have him back: Max to Evelyn Machin, 29 January 1941, private papers.

77 Max had done 'a very pretty piece of work on proteins': W.L. Bragg to R.W. James, 29 December 1941, 78A/47 Bragg archive, Royal Institution.

Max's paper, published in *Nature*: M.F. Perutz, 'X-ray analysis of haemoglobin', *Nature*, 149, 491–6, 1942.

78 'It's quite an important [paper]': Max to Lotte Perutz, 16 March 1942, private papers.

'I had a good idea that the attitude': Interview with Max Perutz, Oral history project on Britain's Response to the Refugee Problem, Imperial War Museum Sound Archive, IWM Collection.

79 Max described as 'the only internee with whom I formed a close, affectionate bond': Max Perutz, unpublished memoir dictated January 2002, private papers.

[H]e would be a great scientist one day, 'which had never occurred to me': Max to Gisela Perutz, 25 May 1973, private papers.

'It was really terrible': Max Perutz, unpublished memoir dictated January 2002, private papers.

80 'We were too different in temperament': Max to Lotte Perutz, 16 November 1941, private papers.

[H]e was 'received by a gorgeous girl': Max Perutz, unpublished memoir dictated January 2002, private papers.

80 'I was very impressed by the letter Max wrote': Gisela Perutz, interview with the author, 17 February 2003.

81 'She has lovely features': Max to Lotte Perutz, 16 November 1941, private papers.

'He [is] . . . not very tall': Gisela Peiser to her parents, 16 November 1941, private papers.

82 'Father likes Gisela a lot': Max to Lotte Perutz, 5 January 1942, private papers.

'I always belittled the effects': Max to Lotte Perutz, 16 March 1942, private papers.

83 'It is frightful to watch her': Max to Evelyn Machin, 20 December 1941, private papers.

Annemarie dismissed Dely as a 'silly woman': Annemarie Wendriner to the Peiser parents, November 1941, private papers.

84 'It is really impossible for us': Max to Evelyn Machin, undated, probably January 1942, private papers.

'Everybody, including the parents': Max to Lotte Perutz, 12 April 1942, private papers.

'[F]or the sake of her decent looks': Annemarie Wendriner to the Peiser parents, April 1942, private papers.

'[Y]ou have witnessed changes': Copy of speech by Herbert Peiser read by Steffen Peiser at Max and Gisela's wedding party, 28 March 1942, private papers.

85 'Gis and I are very happy': Max to Evelyn Machin, 28 April 1942, private papers.

[A] four-page letter: Max to Peiser parents, 17 May 1942, private papers.

[T]he happy home she gave him 'was the ground': Max Perutz, dedication to *SINAQL*.

'Gis and I are happy together': Max to Lotte Perutz, 12 April 1942, private papers.

86 'All the lines are crooked': Max to Evelyn Machin, 21 March 1943, private papers.

[A]s he reported to his paymasters: Max to H.M. Miller, 2 August 43, folder 564, box 43, series 401, RG 1.1, Rockefeller Foundation Archives, RAC.

'In my first lecture I introduced lattice theory': Max Perutz, *Protein Structure: New Approaches to Disease and Therapy*, W.H. Freeman, 1992.

87 A Special Branch policeman . . . appeared one evening: M.F. Perutz, unpublished diary 24 August–17 October 1943, private papers.

'I was excited': ibid.

87 'Everybody was very friendly': ibid.

'Any success I may have had in my war work': Max to Lotte Perutz, 17 September 1943, private papers.

89 'It is not easy in our minute little flat': Max to Franz Perutz, 27 February 1945, private papers.

'Mother put all her jewels on': Max to Lotte and Franz Perutz, 15 April 1945, private papers.

90 'I am very happy here': Max to Felix Haurowitz, 21 May 1945, Haurowitz Papers, Lilly Library, Indiana University, Bloomington, Indiana.

'As I walked across Trinity Great Court': Max to Lotte Perutz, 25 December 1945, private papers.

Chapter 5 Mountains and Mahomet

91 'My own prime purpose': Max to Evelyn Machin, 20 July 1946, private papers.

'Bragg was a little staggered': Max to Franz and Senta Perutz, 11 April 1946, private papers.

92 'Everyone goes about in new clothes': Max to Evelyn Machin, 20 July 1946, private papers.

'Gis had a terrible time in Zurich': Max to Lotte Perutz, 6 June 1946, private papers.

[H]aving lost the Norwegian hickory wood skis: He had left these with his schoolfriend René Jaeger, now a Swiss diplomat like his father, in Bern. A contemporary letter to his parents says that when he went to look for them in 1946 the skis had disappeared, and that he replaced them with a second-hand pair. Much later, he recorded that he had found the original pair exactly where he left them, and that the loss of the skis had featured in a 'recurring nightmare' he suffered from throughout the war. Either he found them subsequently, or he could not bring himself to deal with his nightmare coming true and bestowed the identity of the original skis on their replacements.

93 'I am not good enough really either on foot or on skis': Gisela Perutz to her parents, 6 June 1946, private papers.

'Gisela likes the views': Max to Evelyn Machin, 20 July 1946, private papers.

94 'It will help me to consolidate my recovered physical strength': Max to Peiser parents, 13 June 1946, private papers.

'There is something about the mountains and the flowers': Max to Evelyn Machin, 20 July 1946, private papers.

95 A mountain diary: Max Perutz, climbing diary 1926–47, private papers.

95 'We . . . had four days of creeping about glaciers': Max to Evelyn Machin,
6 September 1937, private papers.

The Jungfraujoch Research Station: see www.ifjungo.ch/jungfraujoch/

'luxurious . . . I have got a pleasant room to myself': Max to Evelyn Machin,
19 June 1938, private papers.

96 'It looks very romantic and blue': Max to Evelyn Machin, 19 June 1938,
private papers.

97 'Notwithstanding the destruction of our apparatus': M.F. Perutz and Gerald
Seligman, 'A crystallographic investigation of glacier structure and the
mechanism of glacier flow', *Proceedings of the Royal Society of London* (A),
172, 335–60, 1939.

Bernal pronounced himself 'very excited': J.D. Bernal to Max, 18 July 1938,
J. 178, Cambridge University Library MS.Add.8287, John Desmond
Bernal Papers.

'This taught me to be humble in my approach to Nature': Max Perutz,
SINAQL, p. 601.

[A] short paper in *Nature*: T.P. Hughes, M.F. Perutz and Gerald Seligman,
'Glaciological results of the Jungfraujoch research party, *Nature* 143, 159,
1939.

[A] much longer and more detailed [paper]: M.F. Perutz and Gerald Seligman,
'A crystallographic investigation of glacier structure and the mechanism
of glacier flow', *Proceedings of the Royal Society of London* (A), 172, 335–60,
1939.

98 'A gaunt figure with a long, sallow face': Max Perutz, *IWIMYAE*, p. 82.

99 'The cover name for this second project': Geoffrey Pyke to Lord Louis
Mountbatten, 23 September 1942, ADM 116/4818, National Archives.

He misspelled the name: Some accounts have blamed the typing error on
an Admiralty clerk, but Pyke's proposal was sent from New York and
presumably typed there.

100 Mountbatten was delighted: Memo from Lord Louis Mountbatten,
10 October 1942, ADM 116/4825, National Archives.

'I attach the greatest importance': Memo from Winston Churchill to Chiefs
of Staff Committee, 7 December 1942, ADM 116/4825, National
Archives.

'The scheme is only possible': ibid.

[A] further document: Document signed 'B', and numbered H/39, ADM
116/4822, National Archives.

101 '[He] told me, with the air of one great man': Max Perutz, *IWIMYAE*,
p. 82.

Lieutenant Douglas Grant . . . first suggested: Note from Douglas Grant to

CMSF (Sir Harold Wernher), 31 March 1943, H/664, ADM 116/4822, National Archives.

102 'I am engaged on war-work now': Max to Evelyn Machin, 21 March 1943, private papers.

'Accordingly I took the bull by the horns': Douglas Grant, letter to Geoffrey Pyke, 22 February 1943, H/361, ADM 116/4821, National Archives.

'Speaking of never meeting': Max to Gisela Perutz, undated 1943, private papers.

103 [A] series of long and chatty 'News Flashes': See, for example, ADM 116/4821 and 4822, National Archives.

104 'It would greatly assist my work': Max to Geoffrey Pyke, 25 March 1943, H/632, ADM 116/4822, National Archives.

'About 5 miles outside Cambridge': Max Perutz, report on progress, 20 April 1943, ADM 116/4822, National Archives.

105 No one ever satisfactorily solved the problem: Max Perutz, *IWIMYAE*, p. 90.

106 Mountbatten got Habbakuk onto the agenda: Andrew Brown, *J.D. Bernal: The Sage of Science*, Oxford University Press, 2005, p. 232.

Grant reported that he himself: M.F. Perutz, unpublished diary, 24 August – 17 October 1943, private papers.

'Yesterday the famous cable': ibid.

107 'He has done his bit, of course': ibid.

After the war Pyke found no further outlets: M.F. Perutz, 'An inventor of supreme imagination', *Discovery*, May 1948, pp. 149–51.

'Dawn found us travelling over a sea of fleecy clouds': ibid.

109 'The unpleasant side of American life': Max to Gisela Perutz, 22 September 1943, private papers.

'[A] quiet and very pleasant man: Max to Gisela Perutz, 3 October 1943, private papers.

'The people here are far more good-looking', Max to Gisela Perutz, 10 October 1943, private papers.

110 'It is strongly recommended': Memo from Chair of US Habbakuk Board (Admiral Moreell) to Chief of Naval Ops, 23 October 1943, ADM 116/4820, National Archives.

'After six days we steamed up the Firth of Clyde': Max Perutz, *IWIMYAE*, p. 94.

111 'I had a very interesting time': Max to Evelyn Machin, 3 June 1944, private papers.

[T]he treatment 'got me back on my feet within 48 hours': ibid.

112 Max wrote up his own investigations: M.F. Perutz, 'A description of the

iceberg aircraft carrier and the bearing of the mechanical properties of frozen wood pulp on some problems of glacier flow', *Journal of Glaciology* 1, 95–104, 1947.

112 'Gerrard: What happens if the pipe goes down': Max Perutz, 'A report on the life and work of the Jungfraujoch research party', unpublished manuscript, private papers.

113 [H]e wrote it all up as a comedy drama: ibid.

115 Max's alpine adventure provided the first experimental means: J.A.F. Gerrard, M.F. Perutz and A. Roch, 'Measurement of the velocity distribution along a vertical line through a glacier', *Proceedings of the Royal Society of London (A)*, 213, 546–58, 1952.

116 [T]he Antarctic Place-Names Committee named a glacier: see Durward Cruickshank, 'The Crystallographic Tourist in the Antarctic', *Crystallography News*, March 1999.

Roger Hanna later remembered: Roger Hanna, interview with the author, October 2004.

117 Max wrote in his diary: Max Perutz, climbing diary, 1926–47, private papers.

'My "Course in Alpine skiing" was a roaring success': Max to Franz and Senta Perutz, 19 April 1947, private papers.

118 'This has been the most successful holiday ever': Max to Lotte Perutz, 28 July 1985, private papers.

'You emerged from the magic snow-covered woods': Max Perutz, memoir dictated January 2002, private papers.

Chapter 6 How haemoglobin was not solved

122 'I have just completed one of the worst and most tedious jobs': Max to Franz and Senta Perutz, 11 April 1946, private papers.

123 'I should have preferred a graduate': Max to Blair Hanson, 1 August 44, folder 565, box 44, series 401, RG 1.1, Rockefeller Foundation Archives, RAC.

Olga Weisz . . . had a similarly disheartening experience: Olga Kennard, interview with the author, 16 June 2005.

[I]t led to a joint publication: M.F. Perutz and O. Weisz, 'The crystal structure of tribromo(trimethylphosphine)gold', *Journal of the Chemical Society*, 438–42, 1946.

124 [T]heir joint preliminary study: M.F. Perutz and O. Weisz, 'Crystal structure of human carboxyhaemoglobin', *Nature* 160, 786–8, 1947.

The first thing she did there was to develop and publish: Olga Weisz and W.F. Cole, 'An improved technique for setting single crystals from zero layer-line photographs', *Journal of Scientific Instruments* 25, 213–14, 1948.

124 'She has just been elected FRS: Max to Lotte Perutz, 15 April 1945, private
papers.

Gutfreund had attended Max's lectures: Herbert Gutfreund, interview with
the author, 25 June 2004.

125 [H]e was 'more useful to him and Gisela as a babysitter': ibid.

His name was John Kendrew: For John Kendrew's life, see Kenneth Holmes,
'Sir John Cowdery Kendrew', *Biographical Memoirs of the Fellows of the
Royal Society* 47, 315–32, 2001.

[H]is whole air said 'I know what I'm doing': Olga Kennard, interview with
the author, 16 June 2005.

Among other things, they were testing: Ritchie Calder, 'Bernal at war', in
Brenda Swann and Francis Aprahamian (eds), *J.D. Bernal: A life in science
and politics*, Verso, 1999, p. 189.

126 'He knew no X-ray crystallography': Max to Peiser parents, 10 August 1946,
private papers.

'[P]robably one of the brightest chaps': Gerald Pomerat, officer's diary,
11 July 1946, folder 566, box 44, series 401, RG 1.1, Rockefeller
Foundation Archives, RAC.

The only conclusion he and Max could draw: J.C. Kendrew and M.F.
Perutz, 'A comparative X-ray study of foetal and adult sheep haemoglobins', *Proceedings of the Royal Society of London* (A), 194, 375–98,
1948.

127 'I am still very anxious to extend the work': M.F. Perutz, research report,
1 July 43–1 August 44, folder 565, box 44, series 401, RG 1.1, Rockefeller
Foundation Archives, RAC.

'The structure of the protein': J.C. Kendrew, report on research, C.27,
J.C. Kendrew papers, Bodleian Library, Oxford.

128 'If I am going to continue my work here': Draft letter from Max to
W.L. Bragg, 3 July 1944, private papers.

'Owing to the time I lost:' ibid.

129 [A] despairing Max confided in David Keilin: Max Perutz, 'Origins of molecular biology', *New Scientist*, 31 January 1980, 326–9.

'The unit exists already': D. Keilin to E. Mellanby, 31 May 1947, FD1 426,
MRC archives, National Archives.

'[W]hen it came to a matter of subsidising foreigners of his own race': Note
of a meeting between E. Mellanby and W.L. Bragg, 2 June 1947, ibid.

Mellanby wrote that 'Rather to [his] surprise': E. Mellanby to W.L. Bragg,
20 October 1947, ibid.

130 'Perutz is very clever and a beautiful worker': W.L. Bragg to E. Mellanby,
22 October 1947, ibid.

130 Max and Kendrew chose the cumbersome handle: Note of a meeting, 30 October 1947, ibid.

'In the English way': Horace Freeland Judson, *The Eighth Day of Creation*, Jonathan Cape, 1979.

'I have been made head of a research unit': Max to Felix Haurowitz, 20 January 1948, Haurowitz archive, Lilly Library, University of Indiana, Bloomington, Indiana.

Mellanby's advice that he should 'get Englishmen': Note of a meeting, 30 October 1947, FD1 426, MRC archives, National Archives.

The next recruit was Hugh Huxley: Hugh Huxley, interview with the author, 25 May 2004.

131 The first had appeared: J. Boyes-Watson, E. Davidson and M.F. Perutz, 'An X-ray study of horse methaemoglobin. I.', *Proceedings of the Royal Society of London* (A), 191, 83–132, 1947.

133 'I have shown my old work a hundred times': Max to J.D. Bernal, 28 May 1947, J. 178 Cambridge University Library MS.Add.8287, John Desmond Bernal Papers.

[C]onference on haemoglobin in [Barcroft's] memory: F.J.W. Roughton and J.C. Kendrew (eds), *Haemoglobin*, Butterworths, 1949.

Since his move to the University of Istanbul, Haurowitz: For a life of Haurowitz, see Frank W. Putnam, 'Felix Haurowitz March 1, 1896– December 2, 1987', *Biographical Memoirs of the National Academy of Sciences* 64, 134–63, 1994.

'Our work here has more than justified': Max to Felix Haurowitz, 20 January 1948, Haurowitz archive, Lilly Library, University of Indiana, Bloomington, Indiana.

134 Haurowitz . . . told him it was 'the most important contribution': F. Haurowitz to Max, 19 February 1948, Haurowitz archive, Lilly Library, University of Indiana, Bloomington, Indiana.

'It would be very important to know': F. Haurowitz to Max, 4 February 1948, ibid.

He submitted his manuscript in June 1948: M.F. Perutz, 'An X-ray study of horse methaemoglobin. II', *Proceedings of the Royal Society of London* (A), 195, 474–99, 1949.

Crick was the son: For more on Crick's life, see Francis Crick, *What Mad Pursuit: A personal view of scientific discovery*, Basic Books, 1988; Matt Ridley, *Francis Crick: Discoverer of the genetic code*, Harper Press, 2006; and Robert Olby, *Francis Crick*, Cold Spring Harbor Laboratory Press, forthcoming.

135 Mellanby asked him if he thought the MRC should support 'these foreigners': Francis Crick, interview with the author, 3 June 2004.

135 According to Max, Kreisel cryptically asked: Horace Freeland Judson, *The Eighth Day of Creation*, Jonathan Cape, 1979, p. 110.

Crick has denied that he sent Kreisel: Francis Crick, interview with the author, 3 June 2004.

'I impressed Perutz sufficiently': Francis Crick, *What Mad Pursuit: A personal view of scientific discovery*, Basic Books, 1988, p. 43.

'He just came and we talked together': Horace Freeland Judson, *The Eighth Day of Creation*, Jonathan Cape, 1979, p. 110.

136 'Even with your generous offer': Max to Peiser parents, 22 May 1949, private papers.

'Gisela gets all the money I earn': Max to Peiser parents, 22 May 1949, private papers.

137 '[S]econd children are luckier than first': Gisela Perutz to Evelyn Machin, January 1950, private papers.

[I]t was 'terribly churchy': Max to Lotte Perutz, 25 January 1950.

'I soon found I could see the answer': Francis Crick, *What Mad Pursuit: A personal view of scientific discovery*, Basic Books, 1988, p. 45.

138 '[M]ost of what I had to say was new to them': Max to Gisela Perutz, April 1950, private papers.

139 'There is a sharp contrast': ibid.

'Harker had kept quiet during the conference': ibid.

Two other American crystallographers: See nobelprize.org/nobel_prizes/chemistry/laureates/1985/

[I]t caused a 'sensation': Max to Gisela Perutz, April 1950, private papers.

140 Hugh Huxley was moaning: Hugh Huxley, interview with the author, 25 May 2004.

[T]he program, which they eventually published: J.M. Bennett and J.C. Kendrew, 'The Computation of Fourier Synthesis with a Digital Electronic Calculating Machine,' *Acta Crystallographica* 5, 109–16, 1952.

Max told the MRC: M.F. Perutz, progress report 1949–1950, FD1 426, MRC archives, National Archives.

141 'I have had work crowding in on me from all sides': Max to Lotte Perutz, 30 May 1950, private papers.

'As a result of my salary increase': Max to Lotte Perutz, 17 December 1950, private papers.

Chapter 7 Annus mirabilis

144 [T]hey published the paper anyway: Sir Lawrence Bragg, J.C. Kendrew and M.F. Perutz, 'Polypeptide chain configurations in crystalline proteins', *Proceedings of the Royal Society of London (A)*, 203, 321–57, 1950.

144 'It was one of those papers you publish': Max Perutz, quoted in Horace Freeland Judson, *The Eighth Day of Creation*, Jonathan Cape, 1979.

'[T]he most ill-planned and abortive': W.L. Bragg, 'First stages in the X-ray analysis of proteins,' *Reports on Progress in Physics* 28, 1–14, 1965.

145 Pauling had . . . a 'vertiginous moment': Thomas Hager, *Force of Nature: the life of Linus Pauling*, Simon & Schuster, 1995.

'The structure looked dead right': Max Perutz, IWIMYAE, p. 190.

146 [T]he volume of essays on science and scientists: Max Perutz, IWIMYAE.

[A] letter to *Nature*: M.F. Perutz, 'New X-ray evidence on the configuration of polypeptide chains', *Nature* 167, 1054, 1951.

147 'I have made a discovery of decisive importance': Max to Peiser parents, 17 June 1951, private papers.

Max wrote to Pauling: Max to Linus Pauling, 17 August 1951, Ava Helen and Linus Pauling papers, Oregon State University Special Collections, box 01.304.1.

Pauling graciously told him: Linus Pauling to Max, 29 August 1951, ibid.

This, according to Todd's account: A. Todd, 'A recollection of Sir Lawrence Bragg', in J.M. Thomas and D.C. Phillips (eds), *Selections and Reflections: the legacy of Sir Lawrence Bragg*, Science Reviews Ltd, 1990.

148 Max afterwards kicked himself: Max Perutz, SINAQL, p. 40.

He had . . . asked [Kendrew] what title he should give his presentation: Francis Crick, *What Mad Pursuit: A personal view of scientific discovery*, Basic Books 1988, p. 50.

'Broadly speaking . . . they were all wasting their time': ibid, p. 49.

'Max proposed a structure': David Davies, interview with the author, December 2003.

149 'Rough calculations show that the vector rods': F.H.C. Crick, 'The height of the vector rods in the three-dimensional Patterson of haemoglobin', *Acta Crystallographica* 5, 381–6, 1952.

'Perhaps I had suspected the weakness': Max Perutz, SINAQL, p. 41.

'He was extraordinarily clever at calculating things': Hugh Huxley, interview with the author, 25 May 2004.

150 [Crick] published a general theory of diffraction by helices: W. Cochran, F. H. Crick and V. Vand, 'The structure of synthetic polypeptides. I. The transform of atoms on a helix', *Acta Crystallographica* 5, 581–6, 1952.

The story of the DNA double helix: See for example James Watson, *The Double Helix*, Athenaeum, 1968; Robert Olby, *The Path to the Double Helix*, Dover, revised edition 1994; Aaron Klug, 'The Discovery of the DNA Double Helix', *Journal of Molecular Biology*, 335, 3–26.

150 [H]e found Max 'very pleasant': James Watson to Betty Watson, 13 September 1951, box 45, folder 16, Watson Archive, Cold Spring Harbor Laboratory.

152 'They used . . . a certain amount of unpublished X ray data': Max to Harold Himsworth, 6 April 1953, FD1 426, MRC archives, National Archives.

153 [T]heir one-page paper: J.D. Watson and F.H.C. Crick, 'A structure for deoxyribose nucleic acid', *Nature* 171, 737–8, 1953.

A form of words suggested by Wilkins: See Maurice Wilkins quoted in Robert Olby, *The Path to the Double Helix*, Dover, revised edition 1994, p. 418.

According to Franklin's biographer: Brenda Maddox, *Rosalind Franklin: The Dark Lady of DNA*, HarperCollins, 2003.

[Wilkins] responded with surprising sang-froid: see Robert Olby, *The Path to the Double Helix*, Dover, revised edition 1994, p. 417.

154 '[I]n a long molecule many different permutations are possible': J.D. Watson and F.H.C. Crick, 'Genetical implications of the structure of deoxyribonucleic acid', *Nature* 171, 964–7, 1953.

[A]s Bragg's biographer Graeme Hunter noted: Graeme Hunter, *Light is a Messenger*, Oxford University Press, 2004.

155 Bragg used the information to calculate: W.L. Bragg and M.F. Perutz, 'The external form of the haemoglobin molecule. I.', *Acta Crystallographica* 5, 277–83, 1952.

In a further paper he went on to show: W.L. Bragg and M.F. Perutz, 'The external form of the haemoglobin molecule. II.', *Acta Crystallographica* 5, 323–8, 1952.

'The papers on these findings, authored by Bragg and myself': Max Perutz, *SINAQL*, pp. 44–5.

A third 1952 paper looked at the internal structure: W.L. Bragg, E.R. Howells and M.F. Perutz, 'Arrangement of polypeptide chains in horse methaemoglobin', *Acta Crystallographica* 5, 136–41, 1952.

156 [E]ven earlier than that [Bernal] was encouraging Dorothy Hodgkin: Georgina Ferry, *Dorothy Hodgkin: A Life*, Granta, 1999, p. 111.

157 'Even the heaviest of heavy atoms would make a negligible contribution': J.C. Kendrew and M.F. Perutz, 'The application of X-ray crystallography to the study of biological macromolecules', in F.J.W. Roughton and J.C. Kendrew (eds), *Haemoglobin*, Butterworths, 1949.

'No heavy atom could stand out in such a crowd': W.L. Bragg, 'Giant molecules', *Nature* 164, 7–10, 1949.

'Essentially the fraction from the atoms in the protein': Francis Crick, interview with the author, 3 June 2004.

[H]e was surprised to discover that the fraction of the X-ray beam: Horace

Freeland Judson, *The Eighth Day of Creation*, Jonathan Cape, 1979, pp. 552–3.

158 Vernon Ingram was born in Breslau: Vernon Ingram, interview with the author, 8 April 2005.

'I said I'm not a green finger biochemist': Herbert Gutfreund, interview with the author, 25 June 2004.

'Perutz says that I[ngram]'s training': Gerard Pomerat, officer's diary, 1 April 1953, folder 567, box 44, series 401, RG 1.1, Rockefeller Foundation Archives, RAC.

159 'I jumped when I saw that': Horace Freeland Judson, *The Eighth Day of Creation*, Jonathan Cape, 1979, p. 552.

'I found the exact intensity changes': Max Perutz, SINAQL, p. 66.

Horace Judson commented on the 'buttery glow of satisfaction': Horace Freeland Judson, *The Eighth Day of Creation*, Jonathan Cape, 1979, p. 555.

160 [I]somorphous replacement in proteins was 'My idea and my discovery': Max Perutz, transcript of interview with Andrew Brown.

'It was a triumph to find that there were no inconsistencies': D.W. Green, V.M. Ingram and M.F. Perutz, 'The structure of haemoglobin. IV. Sign determination by the isomorphous replacement method', *Proceedings of the Royal Society of London* (A), 225, 308–14, 1954.

161 'One of our most exciting events lately': Max Perutz, progress report, 23 February 1953, folder 567, box 44, series 401, RG 1.1, Rockefeller Foundation Archives, RAC.

164 [H]is letters relate the advances: Max to Gisela Perutz, 22 August – 8 September 1953, private papers.

'[Y]ou weren't going to get anywhere with theories': Hugh Huxley, interview with the author.

'I gather that we shall have a hard job': Max to Gisela Perutz, September 1953, private papers.

[I]t was 'most interesting and certainly not a thing': Max to Gisela Perutz, 22 September 1953, private papers.

'My paper was well received': Max to Gisela Perutz, 25 September 1953, private papers.

Chapter 8 In search of solutions

167 Himsworth was a brilliant clinical researcher: Christopher Booth, 'Sir Harold Himsworth (19 May 1905–1 November 1993)', *Proceedings of the American Philosophical Society* 141, 84–7, 1997.

[H]e replied that Max's work: Correspondence and notes of meetings between Bragg and Himsworth January–March 1950, FD1 426, MRC archives, National Archives.

167 'I would prefer on the whole to remain on the MRC staff': Max to Lotte Perutz, 29 June 1952, private papers.

'[P]oor Perutz has rather had to take a financial beating': Gerard Pomerat, officer's diary, 1 April 1953, folder 567, box 44, series 401, RG 1.1, Rockefeller Foundation Archives, RAC.

168 Cambridge appointed Nevill Mott: For more on Mott, see nobelprize.org/nobel_prizes/physics/laureates/1977/mott-bio.html

Max told Himsworth that 'Everyone here is very pleased': Max to H. Himsworth, 9 December 1953, LMB archives.

[Bragg] told Himsworth that he thought Max ought to stay: W. L. Bragg to H. Himsworth, 14 May 1953 and subsequent correspondence, FD1 8687, MRC archives, National Archives.

[I]t was 'fair to say that the protein research was pretty well stuck': W. L. Bragg to H. Himsworth, 29 August 1953, FD1 8687, MRC archives, National Archives.

169 '[E]ven if the team had to be split': Note of W. Weaver's meeting with H. Himsworth, 15 March 1954, FD1/8687, MRC archives, National Archives.

'This attempt to set himself up as Director of Research': Max to Gisela Perutz, April 1954, private papers.

Bernal . . . had started trying to enlist Bragg's support: Max to J.D. Bernal, 22 October 1952, J.178, Cambridge University Library MS.Add.8287, John Desmond Bernal Papers.

'When I see the FRS on the envelopes': Max to Gisela Perutz, April 1954, private papers.

[H]e said : 'Aber dafür zahlen sie dir doch nichts.': Max to H. Gutfreund, 8 March 1981, copy in Perutz archive, Venter Institute.

'You can play such a vital part': W.L. Bragg to J. Kendrew, 5 May 1954, J.117, J.C.Kendrew papers, Bodleian Library, Oxford.

170 Kendrew made it a condition of his acceptance: J. Kendrew to W.L. Bragg, 20 July 1954, J. 117, J.C. Kendrew papers, Bodleian Library, Oxford.

'The only good reason I can think of for going': J. Kendrew to J. Watson, 2 October 1954, box 23, folder 2, Watson archive, Cold Spring Harbor Laboratory.

'John was exceedingly well organised': Vernon Ingram, interview with the author, 8 April 2005.

Max . . . told the writer Horace Judson: Max Perutz, transcript of interview with Horace Judson, 19 December 1974, folder 2, Judson archive, American Philosophical Society Library.

170 'All the members of this Unit very much dislike': Max to H. Himsworth, 26 April 1954, LMB archives.

172 Max discharged himself after ten days: Letters to Gisela Perutz, December 1954, private papers.

[A] letter to the Rockefeller Foundation in March: Max to Rockefeller Foundation, 2 March 1955, folder 568, box 44, series 401, RG 1.1, Rockefeller Foundation Archives, RAC.

[T]here was flu going round: H. Judson, *The Eighth Day of Creation*, Jonathan Cape, 1979, p. 311.

'Max Perutz's psychosomatic illness is back:' J. Watson to Christa Mayr, 15 October 1955, box 46, folder 2, Watson archive, Cold Spring Harbor Laboratory.

'Bragg's departure was the beginning': Max Perutz, transcript of interview with Horace Judson, 4–5 December 1970, folder 1, Judson archive, American Philosophical Society Library.

'[T]hinking naïvely, as we were both Fellows of the Royal Society': Max Perutz, memoir dictated January 2002, private papers.

Max consulted . . . Werner Jacobson: ibid.

173 Ann Cullis . . . made a considerable impact: Ann Kennedy, interview with the author, 20 February 2004.

Max's report to the Rockefeller Foundation: Progress report 53–56, folder 569, box 44, series 401, RG 1.1, Rockefeller Foundation Archives, RAC.

174 'No panel, no referees, no interview': Max Perutz, 'Origins of molecular biology,' *New Scientist*, 31 January, pp. 326–9, 1980.

'The old name "Molecular Structure of Biological Systems"': Max to H. Himsworth 27 September 1957, LMB archives.

'I thought that Perutz probably exercised a great deal': Note of visit to Unit on 2 December 1957 by Dr Norton, FD12/291, MRC archives, National Archives.

'When I took them to the hut': Max Perutz, 'Origins of molecular biology', *New Scientist*, 31 January, pp. 326–9, 1980.

175 'I am unhappy about the idea of being indefinit[e]ly penalised': Max to H. Himsworth, 26 March 1956, LMB archives.

'I believe that Kendrew, with his various part-time activities': Max to H. Himsworth, 21 May 1958, LMB archives.

176 His discovery that the sickle cell mutation caused a single amino acid difference: See Vernon M. Ingram, 'Sickle cell anemia hemoglobin: the

molecular biology of the first "molecular disease" – the crucial importance of serendipity', *Genetics* 167, 1–7, 2004.

176 'One of the things that I learned from Max': Vernon Ingram, interview with the author, 8 April 2005.

'Max's great talent was that he got everybody happily working together': Francis Crick, interview with the author, 3 June 2004.

177 Ann Cullis (now Kennedy) recalls that there were 'lots of parties': Ann Kennedy, interview with the author, 1 March 2004.

[H]e heard about the party afterwards: ibid.

'Max was very friendly to me at all times': ibid.

178 [C]aptured by James Watson in his autobiographical books: See James Watson, *The Double Helix* and *Genes, Girls and Gamow*, Oxford University Press, 2001.

Fred Sanger: For more on Sanger's life and work, see nobelprize.org/nobel_ prizes/chemistry/laureates/1980/sanger-autobio.html

Until Vernon Ingram: Francis Crick, *What Mad Pursuit: A personal view of scientific discovery*, Basic Books, 1988.

Crick invited [Sanger] to attend a series of weekly meetings: ibid; see also Soraya de Chadarevian, *Designs for Life*, Cambridge University Press, 2002.

179 [Sanger] suggested that they might jointly apply to the MRC: Max to H. Himsworth, 27 June 1957, FD12/292, MRC archives, National Archives.

Himsworth assured Max that he regarded his Unit as 'holding the key position in this matter': H. Himsworth to Max, 2 July 1957, FD12/292, MRC archives, National Archives.

His statement: M.F. Perutz, 'Some recent advances in molecular biology', *Endeavour*, 17, 190–203, 1958.

'If such a proposal had been put before the Council': Max Perutz, 'Molecular biology', talk given to MRC on 18 April 1958, selected correspondence, Perutz papers, Venter Institute.

[S]everal members of the Council had told him: Max Perutz, 'Origins of molecular biology', *New Scientist*, 31 January, pp. 326–9, 1980.

Himsworth wrote to Bragg after the meeting: H. Himsworth to W.L. Bragg, 22 May 1958, FD12/292, MRC archive, National Archives.

180 The head of biochemistry, F.G. Young: Soraya de Chadarevian, *Designs for Life*, Cambridge University Press, 2002, pp. 220–21.

He could feel 'safely at home': Miss Potts to Max Perutz, 24 June 1959, Himsworth file, LMB archive.

181 'I got quite desperate about it': Max Perutz, transcript of interview with

H. Judson, 4–5 December 1970, Judson archive, American Philosophical Society Library.

181 In 1957 a Norwegian visiting scientist: Ann Kennedy, letter to the author, 9 February 2007.

In the late summer of 1957: Max to Gerard Pomerat, 16 September 1957, folder 568, box 44, series 401, RG 1.1, Rockefeller Foundation Archives, RAC; also see nobelprize.org/nobel_prizes/chemistry/laureates/1962/kendrew-lecture.html

182 '[I]t looked like nothing so much as 'an anatomical model': Note of visit to Unit on 2 December 1957 by Dr Norton, FD12/291, MRC archives, National Archives.

An unnamed cartoonist: see Richard E. Dickerson, *Present at the Flood: How Structural Biology Came About*, Sinauer, 2005, p. 215.

'Perhaps the most remarkable features': J.C. Kendrew et al., 'A three-dimensional model of the myoglobin molecule obtained by x-ray analysis', *Nature* 181, 662–6, 1958.

'[W]e are all very thrilled': Max to Gerald Pomerat, 16 September 1957, folder 568, box 44, series 401, RG 1.1, Rockefeller Foundation Archives, RAC.

'He may have been depressed at home': Anne Kennedy, interview with the author, 1 March 2004.

183 At last they had what they needed: A.F. Cullis, H.M. Dintzis and M.F. Perutz, 'X-ray analysis of haemoglobin', *National Academy of Sciences Conference on Hemoglobin*, 50–65, 1958.

Michael Rossmann was born in Frankfurt: Details of Rossmann's life and career from interview with the author, 16 April 2005.

'I had no idea what the molecule ought to look like': Max Perutz, *SINAQL*, p. 71.

184 'I felt, probably wrongly, that Max didn't have much trust in me': Michael Rossmann, interview with the author, 16 April 2005.

185 'I need not have worried so much': Max Perutz, *SINAQL*, p. 157.

186 Bragg had described in 1952, 'like Spanish chestnuts': W.L. Bragg to H. Himsworth, 9 October 1959, FD1/9033, MRC archives, National Archives.

'If the myoglobin of a whale is like the haemoglobin of a horse': Max to H. Himsworth, 1 October 59, FD1/9033, MRC archives, National Archives.

187 'There was an alpha helix': David Davies, interview, 16 November 2002.

Max and Kendrew had each quickly written a short paper for *Nature*: M.F. Perutz et al., 'Structure of haemoglobin. A three-dimensional Fourier

synthesis at 5.5Å resolution, obtained by X-ray analysis', *Nature* 185,
416–22, 1960; J. C. Kendrew et al., 'Structure of myoglobin. A three-
dimensional Fourier synthesis at 2Å resolution', *Nature* 185, 422–7, 1960.

187 'When I came back': Max Perutz, 2001 interview for the National Life
Story Collection, British Library.

'MFP settled in empty carriage': Max to his family, July 1962, private papers.

188 Sir Lawrence Bragg had begun to canvass support: 14 D/83, Bragg archive,
Royal Institution.

189 [H]e received a call: Max Perutz, account of the Nobel ceremony written
for his relatives, private papers.

The avalanche of congratulations: File of Nobel congratulations to Max,
Perutz papers, Churchill College, Cambridge.

190 [I]t seemed 'only a few months ago': J.D. Bernal to Max, 5 November 1962,
Perutz papers, Churchill College, Cambridge.

'You know better than I do': ibid.

'I want to say that I have a great admiration': F. Haurowitz to Max,
4 November 1962, Haurowitz papers, Lilly Library, University of Indiana,
Bloomington, Indiana.

191 Max received a toy horse and a bottle of blood: Max to Gertrude Perlman,
2 November 1962, Perutz papers, Churchill College. Cambridge.

'We were, as always, the last to leave the plane': Everything that follows
is based on an account of the Nobel ceremony written for his relatives
by Max, private papers.

192 'Max was in an absolute mood of euphoria': John Kendrew, transcript of
interview with Horace Judson, Judson archive, American Philosophical
Society Library.

193 'If I were titled this would put me on a pedestal': Max to Prime Minister's
permanent private secretary, 24 November 1969, copy among private
papers.

He told friends he thought a title: Max Perutz, transcript of interview with
Horace Judson, 18 November 1974, folder 2, Judson archive, American
Philosophical Society Library.

194 Dorothy Hodgkin: see Georgina Ferry, *Dorothy Hodgkin: A life*, Granta
Books, 1999.

Chapter 9 A structure for science – the LMB

195 Himsworth's diary note of the meeting: 27 August 1957, FD12/292, MRC
archive, National Archives.

Max might rely on 'propinquity and joint round-table meetings': H. Himsworth
to Max, 16 June 59, FD7/1040, MRC archive, National Archives.

196 'We could approach the MRC': Francis Crick, interview with the author, 3 June 2004.

The MRC's Council . . . accepted it on condition: Council minute, 26 May 1961, FD7/1040, MRC archive, National Archives.

The Queen was so interested in everything: Max Perutz, 'Origins of molecular biology', *New Scientist*, 31 January, pp. 326–9, 1980.

197 He had recruited Michael Fuller: Note on file of 11 October 1961, FD7/1040, MRC archive, National Archives.

Fuller's interview with Max, John Kendrew and Tony Broad had been perfunctory: Michael Fuller, interview with the author, 16 February 2004.

Fuller produced a minutely detailed master plan: Soraya de Chadarevian, *Designs for Life*, Cambridge University Press, 2002.

'Once he realised that you could do something': Michael Fuller, interview with the author, 16 February 2004.

198 MRC officers . . . commented that they had had a good lunch: Note of visit, 17 April 1962, FD12/291, MRC archives, National Archives.

199 Michael Fuller arranged to have everyone photographed: Michael Fuller, interview with the author, 16 February 2004.

'It became necessary': Gisela Perutz, interview, 18 February 2003.

200 Newly-arrived young researchers could be misled: Graeme Mitchison, interview with the author, 17 February 2004.

John Kilmartin, who joined the lab as a post doc in 1965, remembers: John Kilmartin, interview with the author, 16 February 2004.

201 'The slightly anarchic way': Notes of visits, 22 October 1965 and 2 April 1968, FD12/291, MRC archives, National Archives.

202 In 1965 [Brenner] decided the time was right: For the history of Brenner's work on C. *elegans*, see Andrew Brown, *In the Beginning was the Worm*, Simon & Schuster, 2003.

Brenner, Sulston and Horvitz were awarded the Nobel prize: nobelprize.org/nobel_prizes/medicine/laureates/2002/

203 Sanger won his second Nobel prize: nobelprize.org/nobel_prizes/chemistry/laureates/1980

The Sanger Centre . . . contributed one-third of the human DNA sequence: See John Sulston and Georgina Ferry, *The Common Thread: A story of science, politics, ethics and the human genome* (Bantam, 2002).

The Nobel committee awarded [Milstein and Köhler]: nobelprize.org/nobel_prizes/medicine/laureates/1984

[T]he National Enterprise Board formed a company: Soraya de Chadarevian, *Designs for Life*, Cambridge University Press, 2002, p. 354.

204 For this achievement [Klug] won the Nobel prize in 1982: nobelprize.org/
   nobel_prizes/chemistry/laureates/1982

   'In 1965 I introduced the use of optical diffraction': Sir Aaron Klug, interview with the author, 17 February 2004.

   Max later thought he had made a tactical error: Hugh Huxley, email to
   the author, 8 February 2007.

   [A] young scientist called John Gurdon . . . had successfully cloned adult
   frogs: J.B. Gurdon, 'The developmental capacity of nuclei taken from
   intestinal epithelium cells of feeding tadpoles', *Journal of Embryology and
   Experimental Morphology* 34, 93–112, 1962.

205 Gurdon has his own institute for developmental biology and cancer research:
   www.gurdon.cam.ac.uk

   He tries to emulate Max's style: Sir John Gurdon, interview with the author,
   22 September 2004.

   Rossmann called the technique 'molecular replacement': M.G. Rossmann
   and D.M. Blow, 'The detection of sub-units within the crystallographic
   asymmetric unit', *Acta Crystallographica* 15, 24–31, 1962.

   [H]e had solved the structure of a large enzyme: M. J. Adams et al.,
   'Structure of lactate dehydrogenase at 2.8 Å resolution', *Nature* 227,
   1098–1103, 1970.

   [Rossmann] became the first to solve the structure of a rhinovirus:
   M.G. Rossmann et al., 'Structure of a human common cold virus and
   functional relationship to other picorna viruses', *Nature* 317, 145–53
   1985.

206 'It's easy to overlook how influential he was': Sir John Gurdon, interview
   with the author, 22 September 2004.

   'I feel tempted to draw their attention': Max Perutz, *IWIMYAE*, p. ix.

   'Max was without any doubt the spiritual leader': Sir Aaron Klug, interview with the author, 17 February 2004.

207 Leo Szilard . . . responded: Max Perutz, notes on the foundation of EMBO,
   Perutz papers, Churchill College, Cambridge.

   [A] meeting at Ravello in Italy: folder F.4, J.C. Kendrew papers, Bodleian
   Library, Oxford. All Max's correspondence on the foundation of EMBO
   is in the Kendrew papers.

   As Max later told the *New Scientist*: Max Perutz in discussion with Graham
   Chedd, 'EMBO – the year of decision', *New Scientist*, 29 February 1968,
   p. 458.

   [A]ccording to Sydney Brenner 'he was a bit cornered': Sydney Brenner
   to David Blow, 6 January 2003, private papers.

   Max went to meet [the Foundation's] officers: Details of negotiations with

Volkswagen Foundation in folder F.43, J.C. Kendrew papers, Bodleian Library, Oxford.

207 Appleyard was immediately impressed: Raymond Appleyard, interview with the author, 17 July 2006.

'For my money, the key figure in the whole thing was Max': ibid.

208 Both EMBO and EMBL continue to flourish: www.embo.org and www.embl.de

EMBL ... 'stands as [Kendrew's] monument': M.F. Perutz, 'Sir John Kendrew (1917–1997) – Obituary', *Protein Science* 6, 2684–5, 1997.

Frederic de Hoffmann, 'went to great efforts to tempt [him]: Francis Crick, *What Mad Pursuit: A personal view of scientific discovery*, Basic Books 1988, p. 145.

'Max wasn't a particularly quick thinker': Francis Crick, interview with the author, 3 June 2004.

209 A year later the Council accepted that it was 'inconceivable': Report to MRC Council meeting of November 1974, folder N.22, D.C. Phillips papers, Bodleian Library, Oxford.

'I certainly find it difficult to conceive': Note of visit by Dr Julie Neale, 21 and 22 March 1974, FD12/291, MRC archive, National Archives.

The Phillips committee ... picked up 'several hints': Report of 12 January 1976 visit by Phillips Committee, folder N.28, D.C. Phillips papers, Bodleian Library, Oxford.

210 'We are very satisfied': Submission by LMB technical staff to Phillips committee, 12 January 1976, folder N.28, D.C. Phillips papers, Bodleian Library, Oxford.

'Contrary to widespread belief': Max Perutz, 'Perutz not retired', *Nature* 273, p. 334, 1978.

Brenner took over the financial side: Sydney Brenner, interview with the author, 21 September 2004.

211 John Walker won the prize: nobelprize.org/nobel_prizes/chemistry/laureates/1997

Andrew Fire and Roger Kornberg ... won the 2006 prizes: nobelprize.org/nobel_prizes/medicine/laureates/2006 and nobelprize.org/nobel_prizes/chemistry/laureates/2006

212 The first beneficiary was: Unpublished memoir by Marie-Alda Gilles-Gonzalez, February 2002.

'You have what we always search out': James Watson, contribution to retirement album for Max, copy in folder 25, box 34, Watson archive, Cold Spring Harbor Laboratory.

212 'In recognition of his distinguished contributions': Royal Society, Copley Medal citation 1979, www.royalsoc.ac.uk/page.asp?tip=1&id=1736

Chapter 10 The breathing molecule

213 He astonished a visiting MRC officer: Note of visit, 22 October 1965, FD12/291, MRC archive, National Archives.

214 Felix Haurowitz had written from Prague: Felix Haurowitz to Max Perutz, 6 April 1938, Haurowitz papers, Lilly Library, University of Indiana, Bloomington, Indiana; F. Haurowitz, *Hoppe-Seyler Zeitung* 254, 266, 1938.

215 Coryell found oxyhaemoglobin to be diamagnetic: L. Pauling and C. Coryell, *Proceedings of the National Academy of Sciences USA* 22, 210–16, 1936.

216 'Please forgive me for presenting': Max Perutz, 'X-ray analysis of haemoglobin', Les Prix Nobel 1962, Nobel Foundation 1963. Reproduced in Max Perutz, *SINAQL*, p. 224.

[T]his term meant 'nothing more complicated': Max Perutz, 'The hemoglobin molecule', *Scientific American* 211, 64–76, 1964.

He invented the term 'allosteric': J. Monod, J.-P. Changeux and F. Jacob, 'Allosteric proteins and cellular control systems', *Journal of Molecular Biology* 6, 306–9, 1963.

217 Hilary Muirhead and Max jointly published a long paper in *Nature*: H. Muirhead and M.F. Perutz, 'Structure of haemoglobin. A three-dimensional Fourier synthesis of reduced human haemoglobin at 5.5Å resolution', *Nature* 199, 633–9.

[H]e announced to his colleagues that he had discovered 'the second secret of life': Horace Judson, *The Eighth Day of Creation*, Jonathan Cape, 1979, p. 576.

[Monod] published a classic paper in 1965: J. Monod, J. Wyman and J.-P. Changeux, 'On the nature of allosteric transitions: a plausible model', *Journal of Molecular Biology* 12, 88–118.

[A] 'molecular lung': M.F. Perutz, 'Haemoglobin: the molecular lung', *New Scientist and Science Journal*, 17 June, 762–5, 1971.

218 Arndt was delighted to return to Cambridge: Uli Arndt, interview with the author, 17 February 2004.

[T]he American post doc ... did not align the diffractometer correctly: Max Perutz, *SINAQL*, p. 159.

'Arndt and Mallett helped me to keep [the diffractometer] running': Max Perutz, *SINAQL*, p. 159.

[T]he model of haemoglobin that Max and Muirhead built: M.F. Perutz, H. Muirhead, J.M. Cox and L.C.G. Goaman, 'Three-dimensional Fourier synthesis of horse oxyhaemoglobin at 2.8Å resolution: II – the atomic model', *Nature* 219, 131–9, 1968.

219 [H]igh-resolution structures of human and horse deoxyhaemoglobin: H. Muirhead and J. Greer, 'Three-dimensional Fourier synthesis of human deoxyhaemoglobin at 3.5Å resolution', *Nature* 228, 516–19, 1970; W. Bolton and M.F. Perutz, 'Three dimensional Fourier synthesis of horse deoxyhaemoglobin at 2.8Å resolution', *Nature* 228, 551–2, 1970.

220 Bob Williams discovered . . . that spin state changes would affect the lengths of the bonds: R.J.P Williams, *Chemical Reviews* 56, 299–337, 1956.

In the course of his talk, Williams proposed: R.J.P. Williams, 'Nature and properties of metal ions of biological interest', *Federation Proceedings* 20 (Suppl. 10), 5–14, 1961.

Williams . . . found that Max appeared to be reluctant to accept the idea: R.J.P. Williams, personal communication.

J. Lynn Hoard . . . discovered that the spin state did indeed affect: J.L. Hoard, in A. Rich and N. Davidson (eds), *Structural Chemistry and Molecular Biology*, Freeman, 1968.

'I realised then . . . that this difference of about half an angstrom: Max Perutz, interview with Horace Judson, 4–5 December 1970, Judson archive, American Philosophical Society.

221 [A] series of papers for *Nature*, the most important of which proposed a mechanism: M.F. Perutz, 'Stereochemistry of cooperative effects in haemoglobin', *Nature* 228, 726–39, 1970.

'Four fleas that make an elephant jump': Max Perutz, 'Haemoglobin: the molecular lung', *New Scientist and Science Journal*, 17 June, 762–5, 1971.

222 'The cooperative mechanism outlined here': *Nature* 228, 726–39, 1970.

Aaron Klug . . . describes it as 'one of the most important [biological] mechanisms: Aaron Klug, interview with the author, 8 March 2004.

'Max was the first in history to allow us to look at a molecule as a machine': Guy Dodson, personal communication, 2006.

David Blow described the paper as 'the pinnacle of Perutz' achievements': D.M. Blow, 'Max Ferdinand Perutz', *Biographical Memoirs of the Fellows of the Royal Society* 50, 227–56, 2004.

Jacques Monod once said that he arrived at his allosteric theory: F. Jacob, 'L'imagination en art et en science', *Actes des colloques du Bicentaire de l'Institut de France 1975–1995*, 69–76, 1995.

'When reviewing scientific work': M.F. Perutz, 'How Lawrence Bragg invented X-ray analysis', *Proceedings of the Royal Institution of Great Britain* 62, 183–98, 1990; reproduced in *IWIMYAE*, p. 339.

223 'The evidence for the stereochemistry of the cooperative mechanism': Max Perutz, *SINAQL*, p. 321.

'[T]he Haemoglobin Battles': ibid, chapter 5.

'After returning home, I complained to Bragg': Max Perutz, *SINAQL*, p. 322.

224 'I have found so often that failure': Max to Robin Perutz, 8 August 1977, private papers.

'By always thinking': Isaac Newton, quoted by Cyril Hinshelwood in *Nature* 207, 1057; and in Max's commonplace book, *IWIMYAE*, p. 426.

225 Shulman now had evidence: P. Eisenberger et al., 'Extended X-ray absorption fine structure determination of iron nitrogen distances in haemoglobin', *Nature* 274, 30–4, 1978.

Max decided that 'the only way to prove [Shulman] wrong': Max Perutz, *SINAQL*, p. 323.

Max found that his EXAFS measurements: M.F. Perutz, S.S. Hasnain, P.J. Duke, J.L. Sessler and H.E. Hahn, 'Stereochemistry of iron in deoxyhaemoglobin', *Nature* 295, 535–8, 1982.

226 'In exhausting 24-hour shifts': Max Perutz, *SINAQL*, p. 325.

The result confirmed that the iron atom was displaced: M.F. Perutz, G. Fermi, B. Luisi, B. Shaanan, and R.C. Liddington, 'Stereochemistry of cooperative mechanisms in hemoglobin', *Accounts of Chemical Research* 20, 309, 1987.

Max and Shulman published a joint paper: G. Fermi, M.F. Perutz and R.G. Shulman, 'Iron distances in hemoglobin: Comparison of X-ray crystallographic and extended X-ray absorption fine structure studies', *Proceedings of the National Academy of Sciences USA* 84, 6167–8, 1987.

Max, John Kilmartin and several colleagues had published a paper: M.F. Perutz, H. Muirhead, L. Mazzarella, R.A. Crowther, J. Greer and J.V. Kilmartin, 'Identification of residues responsible for the alkaline Bohr effect in haemoglobin', *Nature* 222, 1240–3, 1969.

[O]ne participant remembers: Alan Schechter, interview with the author, April 2005.

227 'I remonstrated that [Ho's] experiment must be wrong': Max to Lotte Perutz, 20 August 1984.

[H]is own colleagues had made a similar error: M.F. Perutz et al., 'The pK_a values of two histidine residues in human haemoglobin, the Bohr effect, and the dipole moments of α-helices', *Journal of Molecular Biology* 183, 491–8, 1985.

Ho and his team eventually resolved that particular issue: M.R. Busch et al., 'Roles of the beta 146 histidyl residue in the molecular basis of the

Bohr effect of hemoglobin: a proton nuclear magnetic resonance study', *Biochemistry* 30, 1865–77, 1991.

227 [H]e even wrote to the editor of a leading journal: See John Edsall to Max, 26 March 1985, Perutz papers, Venter Institute.

'It is really difficult for me to understand': Chien Ho to John Edsall, 9 April 1996, Chien Ho's private papers.

While both Shulman and Ho: see, for example, Robert G. Shulman, 'Spectroscopic contributions to the understanding of hemoglobin function: implications for structural biology', *IUBMB Life* 51, 351–7, 2001; Jonathan A. Lukin and Chien Ho, 'The structure-function relationship of hemoglobin in solution at atomic resolution', *Chemical Reviews* 104, 1219–30, 2004.

A whole chapter is devoted: Max Perutz, 'Chapter 5: The haemoglobin battles', *SINAQL*.

228 Max referred to the difficulty of being frank: See John Edsall to Max, 29 November 1985, Perutz papers, Venter Institute.

'I hate fights and would have been happier without them': Max Perutz, 'Chapter 5: The haemoglobin battles', *SINAQL*.

'Max was truly outstanding': Chien Ho, email to the author, 28 June 2006.

229 Max published his final haemoglobin review paper: M.F. Perutz, A.J. Wilkinson, M. Paoli and G.G. Dodson, 'The stereochemical mechanism of the cooperative effects in hemoglobin revisited', *Annual Reviews of Biophysics and Biomolecular Structure* 27, 1–34, 1998.

'What Max said was that the T-R transition': Guy Dodson, interview with the author, July 2006.

230 'One of Max's great offers to the field': Ken Holmes, interview with the author, 25 May 2004.

Chapter 11 Health and disease

231 Linus Pauling published a paper: L. Pauling, H.A. Itano, S.J. Singer and I.C. Wells, 'Sickle cell anemia: a molecular disease', *Science* 110, 543, 1949.

232 They quickly published their result in *Nature*: M.F. Perutz and J.M. Mitchison, 'State of haemoglobin in sickle-cell anaemia', *Nature* 166, 677–82, 1950.

Pauling reacted . . . with irritation: Linus Pauling to Max, 6 February 1951, Ava Helen and Linus Pauling Papers, Oregon State University Special Collections, box 01. 304.1.

'I am very disappointed that you should have been annoyed': Max to Linus Pauling, 14 December 1950, Ava Helen and Linus Pauling Papers, Oregon State University Special Collections, box 01.304.1.

Max . . . 'left the laboratory in shame': Max Perutz, *SINAQL*, p. 431.

233 Allison had been the first to demonstrate: A. C. Allison, 'Two lessons from the interface of genetics and medicine', *Genetics* 166, 1591–9, 2004.

Max and Francis Crick suggested that Ingram might try: V.M. Ingram, 'Hemoglobin: The Molecular Biology of the First "Molecular Disease" – The Crucial Importance of Serendipity', *Genetics* 167, 1–7, 2004.

'This is the first discovery of a specific chemical difference': Max to Gerald Pomerat, 6 December 1956, folder 569, box 44, series 401, RG 1.1, Rockefeller Foundation Archives, RAC.

[T]he Secretary . . . introduced him to a young German doctor: Max Perutz, *SINAQL*, p. 432.

234 'My day with Lehmann became one of the most interesting': ibid, p. 433.

That 1968 publication in *Nature* was a landmark in the field: M.F. Perutz and H. Lehmann, 'Molecular pathology of human haemoglobin', *Nature* 219, 902–9, 1968.

'[V]ery few new concepts have been established': D. Weatherall, 'Towards molecular medicine: reminiscences of the haemoglobin field 1940–2000', *British Journal of Haematology* 115, 729–38, 2001.

235 'The data presented here do not hold out any hope': M.F. Perutz and H. Lehmann, 'Molecular pathology of human haemoglobin', *Nature* 219, 902–9, 1968.

Beatrice Magdoff-Fairchild . . . obtained the first successful X-ray diffraction pattern: B. Magdoff-Fairchild, P.H. Swerdlow and J.F. Bertles, 'Intermolecular organisation of deoxygenated sickle-haemoglobin determined by X-ray diffraction', *Nature* 239, 217–18.

Max announced that these consisted: J.T. Finch, M.F. Perutz, J.F. Bertles and J. Döbler, 'Structure of sickled erythrocytes and of sickle-cell hemoglobin fibers', *Proceedings of the National Academy of Sciences USA* 70, 718–22, 1973.

Max found it hard to believe: Alan Schechter, interview with the author, April 2005.

236 [Schechter] dates his interest in haemoglobin: ibid.

'Even though the structure was wrong': ibid.

'In research you work at proof of principle': ibid.

In the early 1960s Abraham had worked: Don Abraham, interview with the author, April 2005.

237 [P]erhaps most famously when Alexander Fleming: see, for example, Eric Lax, *The Mould in Dr Florey's Coat*, Little, Brown, 2004.

'My spirits rose to the heavens': Don Abraham, reminiscences, www.news.vcu. edu/vcu_view/pages.aspx?nid=1094

These studies of how small molecules bind to proteins: M. F. Perutz,

G. Fermi, D. J. Abraham, C. Poyart and E. Burseaux, 'Hemoglobin as a receptor of drugs and peptides: X-ray studies of the stereochemistry of binding', *Journal of the American Chemical Society* 108, 1064–78, 1986.

237 Max and a French colleague, Claude Poyart, discovered: M.F. Perutz and C. Poyart, 'Bezafibrate lowers oxygen affinity of haemoglobin', *The Lancet*, October 15, 1983.

238 'Don Abraham, my long-time collaborator': Max to Robin Perutz, 1 May 1995, private papers.

[H]e states unequivocally that without Max's collaboration: Don Abraham, interview with the author, April 2005.

239 Alan Schechter believes Max's work . . . underlies advances in many other fields: Alan Schechter, interview with the author, April 2005.

[A] short, introductory book: Max Perutz, *Protein Structure: New Approaches to Disease and Therapy*, Freeman, 1992.

'[W]hose foresight and courage': ibid, dedication.

240 [H]e sat down and wrote a long and detailed review: M.F. Perutz, 'Species adaptation in a protein molecule', *Molecular Biology and Evolution* 1, 1–28, 1983 and *Advances in Protein Chemistry* 36, 213, 1984.

241 A rather acid anonymous referee's report: Quoted by Max in a letter to Robin Perutz, 29 June 1983, private papers.

'I thought what we were doing in Japan was very interesting': Kiyoshi Nagai, interview with the author, 16 February 2004.

[Nagai] devised an ingenious method: K. Nagai and C. Thøgersen, 'Generation of globin by sequence-specific proteolysis of a hybrid protein produced in *Escherichia coli*', *Nature* 309, 810–12,

242 'Max always had very simple ideas': Kiyoshi Nagai, interview with the author, 16 February 2004.

243 [I]f he wanted to do an experiment he would come to Nagai's: ibid.

244 [H]e sat down and wrote a refutation of Popper's argument: Max Perutz, 'A new view of Darwinism', *New Scientist*, 2 October, 36–8, 1986, reprinted in *IWIMYAE*, p. 217.

'In practice, scientific advances often originate from observations': ibid.

245 [T]hey showed that a rearrangement of the amino acids near the haem': I. De Baere et al., 'Polar zipper sequence in the high affinity hemoglobin of *Ascaris suum*: amino acid sequence and structural interpretation', *Proceedings of the National Academy of Sciences USA* 89, 4638–42, 1992.

Max . . . sent off a short paper: M.F. Perutz, R. Staden, L. Moens and I. De Baere, 'Polar zippers', *Current Biology* 3, 249–53, 1993.

[He] thought that was 'the end of the story': Max Perutz, 'Glutamine repeats as polar zippers: their role in inherited neurodegenerative disease', *Molecular Medicine* 1, 718–21, 1995.

245 [A]n international team identified the genetic mutation that causes Huntington's: The Huntington's Disease Collaborative Group, 'A novel gene containing a trinucleotide repeat that is expanded and unstable on Huntington's disease chromosomes', *Cell* 72, 971–83, 1993.

246 Max wrote up the results: M.F. Perutz, T. Johnson, M. Suzuki and J.T. Finch, 'Glutamine repeats as polar zippers: Their possible role in inherited neurodegenerative diseases', *Proceedings of the National Academy of Sciences USA* 91, 5355– 8, 1994.

'Nancy Wexler, the remarkable woman': Max to Robin Perutz, 1 May 1995.

Max and three of his colleagues in Cambridge had discovered: K. Stott, J.M. Blackburn, P.J.G. Butler, and M. Perutz, 'Incorporation of glutamine repeats makes protein oligomerize: Implications for inherited neurodegenerative diseases', *Proceedings of the National Academy of Sciences USA* 92, 6509–13, 1995.

Gillian Bates . . . put part of the human Huntington's disease gene . . . into laboratory mice [and] Stephen Davies . . . looked at their brain cells: S.W. Davies et al., 'Formation of neuronal intranuclear inclusions underlies the neurological dysfunction in mice transgenic for HD mutation', *Cell*, 90, 537–48, 1997.

247 'Your logic, knowledge and keen insight has graced these meetings': Nancy Wexler to Max, 2 February 2002.

Max published a paper in *Nature*, co-authored with Alan Windle: M.F. Perutz and A.H. Windle, 'Cause of neural death in neurodegenerative disease attributed to expansion of glutamine repeats', *Nature* 412, 143–4, 2001.

248 [H]e rushed to complete two papers: M.F. Perutz, J.T. Finch, J. Berriman and A. Lesk, 'Amyloid fibres are water-filled nanotubes', *Proceedings of the National Academy of Sciences USA* 99, 5591–5, 2002; M.F. Perutz, B.J. Pope, D. Owen, E.E. Wanker, and E. Scherzinger, 'Aggregation of proteins with expanded glutamine and alanine repeats of the glutamine-rich and asparagine-rich domains of Sup35 and of the amyloid-peptide of amyloid plaques', *Proceedings of the National Academy of Sciences USA* 99, 5596–5600, 2002.

In 2005, David Eisenberg: Rebecca Nelson, Michael R. Sawaya, Melinda Balbirnie, Anders Ø. Madsen, Christian Riekel, Robert Grothe and David Eisenberg, 'Structure of the cross-beta spine of amyloid-like fibrils', *Nature* 435, 773–8, 2005.

248 Klug now concedes the truth: Aaron Klug, interview with the author, 8 March 2004.

In 1987 its steering committee divided: MRC *Annual Report 1986–1987*.

249 'I suggested other people': Max to Lotte Perutz, 27 March 1987, private papers.

Dwek's colleagues in Oxford have gone on to develop them successfully: T.D. Butters, R.A. Dwek and F.M. Platt, 'Imino sugar inhibitors for the glycosphingolipidoses', *Glycobiology* 15, 43R–52R, 2005.

'I cannot understand the insensitivity and thoughtlessness': Max to *The Times*, 13 July 1990.

250 'Jim, looking more haggard and pock-marked': Max to family, 22 February 1991.

'I wish I were not such a glasshouse plant': Max to Gisela Perutz, 5 January 1960.

251 [H]is doctor referred him to a speech therapist: Max Perutz, unpublished memoir, dictated January 2002, private papers.

'My throat is so sore that I can hardly speak': Max to Robin Perutz, 2 May 1977, private papers.

'I cannot do spectrophotometry': Max to Robin Perutz, 20 May 1977, private papers.

'The *E. coli* . . . have had a marvellous effect': Max to Robin Perutz, 2 June 1977, private papers.

252 'Let me tell you that I am small': Max to P.E. Smith, 7 October 80, copy in private papers.

'My walking was excellent': Max to Lotte Perutz, 23 June 1984, private papers.

'This sort of thing has happened so often': Max to Lotte Perutz, 28 December 1984, private papers.

David Keilin, whose chronic asthma: Max Perutz, 'Keilin and the Molteno', *Cambridge Review*, October, 152–7, 1987, reprinted in *IWIMYAE*, pp. 375–82.

253 'I have been swimming every day': Max to Robin Perutz, 4 April 1979, private papers.

254 'He came running back': Uli Arndt, interview with the author, 16 February 2004.

Under the heading 'The Nobel art of eccentricity': *Observer*, August 1993.

'When I got there . . . I was collared by a man': Max to Lotte, 16 October 1979, private papers.

Aaron Klug . . . admits that: Aaron Klug, interview with the author, 8 March 2004.

255 'I think I get tense only when preparing for a trip': Max to Gisela, 8 June 1973, private papers.

'If ever I get ill again as I did last winter': Max to Gisela, 8 March 1976, private papers.

256 [H]e would anxiously phone his cardiologist: Andrew Grace, personal communication.

257 [T]he warmly sympathetic letter of condolence: Max to Elise Poyart, 25 November 2001, copy kindly supplied by Dr Poyart.

'I have some good and some bad news': Paragraphs typical of letters Max sent to many friends, January 2002.

'The pilgrimage would be sad, I thought': Unpublished memoir of Max by Marie-Alda Gilles-Gonzales, February 2002.

Chapter 12 Truth always wins

259 [Max] kept a commonplace book: Reproduced in Max Perutz, *IWIMYAE*, pp. 425–43.

'[M]any [of my quotations] have become my guiding mottos': Preface to *IWIMYAE*, p. xii.

260 'A volume of the early letters of Hugo van Hofmannsthal': Max to Evelyn Machin, 21 February 1936 (translated from German), private papers.

[H]is mother thought that it was 'too boring and dry': Max to Evelyn Machin, 8 December 1933 (translated from German), private papers.

'[A] witty and very long piece of nonsense': Max to Evelyn Machin, 25 June 1934 (translated from German), private papers.

[H]e completed one entitled 'Proteins': M.F. Perutz, 'Proteins: the machines of life', *Scientific Monthly* 59, 47–55, 1944.

'The stranger is led to the conviction': ibid.

261 'The number of protein molecules': ibid.

[T]his version . . . lacks much of the liveliness: M.F. Perutz, 'Proteins', *Discovery* 5, 326–32, 1944.

'As you can well imagine': Max to Herbert and Nelly Peiser, 17 March 1948, private papers.

262 'I have become more interested in writing as such': Max to Lotte Perutz, 25 January 1950, private papers.

'I have . . . been given a task so fascinating': Max to Herbert and Nelly Peiser, 25 May 1950, private papers.

263 [A]ny means of expressions and display, static or dynamic': Max to Lotte Perutz, 30 May 1950, private papers.

The fee for the 12-page script: Papers on the Festival of Britain, Work25/256, National Archives.

263 He based [the script] around five questions: Papers on the Festival of Britain, Work25/23, National Archives.

[Kathleen Lonsdale and Helen Megaw] worked with a wide range of designers and manufacturers: *Souvenir Book of Crystal Designs*, HMSO, 1951.

264 'I have long been toying with the idea of writing a book': Max to Lotte Perutz, 25 January 1950, private papers.

[H]e found had 'warmth, tact, and a natural dignity': Max Perutz, unpublished account of trip to Cyprus and Israel, 1960, private papers.

'This put our minds into a state of divided loyalty': ibid.

'I wish that the liberal ideals for which [Weizmann] stood': Max Perutz, 'Jewish nationalism and the liberal ideal', *Nature* 302, 781–2, 1983, reprinted in *Is Science Necessary?* as 'Chemist into statesman', pp. 184–90.

'In fact the government's attitude': Max Perutz, unpublished account of trip to Cyprus and Israel, 1960, private papers.

265 [H]e sat down to write his first book: Max Perutz, *Proteins and Nucleic Acids*, Elsevier, 1962.

'The entire edifice of molecular biology': ibid.

'It is absurd that you should not be here': Max to Gisela Perutz, 31 July 1962, private papers.

266 [T]he title of a 1971 *New Scientist* article: Max Perutz, 'Haemoglobin: the molecular lung', *New Scientist and Science Journal*, 17 June, 676–9, 1971.

James Watson's *The Double Helix*: Athenaeum, 1968.

[Max] told Watson that he was 'glad to have it': Max to James Watson, 23 February 1968, Perutz papers, Venter Institute.

267 [H]e heard from Harold Himsworth: Harold Himsworth to Max, 25 March 1968, Perutz papers, Venter Institute.

Hans Krebs almost cut him dead: Max to James Watson, 25 February 1969, ibid.

[O]ne in *Scientific American* by . . . André Lwoff: André Lwoff, 'Truth, truth, what is truth (about how the structure of DNA was discovered)?', *Scientific American* 219, 133–8, 1968.

For over a year drafts circulated: All contained in a file in the Perutz papers held at the Venter Institute.

Himsworth feared that he would make matters worse: Harold Himsworth to Max, 16 December 1968, Perutz papers, Venter Institute.

Randall . . . insisted that even though the MRC report was not marked 'confidential': Randall to Max, 13 January 1969, Perutz papers, Venter Institute.

Watson . . . baldly stated that the idea that the King's researchers: Undated

first draft of letter from James Watson to *Scientific American*, sent to Max
8 January 1969, Perutz papers, Venter Institute.

267 Letters . . . finally appeared in *Science*: *Science* 164 27 June 1969, p. 1537–9.

268 '[A]s a matter of courtesy': ibid.

'I think you will agree that during the period described by [Watson]': Maurice
Wilkins to Max, 20 December 1968, Perutz papers, Venter Institute.

[A] book about molecular biology: Horace F. Judson, *The Eighth Day of
Creation*, Jonathan Cape, 1979.

Judson . . . recorded long conversations: The file of transcripts, held by the
library of the American Philosophical Society, Philadelphia, USA, is notably
thicker than for any other interviewee, of whom there were more than 100.

269 'He plans to write a book': Max to Gisela, 4 April 1970.

'The totem of his intelligence': Horace F. Judson, *The Eighth Day of Creation*,
Jonathan Cape, 1979, p. 599.

The article that appeared in January 1980: Max Perutz, 'Origins of molecular biology', *New Scientist* 85, 326–9, 1980.

'Bragg sounded the University': ibid.

270 [I]t finally appeared in August 1985: Max Perutz, 'That was the war: enemy
alien', *New Yorker*, August 1985; republished as 'Enemy alien' in
IWIMYAE, pp. 73–106.

[Pickersgill] . . . sent copies to Max: Pickersgill correspondence, in 'Enemy
alien' file, Perutz papers, Cambridge.

'A shrunken image of the ideal English gentleman': Max to Lotte Perutz,
14 October 1985, ibid.

271 'So the opposition is very heterogeneous': Max to Robin Perutz, 30 March
1977, private papers.

272 He covered the same material: Max Perutz, 'Why we need science', *New
Scientist*, 19 November, 530–6, 1981.

'You need go back only to your grandmother's early days': ibid.

273 'There is no war that would make the world safe': ibid.

'I am suspicious of scientists': Max Perutz, 'True science', *London Review of
Books*, March 1981; reprinted as 'How to become a scientist' in *Is Science
Necessary?*

Max wrote to Lotte that he was 'very pleased to be asked': Max to Lotte
Perutz, 30 June 1985, private papers.

274 'Feinberg sees the future mainly': Max Perutz, 'Brave new world', *New York
Review of Books*, September 1985; reprinted in *Is Science Necessary?*,
pp. 202–9.

'He is [in his office] day and night': Max to Robin Perutz, 1 May 1995,
private papers.

274 Graeme Mitchison ... often took the opportunity: Graeme Mitchison, interview with the author, 17 February 2004.

275 Vivien and Robin were pressed for their 'uninhibited criticism': Dedication of *IWIMYAE*.

[W]as written in 'a rage of indignation': Max to Gerald Holton, 1 December 1995, Perutz papers, Churchill College, Cambridge.

'Toppling great men from their pedestals': Max Perutz, 'The pioneer defended', *New York Review of Books*, 21 December 1995, reprinted as 'Deconstructing Pasteur' in *IWIMYAE*, pp. 135–46.

[Max] was happy to fire a salvo: 'Pasteur and the culture wars: an exchange', *New York Review of Books*, 4 April 1996.

His review of a life of Fritz Haber: Max Perutz, 'The cabinet of Dr Haber', *New York Review of Books*, 20 June 1996, reprinted as 'Friend or foe of mankind?' in *IWIMYAE*, pp. 3–16.

[L]ectures ... subsequently published as a book: Erwin Schroedinger, *What is Life? The Physical Aspect of the Living Cell*, Cambridge University Press, 1944.

276 'What was true': Max Perutz, 'Erwin Schroedinger's *What is Life?* and molecular biology', in J.W. Kilmister (ed.), *Schroedinger: Centenary Celebration of a Polymath*, Cambridge University Press, 1987, pp. 234–51.

A careful study by ... Edward Yoxen: E.J. Yoxen, 'Where does Schroedinger's "What is Life?" belong in the history of molecular biology?', *History of Science* xvii, 17–52, 1979.

'In science, as in other fields of endeavour': Preface, *Is Science Necessary?*, p. xvi.

277 [H]e 'decided not to sign the contract: Max to Lotte Perutz, 18 November 1983, private papers.

[H]e 'found other scientists' lives more absorbing': Max to Lotte Perutz, 10 June 1989, private papers.

When Jeannine Alton: P. Harper, email to the author, 27 October 2006.

The first, *Science Is Not A Quiet Life*: SINAQL.

278 [H]e put together another collection: *IWIMYAE*.

'My pleasure and gratitude': Max to Chancellor of the Pontifical Academy, 13 May 1981, Pontifical Academy archives.

279 'I felt a bit of a Charlie Chaplin': Max Perutz, 'The challenges of science', in *The challenges of science: A tribute to the memory of Carlos Chagas*, Scripta Varia 103, Pontifical Academy of Sciences, 2002.

'The meetings take place in a beautiful Renaissance villa': Max to Lotte and Franz Perutz, 16 November 1983, private papers.

280 The book containing the proceedings of his workshop finally appeared in

1996: *Resources and population, study week 17–22 November 1991*, Scripta Varia 087, Pontifical Academy of Sciences, 1996.

280 Max continued to express admiration: Max Perutz to Pontifical Academy, 11 March 1998, Pontifical Academy archives.

A recent biography shows that Francis Crick: Matt Ridley, *Francis Crick: discoverer of the genetic code*, HarperPress, 2006.

James Watson too has gone on record: For example in a BBC4 documentary, *The Double Helix*, broadcast on 22 April 2003.

281 Max was embarrassed to have to confess: Max to his family, 18 May 1993, private papers.

He had also joined an (ultimately successful) international campaign: Max to Boris Vainshtein, 4 March 1974, Perutz papers, Venter Institute.

'The people who run this committee': Max to Robin Perutz, 1993, private papers.

'When I was a schoolboy': Max to Keith Carmichael, 22 February 1993, Perutz papers, Churchill College, Cambridge.

His friend the former Peterhouse Master John Meurig Thomas: J.M. Thomas, 'Max Perutz: chemist, molecular biologist, human rights activist', *Chemical Communications*, 3891–4, 2005.

282 The paper was subsequently published: Max Perutz, 'By what right do we invoke human rights?', *Proceedings of the American Philosophical Society* 140, 135–47, 1996, reprinted in *IWIMYAE*, pp. 261–72.

'Scientists the world over are united by a common purpose': ibid.

[H]e wrote despairingly to Robin: Max to Robin Perutz, 23 June 1978, private papers.

He told successive Presidents of the Royal Society: For example, Max to Lord May, 21 January 2002, copy kindly provided by Lord May.

283 [H]is excited announcement . . . had been greeted with 'lukewarm applause': Max to family, 2 March 1991, private papers.

'I am alarmed by the American cries for vengeance': Max to Tony Blair, 14 September 2001, copy in private papers.

He told the Chancellor of the Pontifical Academy: Max to Mgr Sanchez Sorondo, 18 September 2001, Pontifical Academy archives.

'It is one thing': Max Perutz, quoted in Kam Patel, 'Perutz rubbishes Popper and Kuhn', *Times Higher Education Supplement*, 25 November 1994.

284 'Seeing this wonderful picture again': Max Perutz's picture choice, *National Gallery News*, October 2000.

'Nothing that I had read had prepared me': Max to Lord Dacre, 1988, Dacre papers.

284 'I often wonder whether art historians' Max to Lotte Perutz, 11 January 1984, private papers.

'I wish you had all been with me last night': Max to family, 15 April 1979, private papers.

285 'If you want to make money on the stage': Max to Lotte Perutz, 24 November 1984, private papers.

Maurice Cowling ... said they shared: G. Wheatcroft, 'Maurice Cowling' (obituary), *The Guardian*, 6 September 2005.

286 'I usually hate watching the news': Max to Lord Dacre, ? November 1989, Dacre papers.

Klug remembers Max suddenly saying: Aaron Klug, interview with the author, 16 February 2004.

'Gisela and I ... found our friends deeply divided': Max to Lord Dacre, 27 April 1988.

287 'I thoroughly agree with your views about Germany': Max to Lord Dacre.

'I was a chemist but worked in a department of physics': Max Perutz interviewed by Sue Lawley, *Desert Island Discs*, BBC Radio 4, June 2000.

Glossary

Allosteric protein A protein that can adopt one of two conformational states, 'relaxed' and 'tense', one of which is available for interaction with other entities while the other is not. Haemoglobin is an example of an allosteric protein: most others are enzymes.

Alpha helix Conformation adopted by lengths of chain in many proteins, coiling in a helical fashion according to the parameters discovered by Linus Pauling in 1951.

Amino acid Any of around two dozen organic compounds, in which a COOH group and an NH_2 group link to the same carbon atom. They can link to one another via a peptide bond: short chains of amino acids are peptides, and proteins are made up of varying proportions of twenty amino acids.

Angstrom A unit of measurement formerly used to measure interatomic distances, equivalent to a tenth of a nanometre or one ten-millionth of a millimetre. Symbol Å.

Atom The smallest possible unit of a chemical element. It consists of a dense nucleus, containing protons and neutrons, surrounded by a cloud of freely moving electrons. Atoms of different elements have different numbers of protons, neutrons and electrons. The diameter of a typical atom is 20 million times smaller than a millimetre. Yet most of it consists of empty space.

Biochemistry The study of the chemistry of living organisms.

Bohr effect Change in the affinity of the haemoglobin molecule for oxygen with change in the acidity of its surroundings. For example, in the muscles levels of carbon dioxide are relatively high, raising the acidity, so haemoglobin gives up its oxygen more easily.

Cell The fundamental unit of a living organism. Most cells contain a nucleus, within which is packaged the DNA that regulates the cell's activity. Surrounding the nucleus is cytoplasm, which contains many other specialised structures including those for protein manufacture and energy production. Cells are typically between a tenth and a hundredth of a millimetre in diameter. There are many specialised types, such as nerve cells and skin cells: red blood cells are exceptional in that they contain no nucleus.

Compound Chemical entity formed of more than one element.

Creep Gradual deformation of a solid (such as ice) under continuous stress.

Crystal A solid structure that forms when molecules of any substance aggregate in a regular fashion, so that each molecule (or sometimes a group of two or more molecules) is in the same relationship to its neighbours as all the others.

Crystallography The study of the form and structure of crystals.

Diffractometer Device that collects X-ray diffraction data for transfer to a computer without the need for photographic film.

DNA Deoxyribonucleic acid, the genetic material that encodes the instructions to make any living organism. Its molecules take the form of long chains of nucleotides.

Double helix The structure of the DNA molecule, discovered by James Watson and Francis Crick in 1953: two chains of DNA run side by side in opposite directions, forming links between complementary pairs of nucleotides, and twisting as they go like a spiral staircase.

Element Form of matter that cannot be decomposed into simpler substances: examples include carbon, oxygen, iron and sulphur.

Electron density map Contour map of the density of electrons in molecules in a crystal. A two-dimensional map projects the contours on a flat surface: a three-dimensional map traces the density at different levels in the molecule and so builds up a three-dimensional picture. Regions of high electron density show the location of atoms.

Electron microscope Device that uses electrons rather than light to obtain images magnified beyond the range possible with a light microscope.

Enzyme One of a class of proteins that acts as a catalyst in biochemical reactions, facilitating the reaction without itself being changed. There are thousands of enzymes in living things: they are fundamental to life, and so of great interest to science.

Fourier synthesis Mathematical calculation based on the position, intensity and phase of X-ray reflections that reveals the density of electrons throughout the molecules in a crystal.

Gene Unit of inheritance. Physically genes are lengths of DNA that encode particular proteins.

Glutamine One of the twenty amino acids that form proteins. Sequences of 40 or more glutamines that form due to a genetic mutation affecting the protein huntingtin cause the brain damage that is characteristic of Huntington's disease.

Haemoglobin The protein in red blood cells that carries oxygen. It consists of four polypeptide chains, two each of alpha and beta globin, each chain being connected to a haem group – an iron-containing porphyrin. Each cell contains about 1.5 million molecules of haemoglobin. Oxyhaemoglobin is haemoglobin with oxygen attached to the haem groups. Deoxyhaemoglobin is haemoglobin without oxygen. Methaemoglobin is haemoglobin that has been left in air and combined irreversibly with oxygen: its structure is closely similar to oxyhaemoglobin.

Histidine One of the twenty amino acids that form proteins.

Inorganic chemistry The branch of chemistry concerned with compounds other than those containing carbon (compare Organic chemistry).

Isomorphous replacement The technique of adding a heavy atom, such as mercury, to molecules of an organic compound such as a protein, without disturbing the arrangement of the atoms in the molecules. Comparison of the heavy atom compound with the native compound can help to solve the phase problem.

Molecular biology Term coined by Warren Weaver of the Rockefeller Foundation in 1938 to describe the study of the structure and function of large biological molecules, especially proteins and nucleic acids.

Molecule A unit of any chemical entity consisting of two or more atoms. An oxygen molecule contains two atoms of oxygen: a haemoglobin molecule contains 4,800 atoms, not counting the hydrogen atoms. The structure of a molecule is predetermined by its chemical composition.

Myoglobin Protein found in muscle that stores and releases oxygen. One molecule of myoglobin has a structure very similar to each of the four chains in a molecule of haemoglobin.

Nucleotide An organic compound consisting of a nitrogen-containing base linked to a sugar and phosphate backbone. Four nucleotides, adenine, guanine, cytosine and thymine, link together in chains to form DNA: the order of the nucleotides carries the code for the manufacture of proteins.

Organic chemistry The branch of chemistry concerned with compounds that include carbon (compare Inorganic chemistry). Carbon-based molecules are fundamental to life.

Patterson analysis Technique named after its discoverer, Lindo Patterson, for finding the vectors (directions and distances) between key features in a crystal without needing to know the phases of the reflections.

Peptide A compound consisting of two or more amino acids. A polypeptide is a chain of ten or more amino acids.

Phase problem The difficulty of establishing the phases of the waves of X-radiation that form a diffraction pattern after passing through a crystal. For proteins, it was solved by the method of isomorphous replacement (q.v.).

Physical chemistry The study of the physical (e.g. optical, electrical or magnetic) properties of chemical substances.

Plasma The liquid in which blood cells flow in the bloodstream.

Polymer Any substance with large molecules made up of repeated units.

Porphyrin An organic pigment consisting of four linked, nitrogen-containing rings, often with a metal atom in the centre. The haem group in haemoglobin is an iron-linked porphyrin.

Protein Any of a large group of organic compounds found in all living things: examples include insulin, haemoglobin, collagen, pepsin and keratin. They are large molecules consisting of thousands of atoms, arranged in the form of one or more long chains that fold into three-dimensional shapes characteristic of each protein. The chains are made up of dozens to hundreds of amino acids.

Reflection In X-ray crystallography, the term for a spot formed on photographic film by diffracted X-rays.

RNA Ribonucleic acid. Made up of the nucleotides adenine, guanine, cytosine and uracil, different forms of RNA carry the DNA message out of the cell nucleus into the cytoplasm, and construct proteins on the RNA template.

Rotating anode X-ray tube Device for generating intense X-rays that avoids heat damage to the anode by rotating it.

Spectroscopy A range of physical techniques used to measure properties of materials by producing characteristic spectra of absorbed or emitted radiation.

Sickle cell anaemia Also known as sickle cell disease, inherited disorder of the blood mainly affecting people of African origin in which haemoglobin molecules aggregate in the red blood cell and distort its shape. Caused by a single amino acid difference in the haemoglobin sequence.

Unit cell The smallest group of atoms that forms the repeating unit in a crystal.

X-rays Light with wavelengths shorter than a thousand-millionth of a metre, beyond the ultra-violet. Because of their short wavelength they can pass through matter.

X-ray crystallography The study of crystals using X-ray diffraction to locate the positions of atoms inside the crystal.

X-ray diffraction The scattering of X-rays by atoms in a crystal, forming a characteristic pattern of spots on photographic film that is related to the atomic positions.

X-ray tube Device for generating X-rays in which a metal anode is bombarded with electrons in a vacuum, causing X-rays to be emitted.

Index

337